23rd International Conference on Advanced Batteries, Accumulators and Fuel Cells (ABAF 2022)

Journal of Physics: Conference Series Volume 2382

Brno, Czechia
21-24 August 2022

ISBN: 978-1-7138-6488-2
ISSN: 1742-6588

Printed from e-media with permission by:

Curran Associates, Inc.
57 Morehouse Lane
Red Hook, NY 12571

Some format issues inherent in the e-media version may also appear in this print version.

Institute of Physics
Dirac House, Temple Back
Bristol BS1 6BE UK

Phone: 44 1 17 929 7481
Fax: 44 1 17 920 0979

techtracking@iop.org

Additional copies of this publication are available from:

Curran Associates, Inc.
57 Morehouse Lane
Red Hook, NY 12571 USA
Phone: 845-758-0400
Fax: 845-758-2633
Email: curran@proceedings.com
Web: www.proceedings.com

TABLE OF CONTENTS

Preface

Brno University of Technology, Faculty of Electrical Engineering and Communication, has been involved in the research of materials for electrochemical power sources for more than 50 years. In connection with these research activities, the Department of Electrical and Electronic Technology initiated in the year 2000 a series of internationally oriented conferences under the name “Advanced Batteries, Accumulators and Fuel Cells”. They focus on advances in science and technology in the field of electrochemical processes and energy storage (batteries, accumulators, supercapacitors and fuel cells), energy generation from renewable sources, photovoltaic systems, nanomaterials for electrochemical power sources and aspects of corrosion. Every year, the latest discoveries and trends are added to the conference program.

This special issue of *Journal of Physics: Conference Series* contains a selection of 24 full text papers presented at the 23rd International Conference on Advanced Batteries, Accumulators and Fuel Cells, which took place from August 21 to 24, 2022 in Brno, Czech Republic. The book of extended abstracts of all conference lectures and posters is available at www.aba-brno.cz/download/2022-ABAF-Proceeding.pdf .

After several tough years of COVID-19 restrictions, 2022 was the first year without any COVID19 related limitations. Unfortunately, the world was not at peace, and as a gesture of support to our colleagues from Ukraine, all scientists representing Ukrainian scientific institutions could attend our conference for free.

This year's conference also included the Battery 2030+ section, which was designed to exchange knowledge and best practices for long-term battery research in Europe.

We would like to thank our main sponsors, Pragolab and ThermoFisher Scientific and also to all other sponsors, BioLogic, Elektro Recycling Austria, EVC Group, EV Battery and Techsoft Engineering for supporting our conference. We also thank the Battery 2030+ Initiative and the European Energy Research Alliance (EERA) for cooperation in the preparation of this year’s conference and co-sponsoring. Finally, we thank IOP Publishing and Morressier for their assistance in publishing this journal.

The next meeting of the conference is being planned for August 2023 at the Brno University of Technology in the Czech Republic. More information about the conference and its past years is available at the conference website www.aba-brno.cz.

The Organizing Committee
Assoc. Prof. Marie Sedlarikova and Assoc. Prof. Tomas Kazda

Honorary Scientific Committee:
- Petr Vanýsek, *Northern Illinois University, DeKalb, Illinois, USA*
- Arnaldo Visintin, *INIFTA, La Plata, Argentina*
- Günter Fafilek, *TU Wien, Vienna, Austria*
- Elena Shembel, *USCTU, Dnipro, Ukraine*
- Vito Di Noto, *University of Padova, Padova, Italy*
- Harry Hoster, *Universität Duisburg-Essen, Duisburg, Germany*
- Boris Markovsky, *Bar-Ilan University, Tel Aviv, Israel*
- Philipp Adelhelm, *Friedrich Schiller University, Jena, Germany*
- Laurence Hardwick, *University of Liverpool, Liverpool, UK*
- Renata Oriňáková, *UPJŠ, Košice, Slovakia*
- Andrea Fedorková-Straková, *UPJŠ, Košice, Slovakia*
- Grzegorz Lota, *Poznan University of Technology, Poznan, Poland*
- Mariusz Walkowiak, *Institute of Non-Ferrous Metals, Poznan, Poland*

Organisation Committee:
- Marie Sedlaříková, *FEEC BUT, Brno, Czech Republic*
- Tomáš Kazda, *FEEC BUT, Brno, Czech Republic*
- František Klein, *Brno, Czech Republic*
- Vítězslav Novák, *FEEC BUT, Brno, Czech Republic*
- Miroslav Zatloukal, *FEEC BUT, Brno, Czech Republic*
- Jiří Libich, *FEEC BUT, Brno, Czech Republic*
- Josef Máca, *FEEC BUT, Brno, Czech Republic*
- Tomáš Binar, *FEEC BUT, Brno, Czech Republic*
- Jiří Švarc, *University of Defence, Brno, Czech Republic*
- Pavlína Sedlaříková, *Brno, Czech Republic*

Program Committee:
- Marie Sedlaříková, *FEEC BUT, Brno, Czech Republic*
- Arnaldo Visintin, *INIFTA, La Plata, Argentina*
- Mariela Ortiz, *INIFTA, La Plata, Argentina*
- Petr Vanýsek, *Northern Illinois University, DeKalb, Illinois, USA*
- Günter Fafilek, *TU Wien, Vienna, Austria*
- Boris Markovsky, *Bar-Ilan University, Tel Aviv, Israel*
- Harry Hoster, *Universität Duisburg-Essen, Duisburg, Germany*
- Grzegorz Lota, *Poznan University of Technology, Poznan, Poland*
- Mariusz Walkowiak, *Institute of Non-Ferrous Metals, Poznan, Poland*
- Josef Máca, *FEEC BUT, Brno, Czech Republic*
- Vítězslav Novák, *FEEC BUT, Brno, Czech Republic*
- Petr Bača, *FEEC BUT, Brno, Czech Republic*
- Tomáš Kazda, *FEEC BUT, Brno, Czech Republic*
- Helena Polsterová, *FEEC BUT, Brno, Czech Republic*
- Miroslav Zatloukal, *FEEC BUT, Brno, Czech Republic*
- Kristýna Jandová, *FEEC BUT, Brno, Czech Republic*
- Andrea Fedorková-Straková, *UPJŠ, Košice, Slovakia*
- Jiří Libich, *FEEC BUT, Brno, Czech Republic*
- Tomáš Binar, *FEEC BUT, Brno, Czech Republic*

Peer Review Statement

All papers published in this volume have been reviewed through processes administered by the Editors. Reviews were conducted by expert referees to the professional and scientific standards expected of a proceedings journal published by IOP Publishing Publishing.

- **Type of peer review:** Single Anonymous
- **Conference submission management system:** Morressier
- **Number of submissions received:** 32
- **Number of submissions sent for review:** 32
- **Number of submissions accepted:** 24
- **Acceptance Rate (Submissions Accepted / Submissions Received × 100):** 81.3
- **Average number of reviews per paper:** 2
- **Total number of reviewers involved:** 11
- **Contact person for queries:**
  Name: František Klein
  Email: abaf.brno@gmail.com
  Affiliation: Brno University of Technology, Czech Republic

Composite cathode for Li-sulfur system: Effect of inorganic additives on electrochemical performance

M Zukalová[1]*, M Vinarčíková[1], B Pitňa Lásková[1] and L Kavan[1]

[1]J. Heyrovský Institute of Physical Chemistry, Czech Acad. Sci., Dolejškova 3, CZ-18223, Prague 8, Czech Republic

marketa.zukalova@jh-inst.cas.cz

Abstract. The influence of five different inorganic additives with different composition and morphology, viz. nanofibrous rutile/TiO_xN_y, $Li_4Ti_5O_{12}$ (LTS) from Altair, TiO_2-P90 from Evonik, TiO_xN_y-anatase and nano-TiO_2 on the electrochemical performance of the activated carbon/sulfur composite cathode is evaluated by cyclic voltammetry at the scan rate of 0.1 mV s^{-1} and by galvanostatic chronopotentiometry at the 0.1 C rate. The composites are prepared by facile mechanical mixing of activated carbon with the inorganic additive and subsequent melting-diffusion with sulfur. The addition of nano-TiO_2 from titanium isopropoxide, TiO_xN_y-anatase, rutile/TiO_xN_y, and LTS Altair to the carbon/sulfur composite cathode increased the charge capacity of the corresponding Li-sulfur cell. The negative effect of the P90 additive can be attributed to the blocking of its surface by Al_2O_3 and SiO_2, hindering the adsorption of polysulfides. TiO_xN_y-anatase additive exhibits the highest capacity improvement due to its surface-enhanced redox chemistry of conductive and sulphiphilic surface resulting in a 7 % increase of the voltammetric charge capacity. In addition, this additive has a beneficial effect on the cycling stability of the sulfur composite cathode during galvanostatic cycling.

1. Introduction

Due to its high theoretical capacity of 1675 mAh g^{-1}, exceeding substantially the limits of currently used Li-ion batteries, the Li-sulfur battery represents an attractive energy storage system. However, despite the joint effort of researchers, fundamental issues of this system, which have been not fully addressed yet, hinder its commercialization. The cathode architecture represents a scientific challenge. It must contain an electronically conductive additive with special morphology permitting the electrochemical reduction of non-conductive sulfur and buffering volume changes during its reduction to soluble higher polysulfides (PS) and their oxidation back to elemental sulfur. Conductive carbonaceous materials (including carbon nanotubes[1,2], graphene derivatives[3,4], and calcined carbon materials) with various morphology have been demonstrated to enhance the performance of sulfur-based cathode composites. Further improvement of this system represents either heteroatom doping of carbon increasing its adsorption ability[5,6] or using an additional inorganic component for the efficient immobilization of PS. They are used as an additive or micro/nanostructured species in spherical materials[7]. Inorganic oxides[8,5,9,10], suboxides[11], sulfides[5], or nitrides[12] incorporated in the sulfur/carbon cathode composite increased the charge capacity and coulombic efficiency and improved cycling stability of the Li-sulfur battery. Due to its low price and non-toxicity, TiO_2 is frequently used as an efficient PS-immobilizing component together with porous carbon[5], fibrous carbon[5], graphene[13], carbon nanotubes[14], or conductive polymer[15]. Recently, we studied the electrochemical behavior of different porous carbons in nanocrystalline TiO_2/carbon/sulfur composite cathodes[16]. We found that the in-situ sol-gel impregnation of the high surface area carbon (1600 m^2 g^{-1}) with TiO_2 had a negative effect on the performance of the corresponding sulfur composite cathode due to partial blocking of the carbon surface by TiO_2[16]. Hence, in this study, we compare the electrochemical performance of a commercially available activated carbon with different morphologies of TiO_2, TiO_xN_y, and $Li_4Ti_5O_{12}$ powders in the sulfur composite cathode for Li-sulfur battery to reveal any synergy or compatibility in the particular inorganic additive/activated carbon combination leading to its improved electrochemical performance. To provide a representative comparison of the particular components concerning their influence on the charge capacity of the corresponding Li-sulfur battery and to facilitate prospective scale-up, we tested mechanical mixtures of inorganic additives with activated carbon.

2. Materials and Methods

2. 1. Inorganic additives

Nano-TiO_2 was prepared by hydrolysis of titanium isopropoxide (Aldrich) at 0°C. The product was recovered by filtration under suction, washed with deionized water and ethanol, and dried at 50°C. TiO_xN_y-anatase was synthesized by annealing of the phase-pure TiO_2B, (prepared according to the procedure described in Ref. [17]), in the NH_3 flow at 600°C for 6 hours. Nanofibrous rutile/TiO_xN_y was prepared from electrospun TiO_2 anatase nanofibers (Kertak) by annealing in the NH_3 flow. P90 is a commercial highly dispersed TiO_2 (AEROXIDE® TiO_2 P90, Evonik). According to the producer's specification, this material is stabilized by 0.3% of Al_2O_3 and 0.2% of SiO_2. LTS Altair material was commercial $Li_4Ti_5O_{12}$ spinel.

2. 2. Preparation of electrodes

Carbonaceous additive (activated carbon, Fluka (AC)) was mixed in an agate mortar with the respective inorganic component (nano-TiO_2 from titanium isopropoxide, TiO_xN_y-anatase, rutile/TiO_xN_y, P90, LTS Altair) in a mass ratio 4:1. The mixture was blended with sulfur (Aldrich) in a mass ratio 4:1 (S : mixture), mortared, and then treated at 155 °C for 15 hours under Ar atmosphere in a Teflon container. The product was again mortared, mixed with conductive carbon black C65 (Timcal), and a 2 % aqueous solution of carboxymethylcellulose (CMC, Sigma). The mass ratio of the TiO_2-carbon/sulfur composite to C65 and CMC was 7:2:1. The mixture was then diluted with deionized water to a consistency of viscous paste and coated by doctor-blading on Al foil. After drying in air at ambient temperature and 50 °C in vacuum overnight, the coated Al foil was cut into disc electrodes of 14 mm in diameter. The areal sulfur loading was adjusted to 1–2 mg cm^{-2}. The electrodes were stored in a glove box under an Ar atmosphere. The coin cells were assembled with a Li-foil anode (16 mm in diameter), a glass microfiber separator (Whatman), and 20 μl of electrolyte solution. The electrolyte solution consisted of 1.0 M lithium bis-(trifluoromethanesulfonyl) imide (LiTFSI) dissolved in a mixture of 1,3-dioxolane and 1,2-dimethoxyethane (1:1 by volume) with 1.0 wt% $LiNO_3$ as an electrolyte-additive. LiTFSI was dried in vacuum at 130 °C overnight, $LiNO_3$ at 50 °C in vacuum overnight, and the mixture of organic solvents was dried over a molecular sieve 4Å (Aldrich). The electrolyte solution contained 8-12 ppm H_2O as determined by Karl Fischer coulometric titration (Mettler Toledo). Electrolytes, solvents, and other chemicals were of standard quality (p.a. or electrochemical grade) purchased from Aldrich or Merck.

2. 3. Methods

X-ray diffraction was measured on the Bruker D8 Advance diffractometer using CuKα radiation. Adsorption measurements of nitrogen were carried out with ASAP 2020 apparatus (Micromeritics) at 77 K. All samples were degassed before analysis at 250°C in vacuum. The surface area was determined by the BET (Brunauer, Emmett, Teller) equation. The BET surface areas of carbonaceous material containing micropores in the structure have been calculated using the Rouquerol criterion[18,19]. Both the surface area in micropores and the external surface area were determined by t-plot analysis (thickness curve: Carbon Black STSA) [20]. The FE SEM S-4800 microscope (Hitachi) served for acquiring scanning electron microscopy (SEM) images. Electrochemical measurements were carried out in the 2032 coin-type test cells with Li-metal anode and glass microfiber separator in the potential window of 1.7 V/2.9 V vs. Li^+/Li. Cyclic voltammetry was measured with Autolab 302N potentiostat (Metrohm) controlled by Nova and Nova Battery SW. For each sample, two consecutive voltammetric scans were measured at the scan rate of 0.1 mV s^{-1}. Galvanostatic chronopotentiometry at 0.1C was measured by Neware Battery Testing System controlled by BTS 7.6 SW.

3. Results and Discussion

3. 1. Materials Characterization

The crystal structure of our particular inorganic components was analyzed by X-ray diffraction and the results are depicted in Figure 1. The diffractogram of nano-TiO_2 is featureless, confirming just

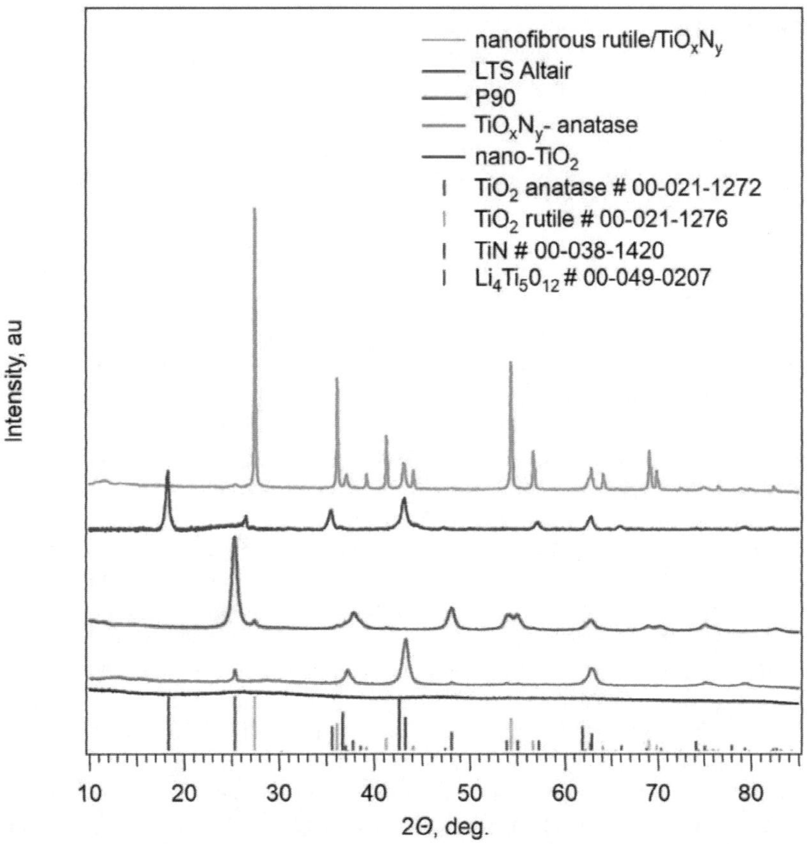

Figure 1. X-ray diffractograms of inorganic additives: nanofibrous rutile/TiO_xN_y (green), LTS Altair (brown), P90 (red), TiO_xN_y-anatase (blue) and nano-TiO_2 (black). The traces are offset for clarity, but the intensity scale is identical for all the diffractograms. JCPDS patterns of all inorganic compounds referred to in the study are marked with sticks at the bottom axis.

short-range ordering in the structure of this disperse material. The diffractogram of TiO_xN_y-anatase is dominating with peaks of TiO_xN_y, only the peak at 25 deg (2Θ) indicates the presence of anatase impurities. For comparison, the JCPDS pattern of TiN-osbornite (# 00-038-1420) is presented at the bottom of Figure 1 (blue sticks). Due to the cubic rock-salt structure of both TiO_xN_y and TiN, they exhibit similar diffraction patterns except for the shift of the peaks of TiO_xN_y to higher 2Θ due to differences in the lattice constants. The lattice constant of TiN is 4.24173 Å (pattern No.# 00-038-1420), whereas the value of 4.1769 Å was calculated from the diffraction pattern of TiO_xN_y (Figure 1). The diffraction pattern of P90 evidences that this material consists of an anatase phase with small rutile content. The diffractogram of LTS Altair exhibits the features of $Li_4Ti_5O_{12}$ spinel with a small peak at 25.3 deg (2Θ) evidencing the presence of anatase impurities. The peaks of TiO_2 rutile are dominating the diffractogram of nanofibrous rutile/TiO_xN_y. Also, in this sample, analogously to the diffractogram of TiO_xN_y-anatase, the value of the lattice constant (4.1958 Å) calculated from the peak at 43.08 deg (2Θ) confirms the presence of TiO_xN_y.

Since the surface area and porosity are essential attributes influencing the PS adsorption ability, the morphology of inorganic additives was characterized by nitrogen adsorption measurements. AC exhibited a BET surface area of 1015 $m^2 g^{-1}$ and a surface area outside micropores (a difference between the BET surface area and the surface area in micropores) of 141 $m^2 g^{-1}$ (Ref. [21]). Considering both the diameter of the micropores <2 nm and the dimensions of the S_8 molecule[1], it is obvious, that they are accessible neither for melted sulfur nor for PS dissolved in the electrolyte solution. Hence, their surface has no impact on the electrochemical behavior of the carbon/composite cathode. Nevertheless, the effects of inorganic additives should be more evident in the composite cathode with carbonaceous additive exhibiting a smaller surface area outside micropores.

Adsorption isotherms of inorganic additives are depicted in Figure 2. The hysteresis observable at relative pressures above 0.4 particularly in the isotherms of nano-TiO_2 and TiO_xN_y-anatase can be explained by capillary condensation of the adsorbate in interparticular spaces in the case of the former and in the channels in the nanofibrous structure of the latter. The high dispersity of nano-TiO_2 results in a network of interconnected spaces surrounded by TiO_2 nanoparticles. Figure 3 presents the BET surface areas of all inorganic additives evaluated in this study.

The TiO_2-containing materials with relatively different particle sizes (BET surface areas) were chosen intentionally to correlate both their crystalline structure and particle size with the electrochemical performance of the resulting composite with sulfur. In addition, inorganic materials listed in Figure 3 differ not only by their particle size but by their morphology as well. Whereas nano-TiO_2, P90, and LTS Altair represent standard dispersed powders with quasi-spherical particles (data not shown), TiO_xN_y-anatase, and nanofibrous rutile/TiO_xN_y exhibit fibrous morphology (Figures 4 a, b). Even though the former contains just crystals of TiO_xN_y and anatase, it still maintains the fibrous morphology of the precursor, i.e. the monoclinic TiO_2B formed by two TiO_6 edge-sharing octahedra linked to the neighboring doublet of octahedra by corners[17]. Fibrous morphology of the latter results from the preparation process (nanofibrous rutile/TiO_xN_y). Rutile and TiO_xN_y crystals arranged in the particular fibers are obvious in Figure 4 b.

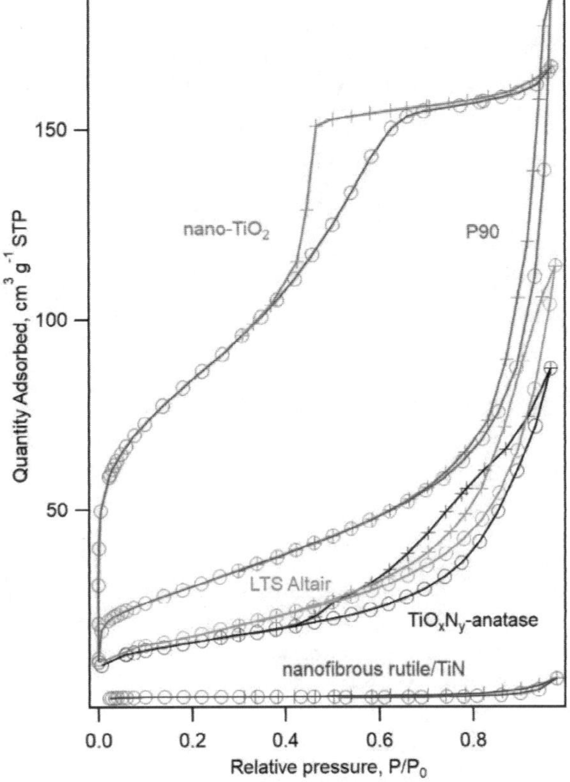

Figure 2. Nitrogen adsorption isotherms at 77 K measured on inorganic additives. The adsorption branch of the isotherm is marked by circles, desorption branch by crosses.

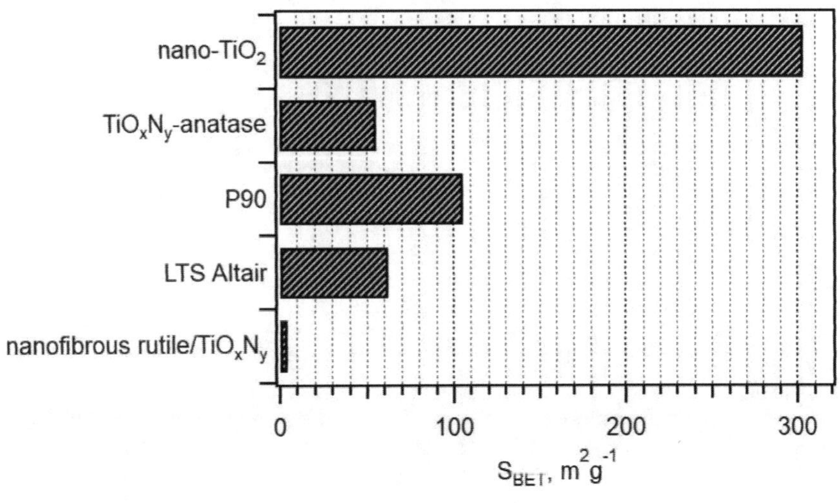

Figure 3. BET surface areas of inorganic additives determined from nitrogen adsorption measurements.

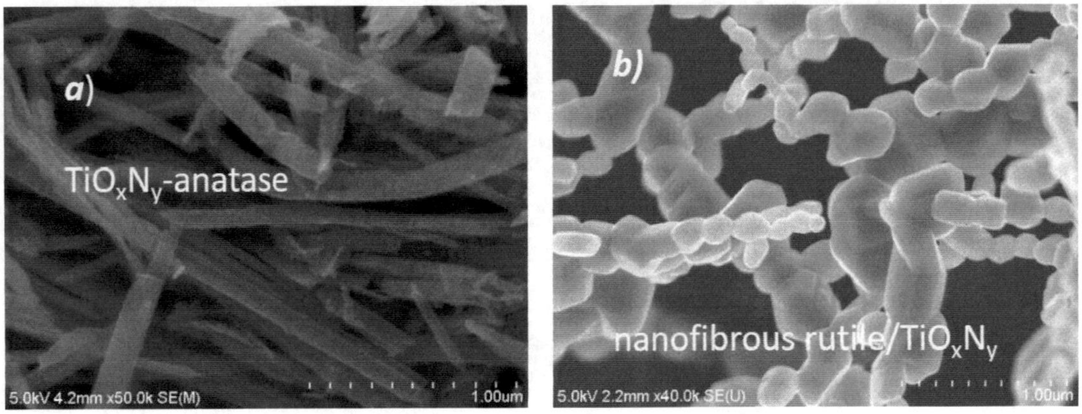

Figure 4. SEM images of TiO_xN_y-anatase (a) and nanofibrous rutile/TiO_xN_y (b). Scale bar 1 μm.

3.2. Electrochemical Measurements

Figure 5 shows cyclic voltammograms measured on coin cells containing the sulfur composite cathode with AC, either pure or modified with an inorganic additive. The cell further contains a Li anode, and a glass microfiber separator. This separator exhibited an improved performance and higher cycling stability in coin cells as compared to the standardly used polypropylene Celgard separator[22]. The overview of the corresponding specific charge capacities calculated from the oxidation (charge) branch of the cyclic voltammogram is presented in Figure 6.

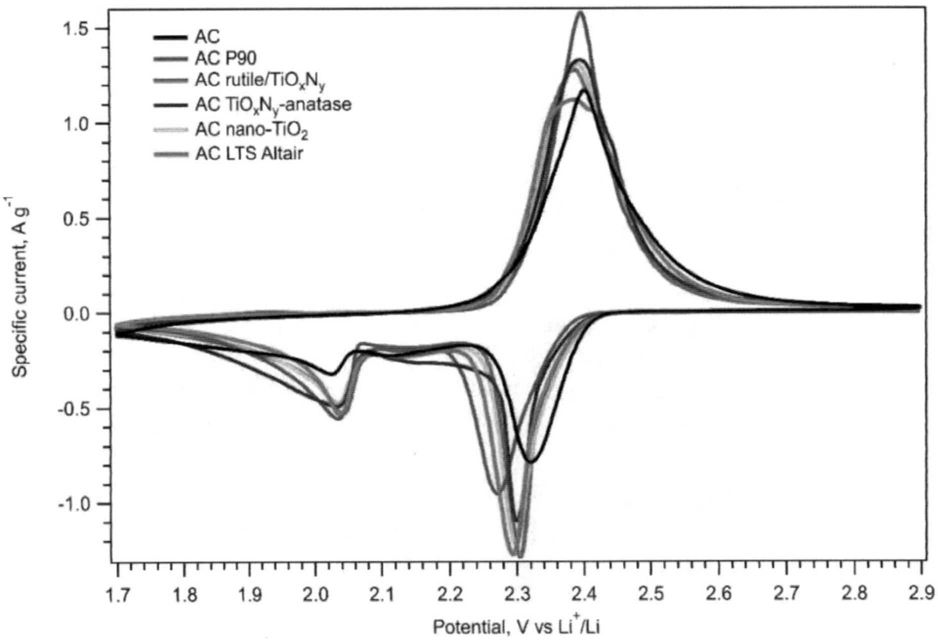

Figure 5. Cyclic voltammograms measured on the cell with sulfur composite cathode with AC, AC P90, AC rutile/TiO_xN_y, AC TiO_xN_y-anatase, AC nano TiO_2, and AC LTS Altair. Currents are normalized to the mass of sulfur.

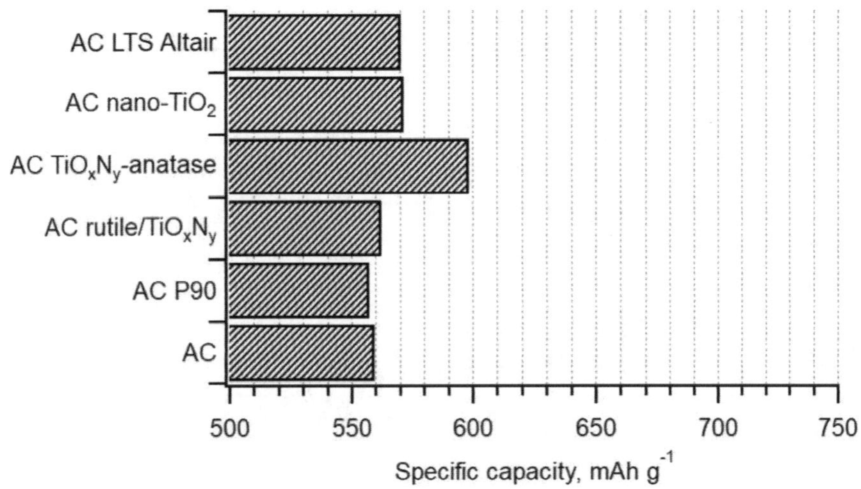

Figure 6. The overview of specific charge capacities calculated from the oxidation branch of the cyclic voltammograms (presented in Figure 5). Specific capacities are normalized to the mass of S in the cathode material.

Except for the P90 additive, the specific capacities of all other composites with inorganic components are enhanced as compared to the parent AC/sulfur composite cathode. Surprisingly, AC/nano-TiO_2 does not exhibit the highest specific charge capacity despite the BET surface area of nano-TiO_2 exceeding 300 $m^2\ g^{-1}$ (Figure 3). Cyclic voltammograms of all the composite cathodes show two reduction peaks at 2.3 V and 2.04 V vs. Li^+/Li. In the case of the cathode containing P90, the potential of the first reduction peak is downshifted by about 30 mV. This is probably an effect of the surface treatment of this commercial material by 0.3% of Al_2O_3 and 0.2% of SiO_2, *vide ultra*. This treatment results in high dispersity of this material, nevertheless, the particular surface blocking manifests itself by lower charge capacity (in addition to the above-discussed peak shift). All the cathodes exhibit just one broad anodic peak in the cyclic voltammogram representing the overlap of the features of the Li_2S/Li_2S_2 oxidation to PS and PS oxidation to elemental S. The AC/TiO_xN_y-anatase composite exhibits the best performance out of all AC-based cathodes despite the small BET surface area of TiO_xN_y-anatase. This behavior of AC/TiO_xN_y-anatase resembles the ability of Magnéli phases to chemisorb PS efficiently[23,24]. Hence, the positive effect of TiO_xN_y-anatase can be ascribed to the surface-enhanced redox chemistry on its conductive and sulphiphilic surface, analogously to the effect of Ti_4O_7 [24]. The soluble PS generated efficiently at the AC-sulfur interface are partially immobilized on the surface of inorganic oxide or oxynitride. This effect confirms the higher PS adsorption ability of inorganic additives as compared to activated carbon. A similar effect can be expected for rutile/TiO_xN_y additive as well, however, this material is handicapped by its small BET surface area.

Figure 7 demonstrates the beneficial effect of TiO_xN_y-anatase additive on the performance of the sulfur composite cathode during galvanostatic charging/discharging. In agreement with cyclic voltammetry, the beneficial effect of inorganic additive is obvious. The AC/sulfur/TiO_xN_y-anatase cathode shows excellent cycling stability within 20 galvanostatic cycles and exhibits even a slight increase of capacity during cycling.

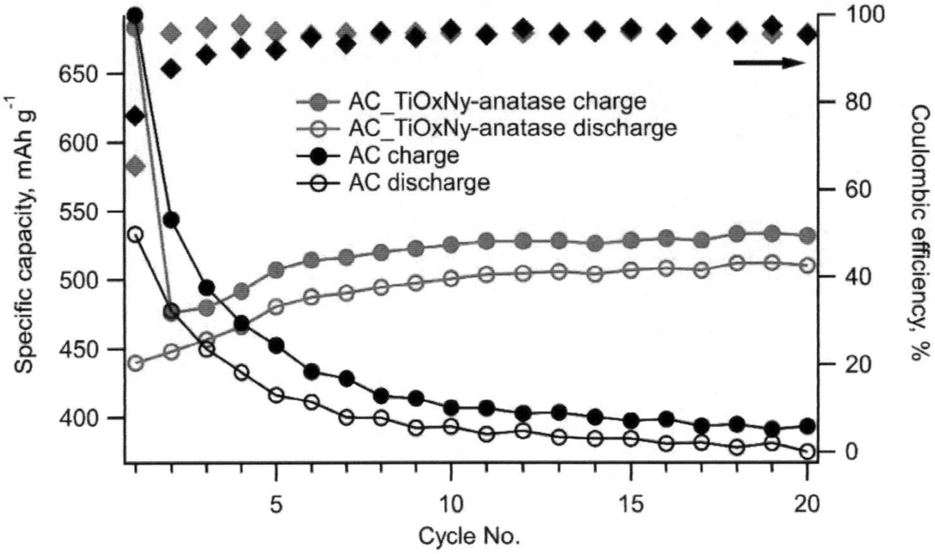

Figure 7. Specific charge/discharge capacities calculated from galvanostatic chronopotentiometry at 0.1 C rate measured on cathode composites containing AC and AC TiO_xN_y-anatase. Specific capacities are normalized to the mass of S in the cathode material. The corresponding coulombic efficiencies are marked with black and red diamonds.

4. Conclusions

The effect of five different inorganic additives on the electrochemical performance of the AC sulfur composite cathode was evaluated by electrochemical measurements. Nano-TiO_2 from Ti-isopropoxide, TiO_xN_y-anatase, rutile/TiO_xN_y, and LTS Altair increased the charge capacity of the corresponding Li-sulfur cell as compared to that of bare carbon/sulfur composite cathode. The negative effect of the P90 additive can be attributed to the blocking of its surface by Al_2O_3 and SiO_2, hindering PS adsorption. TiO_xN_y-anatase additive exhibits the highest capacity improvement due to its surface-enhanced redox chemistry of conductive and sulphiphilic surface resulting in the increase of the voltammetric charge capacity from 559 to 599 mAh g^{-1}. In addition, chronopotentiometric measurements prove that this additive has a beneficial effect on the cycling stability of the sulfur composite cathode during galvanostatic cycling.

Acknowledgments

This research was funded by the Grant Agency of the Czech Republic (contract No. 20-03564S).

References

[1] Xin S, Gu L, Zhao NH, Yin YX, Zhou LJ, Guo YG, Wan LJ 2012 *J Am Chem Soc* **134** 45:18510-18513. doi:10.1021/ja308170k
[2] Huang YP, Sun XG, Wang J, Li X, Chen W, Wei CC, Hu H, Liang GD 2019 *J Alloy Compd* **776** 187-193. doi:10.1016/j.jallcom.2018.10.146
[3] Kim J, Kang Y, Song SW, Suk J 2019 *Electrochim Acta* **299** 27-33. doi:10.1016/j.electacta.2018.12.165
[4] Yi R, Liu C, Zhao Y, Hardwick LJ, Li Y, Geng X, Zhang Q, Yang L, Zhao C 2019 *Electrochim Acta* **299** 479-488. doi:10.1016/j.electacta.2019.01.015
[5] Eftekhari A, Kim D-W 2017 *J Mater Chem A* **5** 34:17734-17776. doi:10.1039/c7ta00799j
[6] Li S, Jin B, Zhai X, Li H, Jiang Q 2018 *ChemistrySelect* **3** 8:2245-2260. doi:10.1002/slct.201703112

[7] Gong Q, Gao T, Hu T, Zhou G 2019 *Nanomaterials* **9** 9. doi:10.3390/nano9091207
[8] Wei Seh Z, Li W, Cha JJ, Zheng G, Yang Y, McDowell MT, Hsu PC, Cui Y 2013 *Nat Commun* **4** 1331. doi:10.1038/ncomms2327
[9] Zeng T, Ji P, Shang B, Peng Q, Hu X, Li G 2018 *Ionics* **24** 10:2973-2982. doi:10.1007/s11581-018-2468-8
[10] Ahn H, Kim Y, Bae J, Kim YK, Kim WB 2020 *Chem Eng J* **401** 126042. doi:10.1016/j.cej.2020.126042
[11] Wei H, Rodriguez EF, Best AS, Hollenkamp AF, Chen D, Caruso RA 2017 *Adv Energy Mater* **7** 4. doi:10.1002/aenm.201601616
[12] Lim WG, Jo C, Cho A, Hwang J, Kim S, Han JW, Lee J 2019 *Adv Mater* **31** 3:e1806547. doi:10.1002/adma.201806547
[13] Wang M, Tan S, Kan S, Wu Y, Sang S, Liu K, Liu H 2020 *J Energy Chem* **49** 316-322. doi:10.1016/j.jechem.2020.03.011
[14] Geng XW, Yi RW, Lin XF, Liu CG, Sun Y, Zhao YC, Li YQ, Mitrovic I, Liu R, Yang L, Zhao CZ 2021 *J Alloy Compd* **851** 156793. doi:10.1016/j.jallcom.2020.156793
[15] Zhao Y, Zhu W, Chen GZ, Cairns EJ 2016 *J Power Sources* **327** 447-456. doi:10.1016/j.jpowsour.2016.07.082
[16] Zukalova M, Vinarcikova M, Bousa M, Kavan L 2021 *Nanomaterials* **11** 2. doi:10.3390/nano11020541
[17] Zukalova M, Kalbac M, Kavan L, Exnar I, Graetzel M 2005 *Chem Mat* **17** 5:1248-1255. doi:10.1021/cm048249t
[18] Rouquerol J, Llewellyn P, Rouquerol F 2007 *Studies in Surface Science and Catalysis* **160** 49-56. doi:10.1016/S0167-2991(07)80008-5
[19] Rouquerol F, Rouquerol J, Sing KSW, Llewellyn P, Maurin G (2014) *Adsorption by Powders and Porous Solids*. Elsevier Ltd. doi:10.1016/c2010-0-66232-8
[20] Lowell S, Shields JE, Thomas MA, Thommes M (2004) *Characterization of Porous Solids and Powders: Surface Area, Pore Size and Density*. Particle Technology Series. Kluwer Academic Publishers. doi:10.1007/978-1-4020-2303-3
[21] Zukalova M, Vinarcikova M, Pitna Laskova B, Kavan L 2021 *ECS Trans* **105** 1:183-189. doi:10.1149/10501.0183ecst
[22] Murray V, Hall DS, Dahn JR 2019 *J Electrochem Soc* **166** 2:A329-A333. doi:10.1149/2.1171902jes
[23] Zubair U, Amici J, Francia C, McNulty D, Bodoardo S, O'Dwyer C 2018 *ChemSusChem* **11** 11:1838-1848. doi:10.1002/cssc.201800484
[24] Pang Q, Kundu D, Cuisinier M, Nazar LF 2014 *Nat Commun* **5** 4759. doi:10.1038/ncomms5759

Lithium-ion battery module-to-cell: disassembly and material analysis

A Pražanová[1,*], M Havlík Míka[2] and V Knap[1]
[1] Department of Electrotechnology, Faculty of Electrical Engineering, Czech Technical University in Prague, Technická 1902/2, 16627 Prague, Czech Republic
[2] Department of Glass and Ceramics, University of Chemistry and Technology, Technická 5, 166 28 Prague, Czech Republic

*Corresponding author mail: anna.prazanova@cvut.cz

Abstract. Lithium-ion batteries (LIBs) are one of the most popular energy storage systems. Due to their excellent performance, they are widely used in portable consumer electronics and electric vehicles (EVs). The ever-increasing requirements for global carbon dioxide CO_2 emission reduction inhibit the production of new combustion vehicles. Thus, the demand for EVs increases, as well as the number of spent LIBs. Due to increases in raw materials saving and reduction in energy and environmental impacts, recycling is one of the most promising solutions for end-of-life (EOL) treatment for spent LIBs. This work describes the first step in recycling the LIBs nickel-manganese-cobalt (NMC) based module from a full battery electric vehicle (BEV) holding its high recycling efficiency and considering the process costs and environmental impact. This paper is devoted to module-to-cell disassembly, discharge state characterization measurements, and material analysis of its components based on x-ray fluorescence (XRF) and diffraction (XRD).

1. Introduction

Lithium-ion batteries (LIBs) currently represent one of the most promising electrochemical energy storage systems. They are widely used in a broad range of applications, from powering small consumer electric devices such as smartphones or laptops to integrating into electric transport systems considering electric bikes or scooters and electric vehicles (EVs). Moreover, LIBs are one of the key technologies for stationary energy storage systems [1,2].

The automotive industry is undergoing a fundamental change. The ever-increasing requirements for global carbon dioxide CO_2 emission reduction led to a decrease in the production of new combustion vehicles and an increase in the demand for electric ones [3]. In Europe, the whole emission-free concept shield within the European Green Deal is supported by Regulation (EU) 2019/63; it sets EU fleet-wide emission performance standards for new passenger cars and new light commercial vehicles (vans) registered in the EU. Moreover, the Fit for 55 legislative package presents EU's ambitious target of achieving net emission reductions of at least 55% by 2030, compared to 1990 levels [4].

The electric vehicle LIBs are generally based on valuable metals such as cobalt (Co), manganese (Mn), nickel (Ni), and lithium (Li), other metals such as iron (Fe) and aluminium (Al) or the non-metallic elements such as phosphorous (P) in the cathode, other elements such as graphite (C) in the anode, and materials such as Al, copper (Cu), and Fe in coverings and electronic components. The electrolyte of

LIBs is typically made up of lithium salt solution such as lithium hexafluorophosphate $LiPF_6$ salt. Additionally, the battery consists of organic polymers, e.g., a separator, binders, and adhesives [5,6].

The constant growth in the EVs market leads to the ever-increasing number of spent LIBs and brings numerous waste-management challenges in end-of-life (EOL) processing. Due to increased requirements on raw materials savings and reduction in energy and environmental impacts of mining, recycling is one of the most promising solutions for battery waste treatment [7]. A proper recycling technique recovers selected materials with high efficiency and purity, reaching more than 90 % [8]. The EU responded to the current situation and issued a proposal for a Regulation of the European Parliament and the Council concerning batteries and waste batteries, repealing Directive 2006/66/EC and amending Regulation (EU) No 2019/1020. The proposal has established, e.g., collection targets for portable waste batteries, which are 70 % by the end of 2025, instead of the current 65 %, and 80 % by the end of 2030; it regards recycling efficiencies for LIBs treatment: 65 % by 2025, 70 % by 2030 [9].

This work represents the first step in recycling the LIB module while maintaining its high efficiency and considering the process's overall cost and environmental impact. It focuses on the disassembly of the LIB module typically used in a full battery electric vehicle (BEV) to its NMC-based cells and provides the material analysis. Electrochemical impedance spectroscopy (EIS) and five cycles of cyclic voltammetry (CV) were measured in terms of safety and complex discharge state characterization of the cell. The detailed analysis of the material composition of the battery cell components was determined using x-ray fluorescence (XRF) and diffractometry (XRD). According to the analysis, the content of valuable metals in the N:M:C ratio was determined as 54%:23%:23%. The difference between the proportions of these metals could affect the recycling process's overall economic and environmental impact and will be studied in more detail in future work.

2. Experimental

This paper presents performed dismantling treatment and the material analysis on the battery module's scale to its cells. The LIB module typically used in the battery electric vehicle (BEV) was chosen for laboratory testing. The module consists of 24 NMC pouch cells in a 3P8S connection, each with 3.65 V and 78 Ah capacity, which create the base for the nominal module voltage of 29.36 V.

2.1. Disassembly procedure

For an evaluation of the process complexity, the module was manually disassembled into individual parts. Before disassembling, the module was fully discharged and short-circuited to ensure safe handling. Firstly, the Al packaging was milled off. It was followed by removing the insulating and thermo-insulating materials, which were later analyzed as commonly used polyurethane and polymer based on a mixture of aryl phosphates and hydrated aluminum oxides, with the presence of rutile (TiO_2) and corundum (Al_2O_3). Secondly, the remaining plastics and metal coverings were manually split off, and finally, the individual cells were separated. The treated module is shown in Figure 1 (a), the removed covering materials are illustrated in Figure 1 (b), and the separated battery cell is in Figure 1 (c).

2.2. Battery cell testing

For a characterization of the discharged state of the cell after short-circuiting the entire module, one of the separated battery cells has undertaken a series of measurements, including the EIS and CV. These measurements are used to describe the migration of charge carriers, characterize the interfacial properties of materials at the electrode surface and identify the electrochemical reactions of active electrode materials.

(a) (b) (c)

Figure 1. Example of (a) treated BEV module, (b) removed coverings, and (c) battery cell.

2.2.1. Electrochemical impedance measurement

The EIS measurement is commonly used for investigating the properties of materials or electrode reactions by measuring the impedance over a range of frequencies. The complex impedance Z^* was recorded with an Autolab PGSTAT30 potentiostat using a harmonic AC field from 1 MHz to 1 mHz at an amplitude of 1 mV. The absolute value of impedance /Z/ and its phase angle φ were plotted in a Bode plot as functions of frequency. For the spectra analysis, the negative imaginary part $-Z_{Im}$ was plotted vs. the real part Z_{Re}, thereby forming a Nyquist plot on which frequency is a function parameter.

An equivalent circuit model (ECM) with electric components whose parameters quantify the physicochemical processes inside the cell under the electric field can be extracted from an impedance spectrum obtained via EIS. Based on the plots shape, a fractional-order ECM for LIB, which is shown in Figure 2, was used [10]. Using the non-least-squares method (NLSM), the impedance data were fitted to the equivalent circuit and the parameters of circuit components were calculated.

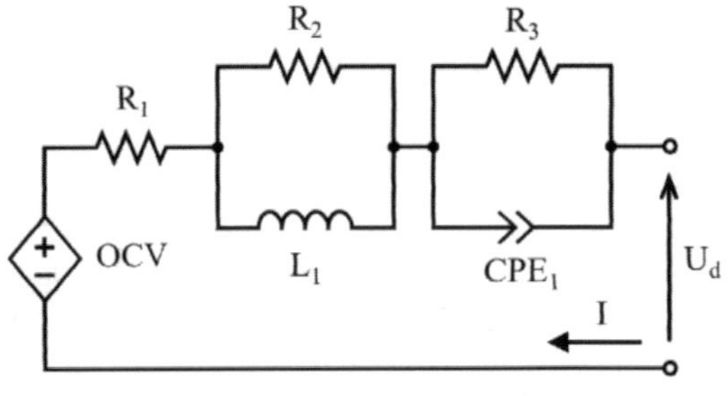

Figure 2. A fractional-order equivalent circuit model for LIB cell.

In the circuit, the resistor R_1 represents the high-frequency resistance of the electrolyte $LiPF_6$ and its impedance Z_1 was expressed as:

$$Z_1 = R_1. \tag{1}$$

Because the battery cell was in a completely discharged state after it had been short-circuited, the value of the original electrolyte resistance might be additionally lowered by the post formation of highly conductive "channels" during this discharge process.

The parallel combination of the resistor R_2 and inductor L_1 represents high-frequency inductive behavior and its impedance Z_2 was determined according to equation (2).

$$Z_2 = \left(\frac{1}{R_2} + \frac{1}{i\omega L_1}\right)^{-1} \tag{2}$$

The inductive response is probably caused by the layered structure of metallic foils inside a battery cell or by the prevailing electronic conductivity of cathode or anode materials.

The parallel combination of the resistor R_3 and constant phase element (CPE) represents the distributed capacitance response to the migration of charge Li^+ carriers and electrochemical reactions in electrode materials. Its impedance Z_3 was calculated as:

$$Z_3 = \left[\frac{1}{R_3} + A(i\omega)^{\alpha}\right]^{-1}, \tag{3}$$

where A is the distributed capacity in an electrode and at the electrode/metal collector surface, and α is a parameter characterizing the distribution. The total impedance of the cell Z was calculated as a sum of the partial impedances above:

$$Z = Z_1 + Z_2 + Z_3. \tag{4}$$

2.2.2. Cyclic voltammetry measurement

The CV is typically used for the analysis of electrochemical redox reactions of active materials by measuring the current as a function of potential. In this work, this was also recorded with an Autolab PGSTAT30 potentiostat in the staircase mode from 0 V to the potential when the current reached the value of 1 A; the potential step was 2.44 mV with scan rate 2 mV/s.

The plots of current I vs. potential E indicate the change of polarization resistance R_p during the potential cycles when a total of five cycles were measured. The derivative of recorded CV curves dI/dE represents the reciprocal polarization resistance $1/R_p$ of the electrochemical reactions inside a battery cell. Moreover, the hysteresis of the curves indicates the difference of R_p due to the charging or discharging process. Such CV measurements are very useful as an EIS complement for precise determining the polarization resistance R_p which could relatively indicate the state of health (SOH) of a spent battery cell.

2.3. Material analysis

For further processing, the battery cell was opened and dismantled to its individual layers of electrode materials soaked with electrolyte. The material of these thin layers on Cu, and Al metal sheets or separator were analyzed using the x-ray fluorescence (XRF) using the Philips PANAlytical spectrometer and x-ray diffraction (XRD) using Philips PANAlytical XPERT-PRO diffractometer. The composition of cathode, anode and electrolyte materials were determined, and the material ratio of valuable metals (Ni:Mn:Co) was discussed. The first cut in a battery cell is shown in Figure 3 (a); the individual layers of the cell's electrode materials are demonstrated in Figure 3 (b).

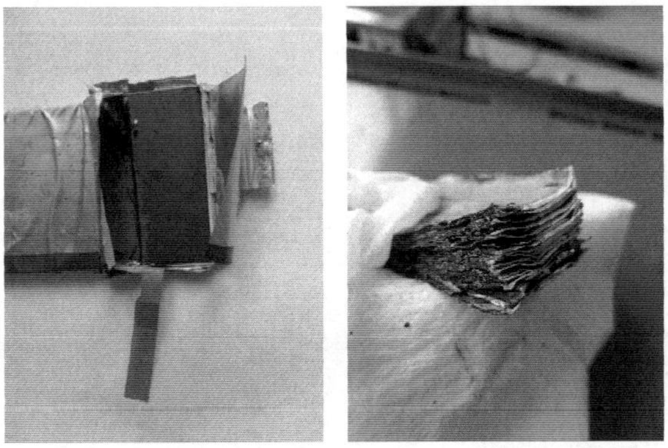

(a) (b)

Figure 3. Example of (a) a cut in battery cell, (b) individual layers of electrode materials.

3. Results and Discussion

The module was disassembled manually using a milling cutter, pliers, and other specialized tools. Due to this was the first processing of this technology, the process was complicated, and the total disassembly time was around 3 hours. The separation of one battery cell, including adhesive cleaning with ethanol, was about 30 minutes on average.

The discharge state of the cell was evaluated using EIS measurements. In the frequency range of 2 kHz to 10 Hz, the impedance exhibited ohmic character with φ close to zero. For the frequencies above 2 kHz, the character was inductive ($\varphi > 0$), and below 10 Hz the character was capacitive ($\varphi < 0$). The distribution of capacity A is characterized by the parameter $0.5 < \alpha < 1$ which indicated the formation of a space charge in the region close to the metallic current collectors of a battery cell. This space charge was disturbed by the migration of charge carriers; when $\alpha = 1$ than the electrochemical system behaves as an ideal capacitor, when $\alpha = 0.5$ than the system response is mainly controlled by the diffusion of ions. The measurements are shown in Nyquist and Bode plots illustrated in Figure 4. The relatively low resistance values for resistors R_1, R_2 and R_3 indicate a high electric conductivity of the electrolyte and electrode materials that form thin layers of the cell. The values could be also affected by the deep discharge process.

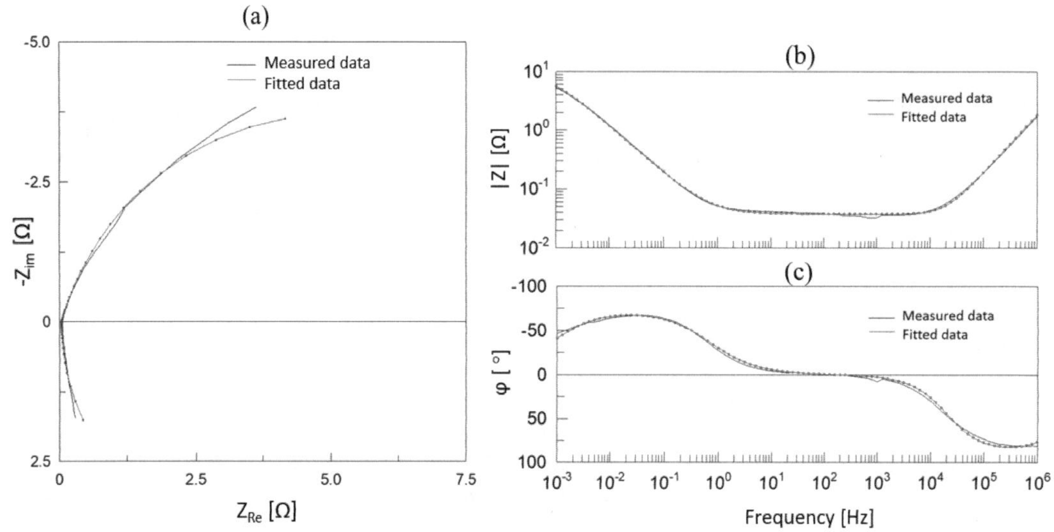

Figure 4. Impedance measurement (a) Nyquist impedance plot; Bode plots of (b) absolute value of impedance /Z/, and (c) its phase angle φ as functions of frequency.

The parameters of the ECM were calculated and are given in Table 1.

Table 1. Established ECM parameters according to the NLSM fit.

Parameter	Fitted value	Error	Relative Error [%]
R_1 [mΩ]	37.68	0.27	0.72
R_2 [mΩ]	8.34	1.06	12.72
R_3 [mΩ]	9.78	0.46	4.69
L_1 [μH]	0.29	0.33	1.12
A [F]	8.05	0.09	1.15
α [-]	0.82	0.01	0.55

The resistance calculated from CV data dynamically changed during the potential cycle in the range of 10 Ω to 1 Ω. The lowest resistance values were achieved for the highest current values at the high and low vertex potentials, which is demonstrated in Figure 5. The observed extent of hysteresis indicates the significantly decreased reversibility and SOH of the spent battery cell in the measured potential window.

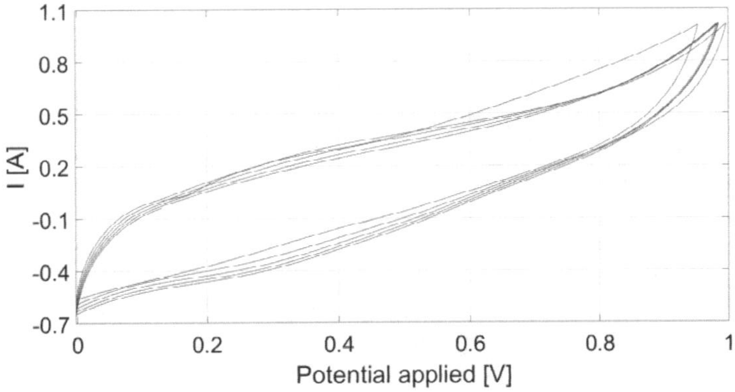

Figure 5. CV plots for five cycles.

A detailed characterization of the cell composition is an initial step to determine the suitable recycling procedure. A proportion overview of individual materials, especially metals, is necessary to evaluate the effectiveness of the selected technique or to define the expected economic return. In this work, the x-ray fluorescence (XRF) and x-ray diffractometry (XRD) analysis for the material composition determination was used; its results are summarized in Table 2.

Table 2. Determined material composition of the battery cell.

Component	Material composition
Cathode	Al sheet, $LiMn_{0.15}Co_{0.15}Ni_{0.7}O_2$
Anode	Cu sheet, C (min. 90 %) mixed with $LiCo_{0.25}Ni_{0.75}O_2$ and CuO or Al_2O_3
Electrolyte	$LiPF_6$
Separator	polymeric organic thin film

Cathode founded on Al sheet is created by mixed oxide of stoichiometry $LiMn_{0.15}Co_{0.15}Ni_{0.7}O_2$ and contains an amorphous phase, which is probably a binder of organic composition. Other impurities can be due to the dissolution of the surroundings and transport by diffusion into the cathode mass. These impurities are low in the amorphous phase. Anode based on Cu sheet consists of more than 90 % graphite with a 1 to 3 wt% admixture of lithium mixed oxide of stoichiometry $LiCo_{0.25}Ni_{0.75}O_2$ and copper oxide CuO and possibly aluminum oxide Al_2O_3. These impurities are probably caused by the partial dissolution of the surrounding components in the electrolyte and their subsequent diffusion into the electrode mass. The electrolyte represents a commonly used lithium hexafluorophosphate $LiPF_6$. The separator is created by a polymeric organic thin film, which has not yet been analyzed in detail. Moreover, the ratio between Ni, Mn, and Co (N:M:C) in the cathode material was determined and was closest to the values of 54%:23%:23%. It can be assumed that this proportion between the main elements is used to achieve the most suitable properties of the battery cell.

4. Conclusions

Recycling is one of the most promising disposal methods for spent LIBs, especially those from EVs, which contain a considerable proportion of valuable metals such as Co, Ni, and Mn. Therefore, it is necessary to optimize current recycling techniques to achieve the highest possible proportion of recovered materials of sufficient quality for further use.

This paper represents the first step in the recycling process of a LIB module based on NMC cells. Short-circuiting the entire module was identified as a sufficient way to discharge individual cells to their safe limit, which is suitable for further handling. The material analysis results determined the proportions of metals and other elements of the battery cell compounds, which are necessary for further processing, including crushing, particle separation, and final recovery of valuable metals. The ratio between Ni, Mn, and Co (N:M:C) in the active cathode material was established and was closest to the values of 54%:23%:23%. In future work, a more detailed analysis of more battery cells will be provided. The results will be included in the final environmental and economic evaluation of the BEV modules, considering the production and waste treatment.

References

[1] Armand M, Axmann P, Bresser D, Copley M, Edström K, Ekberg C, Guyomard D, Lestriez B, Novák P, Petranikova M, Porcher W, Trabesinger S, Wohlfahrt-Mehrens M and Zhang H 2020 Lithium-ion batteries – Current state of the art and anticipated developments *J. Power Sources* **479** 228708

[2] Chen W, Liang J, Yang Z and Li G 2019 A Review of Lithium-Ion Battery for Electric Vehicle Applications and Beyond *Energy Procedia* **158** 4363–8

[3] Lander L, Cleaver T, Rajaeifar M A, Nguyen-Tien V, Elliott R J R, Heidrich O, Kendrick E, Edge J S and Offer G 2021 Financial viability of electric vehicle lithium-ion battery recycling *iScience* **24**

[4] European Commission CO_2 emission performance standards for cars and vans *Clim. Action*

[5] Xu C, Dai Q, Gaines L, Hu M, Tukker A and Steubing B 2020 Future material demand for automotive lithium-based batteries *Commun. Mater.* **1**

[6] Diekmann J, Hanisch C, Froböse L, Schälicke G, Loellhoeffel T, Fölster A-S and Kwade A 2017 Ecological Recycling of Lithium-Ion Batteries from Electric Vehicles with Focus on Mechanical Processes *J. Electrochem. Soc.* **164** A6184–91

[7] Harper G, Sommerville R, Kendrick E, Driscoll L, Slater P, Stolkin R, Walton A, Christensen P, Heidrich O, Lambert S, Abbott A, Ryder K, Gaines L and Anderson P 2019 Recycling lithium-ion batteries from electric vehicles *Nature* **575** 75–86

[8] Pražanová A, Knap V and Stroe D-I 2022 Literature Review, Recycling of Lithium-Ion Batteries from Electric Vehicles, Part I: Recycling Technology *Energies 2022, Vol. 15, Page 1086* **15** 1086

[9] EUR-Lex Proposal for a REGULATION OF THE EUROPEAN PARLIAMENT AND OF THE COUNCIL concerning batteries and waste batteries, repealing Directive 2006/66/EC and amending Regulation (EU) No 2019/1020 *Doc. 52020PC0798*

[10] Lvovich V F 2012 Impedance Spectroscopy: Applications to Electrochemical and Dielectric Phenomena, J.Wiley& Sons New Jersey.

Four-component high entropy spinel oxide as anode material in lithium-ion batteries with excellent cyclability

D Csík[1,2], D Zalka[2,3], K Saksl[1,2], D Capková[4], R Džunda[1]

[1] Department of Materials Science, Faculty of Materials, Metallurgy and Recycling, Technical University of Košice, Letná 9, 042 00, Košice, Slovac Republic
[2] Institute of Materials Research, Slovak Academy of Sciences, Watsonova 47, 040 01, Košice, Slovak Republic
[3] Department of Physics of Condensed Matters, Faculty of Sciences, Pavol Jozef Šafárik University in Košice, Park Angelinum 9, 041 54 Košice, Slovak Republic
[4] Department of Physical Chemistry, Faculty of Sciences, Pavol Jozef Šafárik University in Košice, Moyzesova 11, 041 54, Košice, Slovak Republic

Email: dcsik@saske.sk

Abstract. Nowadays, energy storage technologies are in focus of public interest, especially in the field of the automotive industry. Lithium-ion batteries (LIBs) are evaluated as one of the most advanced energy storage devices because of their high energy density, which can meet rapidly growing energy requirements. Graphite based anode materials in LIBs are reaching their fundamental limits, especially their specific capacities. Recently, it has been demonstrated that high entropy oxides (HEOs) possess promising and unexpected electrochemical properties, such as remarkable reversible capacity and cycle stability due to the high entropy of the system. The highly disordered structure can provide self-healing properties resulting in regeneration of the capacity by applying low current densities. In addition, they can alleviate volume changes during the cycling process, unlike simple oxides. Among the various types of high entropy oxides, spinel-structured HEOs are the most studied because they ensure the three-dimensional transport of lithium ions ensuring high rate capability. Herein, we report a simple method of preparation of high entropy oxide (HEO) with a spinel structure consisting of 4 different elements (Co, Fe, Cr, Ni). The prepared HEO exhibited excellent cycle stability during (116 $mAh.g^{-1}$) 500 cycles at a current density of 500 $mA.g^{-1}$, which confirms their usage as anode active materials in lithium-ion batteries.

1. Introduction

Our planet faces several environmental problems such as global warming and climate change, which are related to emissions of CO_2. To suppress CO_2 emissions and obtain climate neutrality, utilization of green power sources, such as wind power, solar or geothermal energy, is necessary. Lithium-ion batteries play a key component when it comes to energy sustainability. They have huge potential for usage in off-grid power supply systems, e.g., renewable resources of energy. [1]

Anode materials are necessary in LIBs because they host the lithium ions in a fully charged state. Metallic lithium would be an ideal anode material due to its low electrochemical potential (-3.04 V vs. standard hydrogen electrode) and high theoretical capacity (3861 $mAh.g^{-1}$) [2] but suffers from dendrite formation during cycling, which can cause a short circuit and thermal runaway. The two most used commercial anode materials are graphite and LTO ($Li_4Ti_5O_{12}$) which both have relatively low theoretical capacities (372 $mAh.g^{-1}$ [2] and 175 $mAh.g^{-1}$ [2], respectively). The so-called conversion materials based on transitional metal oxides and alloying materials are the ones with higher theoretical capacities. These materials enable multi-electron redox processes while intercalation materials, such as graphite is capable of one electron exchange [2].

Like alloying materials, transitional metal oxides suffer from large volume variation and pulverization, resulting in subsequent severe capacity degradation. To alleviate the disadvantages of intercalation or conversion electrodes, high entropy oxides (HEO) can be used as anode materials that possess excellent electrochemical properties [3–8]. Due to the stabilization effect of the entropy, remarkable reversible Li-in storage capability (1235 $mAh.g^{-1}$ at a current density of 100 $mA.g^{-1}$ [3]) and great cycling stability (500 cycles without significant capacity degradation [7]) can be achieved. Moreover, high entropy oxides with spinel structure enable high rate capability (500 $mAh.g^{-1}$ at current density of 2000 $mA.g^{-1}$ [3]) because they ensure the three-dimensional transport of lithium ions [3,5,6]. In addition, the disordered structure of the HEO provides more space to buffer volume changes during charging and discharging, resulting in improved electrode stability as well as integrity [5,7]. Furthermore, HEO can be used in lithium-sulfur batteries which are anticipated as one of the most promising candidates for next-generation rechargeable batteries. HEOs can act as a polysulfide anchor mitigating the shuttle effect caused by the dissolution of lithium polysulfides in the electrolyte [8].

2. Experimental methods

2.1. Material synthesis

Pure elements (>99.9%) of Co, Fe, Cr, and Ni were used to prepare medium entropy alloy by arc melting under an argon atmosphere using a Mini Arc Melter (Edmund Buhler). After the melting process, the samples were milled by a vibration mill at a frequency of 50 Hz. The powder particles (<40 μm) of the alloy were heat treated in a tube furnace at 1000 °C for 6 h in an oxygen atmosphere. The sintered oxide was milled by a vibration mill at a frequency of 50 Hz. The obtained oxide particles were wet-milled in isopropyl alcohol using a ball mill at 500 rpm for 30 min.

2.1.1. Materials characterization

The crystalline structure of the alloy and the oxide was examined by X-ray diffraction (XRD) (Philips X´Pert Pro) with a Cu Kα radiation source. The diffraction angle was scanned from 15° to 140° (2theta) with a step size of 0.03° and time per step of 60 s. The morphology of the alloy and oxide particles was examined by scanning electron microscopy (Tescan VEGA 3). To analyse the chemical composition of the materials energy dispersive X-ray spectroscopy (Bruker Nano Xflash) was performed.

2.1.2. Electrode preparation and electrochemical measurements

The working electrode was prepared by casting a slurry, made out of 80 wt. % HEO powder, 10 wt. % carbon black Super P and 10 wt. % PVDF dissolved in N-methyl-2-pyrrolidone, onto Cu foil using a casting knife. The film was dried in a vacuum dryer at 120 °C overnight. Electrodes of a diameter of 12 mm were punched. The active material loading was about 2.9 $mg.cm^{-2}$. The electrochemical properties of the high entropy oxide were examined in electrochemical cells (CR2032, 2-electrode Swagelok cell, El-Cell). The half cells were assembled in a glovebox (Jacomex) with a controlled argon atmosphere (O_2, H_2O < 1 ppm). Metallic lithium, Whatman GF/A and 1 M $LiPF_6$ in a mixture of EC/DEC 50/50 were used as counter electrode, separator and electrolyte, respectively. The galvanostatic measurements were performed on a Land Battery Cycler. Cyclic voltammetry (potential window 0.01-3 V vs. Li/Li^+ and scan rate 0.1 mVs^{-1}) and electrochemical impedance spectroscopy (frequency range from 1 MHz to 100 mHz) measurements were carried out on a Biologic SP-150 potentiostat.

3. Results

3.1. Material characterization

The XRD patterns of the prepared alloy CoFeCrNi are shown in **Figure 1a**, which confirms the single-phase cubic structure of the alloy with an Fm-3m space group and lattice parameter of 3.598(9) Å. The powder particles of the alloy after the milling process had irregular morphology with sizes in the ranges of 1-100 μm (**Figure 1b**). Even though the elements were weighed equimolarly, the final chemical composition of the alloy is just quasi-equimolar (**Figure 1c**). The biggest deviation from the equimolar composition had iron and nickel. Despite that, the individual elements were homogeneously distributed.

The equation to calculate the configurational entropy of the prepared alloy is described in [9]. According to that, our four-component alloy belongs to the medium entropy alloys, because the configuration entropy is

$$\Delta S = -R\left[\sum_{i=1}^{n} x_i \ln x_i\right] = -R[0.2475 * \ln(0.2475) + 0.3019 * \ln(0.3019) + 0.2504 * \ln(0.2504) + 0.2003 * \ln(0.2003)] = \mathbf{1.37}\boldsymbol{R} \quad (1)$$

, where x_i is the atomic fraction of the i-th element and R is the universal gas constant (8.314 $J.K^{-1}.mol^{-1}$). The definition of high entropy materials says that the configuration entropy of the material needs to be higher than 1.61R.

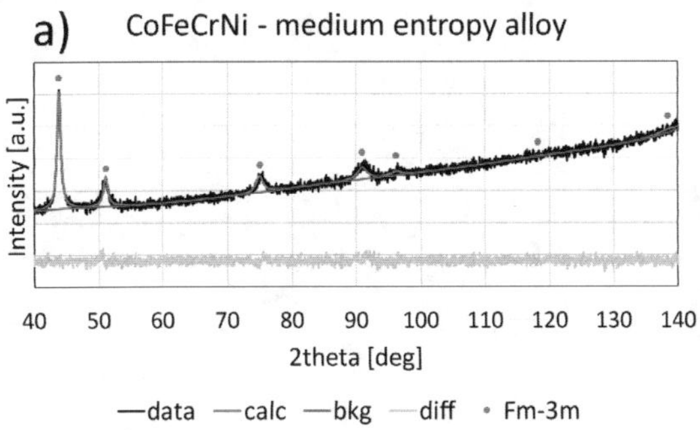

c)

Element	wt.%	σ wt.%	at.%	σ at.%
Chromium	22.88	0.07	24.75	0.08
Iron	29.98	0.17	30.19	0.17
Cobalt	26.24	0.26	25.04	0.25
Nickel	20.90	0.10	20.03	0.10

Figure 1 a) XRD, b) SEM/EDS mapping, c) EDS analysis of CoFeCrNi medium entropy oxide

After the oxidation of the powder particles of the alloy, a dual phase oxide system was formed according to the XRD analysis supplemented by Rietveld refinement (**Figure 2a**). The major phase has a cubic structure with an Fd-3m space group and lattice parameters of 8.319(1) Å belonging to the group of spinel oxides, while the minor phase has a cubic structure with an Fm-3m space group and lattice parameters of 4.199(7) Å belonging to the group of rock salt oxides. The weight fraction of the major phase is 0.95. Generally, dual phase oxides are not suitable for application in lithium-ion batteries, because the two phases usually have different volume expansion coefficients and during the oxidation/reduction processes electrode degradation could occur. After the final milling process the average particle size was around 2.5 μm (**Figure 2b**). The individual elements were distributed relatively homogeneously, the biggest deviation had a nickel. According to works [10,11], the configuration entropy of spinel oxides is higher than simple oxides (e.g., with rock salt structure), because of the fact that in a spinel oxide the cations are positioned on two Wickoff-sites. Apart from that we are not aware of the distribution of the cation types, the configuration entropy of our oxide $(CoFeCrNi)^{II}(CoFeCrNi)^{III}{}_2O_4$ in equimolar composition can be calculated as:

$$EM = S^{conf}.L = -R\left[\frac{\sum a^x \sum f_i^x \ln f_i^x}{\sum a^x}\right].L$$
$$= -R\left[\frac{1 * (4 * 0.25 * \ln(0.25) + 2 * (4 * 0.25 * \ln(0.25) + 4 * 1 * \ln(1)}{1 + 2 + 4}\right] * 3$$
$$= \mathbf{1.782R} \qquad (2)$$

, where a^x is the number of sites on the x sublattice, f_i^x is the fraction of element species randomly distributed on the x sublattice, L is the total number of sublattices. It means that the prepared oxide belongs to the high entropy group.

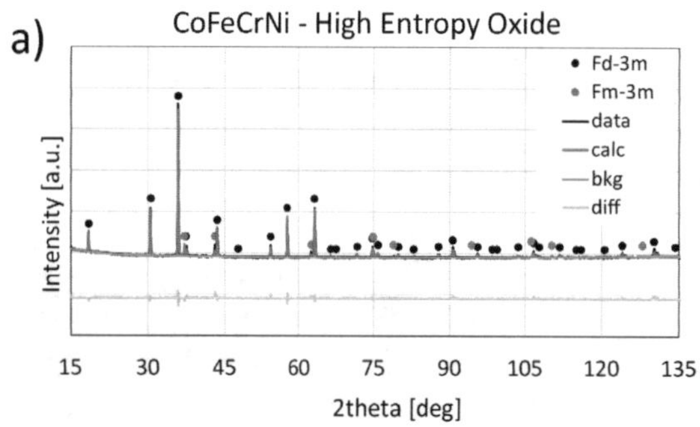

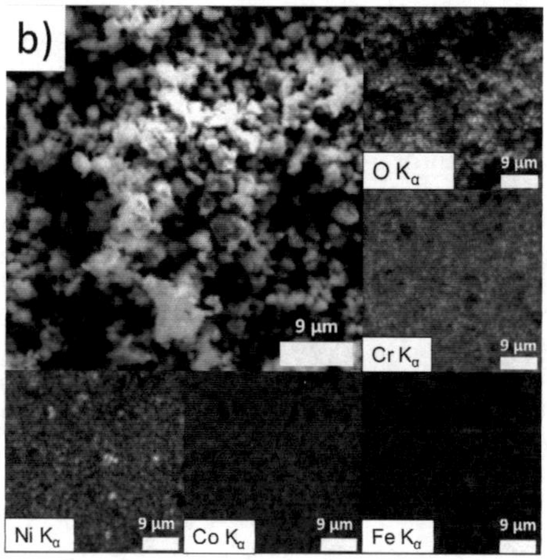

c)

Element	wt.%	σ wt.%	at.%	σ at.%
Chromium	15.13	0.24	9.82	0.26
Iron	22.17	0.27	13.40	0.31
Cobalt	19.23	0.23	11.01	0.26

Figure 2 a) XRD, b) SEM/EDS mapping, c) EDS analysis of CoFeCrNi high entropy oxide

3.2. Electrochemical properties

The cyclic voltammetry (CV) curves revealed reduction and oxidation peaks for the HEO (**Figure 3**). In the first reduction cycle the formation of the solid-electrolyte interphase takes place at 0.1 V vs. Li/Li^+ [3,5,6]. In the oxidation region, the anodic peak is positioned at 1.5 V vs. Li/Li^+ which can be identified as the oxidation of the metallic elements (Co, Fe, Cr, Ni) to the corresponding metal oxides [3,5,6]. The corresponding reduction peaks around 1.5 – 0.7 V vs. Li/Li^+ are represented in the subsequent cycles. The consecutive CV curves (**Figure 3b**) have similar shapes without a sharp reduction peak at 0.1 V vs. Li/Li^+ which indicate that the formation of the SEI layer is less intensive.

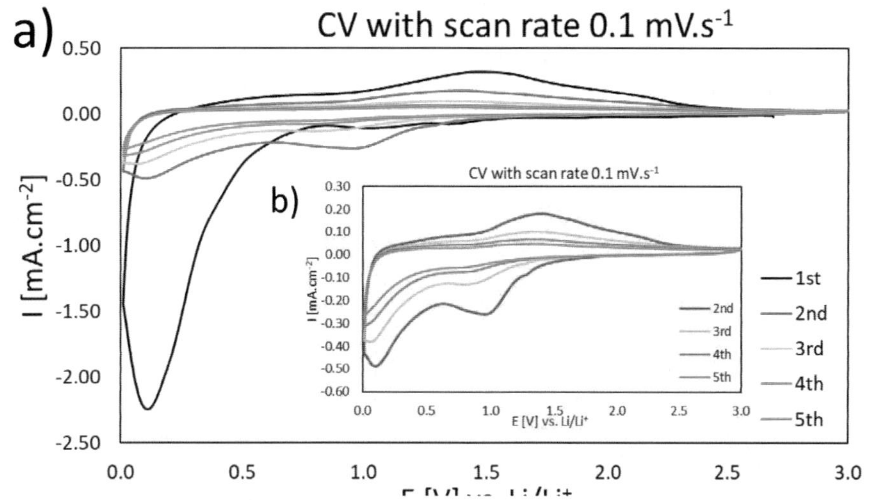

Figure 3 a) Cyclic voltammetry of CoFeCrNi high entropy oxide (from 1st to 5th cycle), b) CV from 2nd to 5th cycle

EIS measurements before cycling and after 5 formation cycles were carried out (**Figure 4**). The high- and middle-frequency regions of the complex-plane impedance plots show a semicircle, which include the ohmic resistance, SEI film resistance and the charge transfer resistances between the Li-electrolyte and the active material-electrode interfaces since the impedance were recorded in a two-electrode cell (**Figure 4b**). A straight sloping line in the low-frequency region can be observed because of the lithium-ion diffusion resistance. The charge transfer resistance before cycling is lower than that after cycling, indicating the increasing polarization resistance of the electrode during cycling. It can be caused by the active material dissolution to the electrolyte and the irreversible, undesired side-reactions (alloying with the Li metal). To confirm these assumptions in operando structural analysis would be performed in the future.

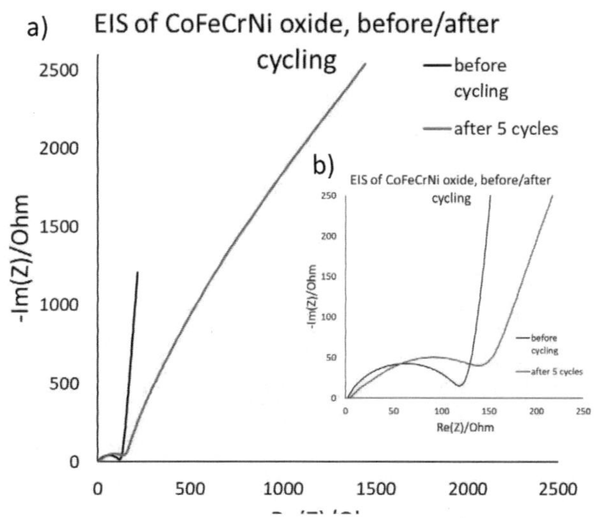

Figure 4 Electrochemical impedance spectroscopy of CoFeCrNi oxide before and after cycling, b) detail

In order to determine the cyclic stability of the prepared high entropy oxide, galvanostatic cycling with potential limitation (GCPL) was performed at a current density of 100 mA.g^{-1} for 300 cycles, as shown in **Figure 5a**. The initial specific discharge and charge capacities were 1011 and 682 mAh.g^{-1}, respectively, corresponding to low initial Coulombic efficiency of 67 %, but comparable with other high entropy spinel oxides (**Table 1**). During the first 40 cycles, the capacity was gradually decreased to 220 mAh.g^{-1}. After the 50th cycle the capacity started to improve and a 308 mAh.g^{-1} capacity was achieved, which was relatively stable during subsequent cycles with the CE being above 98.9 %. It indicates the self-healing properties of the high entropy oxide. Long-term galvanostatic cycling was performed at a current density of 500 mA.g^{-1} (**Figure 5b**). Similar capacity changes occurred during cycling, e.g., capacity dropped to 116 mAh.g^{-1} during the first 50 cycles, which is a much lower value than in other types of high entropy oxides (**Table 1**). After that, excellent cycling stability was achieved during 500 cycles. In the regeneration process (at a current density of 100 mA.g^{-1}) the capacity was increased to 334 mAh.g^{-1}. For rate performance, the high entropy oxide delivered discharge capacities of 380, 251, 149, 73 and 29 mAh.g^{-1} at 100, 200, 500, 1000 and 2000 mA.g^{-1}, respectively (**Figure 5c** and **Table 1**). As current densities were increased, the capacities were decreased due to polarization. These values are a little bit lower than in the cases of other HEOs (**Table 1**) because of the dual phase structure and bigger particle sizes of the CoFeCrNi oxide. However, in the regeneration process at a current density of 100 mA.g^{-1} the capacity was improved in subsequent cycles (up to 280 mAh.g^{-1}) indicating excellent self-healing properties of the electrode material similarly as in other HEOs.

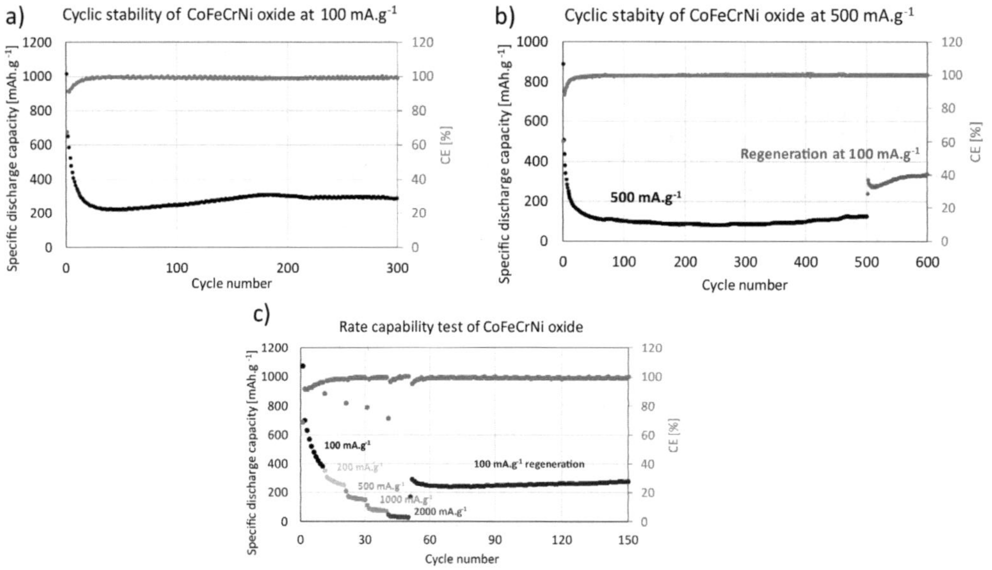

Figure 5 a) Cyclic stability of CoFeCrNi oxide at 100 mA.g^{-1}, b) Cyclic stability of CoFeCrNi oxide at 500 mA.g^{-1}, c) Rate capability test of CoFeCrNi oxide

Table 1. Comparison of electrochemical performances of different high entropy spinel oxides with current work

Composition	1st sp. disch. cap. [mAh.g^{-1}]	1st sp. ch. cap. [mAh.g^{-1}]	CE [%]	Long term cycling			Rate capability test - [mAh.g^{-1}]					Ref.
				Specific capacity [mAh.g^{-1}]	Current density [mA.g^{-1}]	Cycles	100 mA.g^{-1}	500 mA.g^{-1}	1000 mA.g^{-1}	2000 mA.g^{-1}	Reg. at 100 mA.g^{-1}	
$(CrMnFeCoNi)_3O_4$	1869	1245	67	752	500	200	1072	824	649	500	-	[3]
$(FeCoNiCrMn)_3O_4$	1034	680	66	402	500	300	586	361	269	180	483	[5]
$(FeCoNiCrMnZnLi)_3O_4$	1049	706	67	370	500	100	429	220	145	83	389	[6]
CoFeCrNi-oxide	1011	682	67	125	500	500	380	149	73	29	280	current work

4. Conclusion

In this work, a four-component high entropy spinel oxide was prepared by a relatively simple two-step method using arc melting of pure elements and oxidizing it in an oxygen atmosphere. The formed oxide has a dual phase structure consisting of spinel and rock salt phases. The weight fraction of the secondary phase was around 0.05. The EDS mapping revealed that all the elements were homogeneously distributed in the material, and no segregation was observed. CV showed the reduction and oxidation regions of the material, while EIS performed before and after the cycling process unveiled that the charge transfer resistance of the electrode was higher after cycling. The initial discharge capacity was 1011 mAh.g^{-1} with a Coulombic efficiency of 67 % in the first cycle. Even though, the capacity was decreasing in the first 40 cycles, after that, excellent cycling stability was observed even at higher current density. These results suggest that high entropy oxides have greatly improved cyclic stability due to the entropy stabilization effect. However, an optimization of the prepared method of the materials and electrode is required.

Acknowledgments

This work was realized within the frame of the project "Research centre for progressive materials and technologies for present and future applications "PROMATECH", ITMS 26220220186"; by the projects of Slovak Research and Development Agency APVV-20-0138, APVV-20-0205, APVV-21-0274 and of Slovak Grant Agency VEGA 2/0039/22.

References

[1] Eftekhari A 2019 *Future Lithium-ion Batteries* (London: Royal Society of Chemistry)
[2] Wu F, Maier J and Yu Y 2020 *Chemical Society Reviews* **49** 1569–614
[3] Nguyen T X, Patra J, Chang J K and Ting J M 2020 *J Mater Chem A Mater* **8** 18963–73
[4] Sarkar A et al. 2018 *Nat Commun* **9** 3400
[5] Wang D, Jiang S, Duan C, Mao J, Dong Y, Dong K, Wang Z, Luo S, Liu Y and Qi X 2020 *J Alloys Compd* **844** 156158
[6] Duan CQ, Tian K, Li X, Wang D, Sun H, Zheng R, Wang Z and Liu Y 2021 *Ceram Int* **47** 32025–32
[7] Qiu N, Chen H, Yang Z, Sun S, Wang Y and Cui Y 2019 *J Alloys Compd 777* 767–74.
[8] Zheng Y, Yi Y, Fan M, Liu H, Li X, Zhang R, Li M and Qiao Z 2019 *Energy Storage Mater* **23** 678–83
[9] Zhang Y 2019 *High-Entropy Materials (*Singapore: Springer Singapore)
[10] Miracle D B and Senkov O N 2017 *Acta Materialia* **122** 448–511
[11] Dippo O F and Vecchio K S 2021 *Scr Mater* **201** 113974

A comprehensive heat generation study of lithium titanate oxide-based lithium-ion batteries

Seyed Saeed Madani[1,2 *], Erik Schaltz[2], Søren Knudsen Kær[2] and Carlos Ziebert[1]

[1] Institute of Applied Materials-Applied Materials Physics, Karlsruhe Institute of Technology, Hermann-von-Helmholtz-Platz 1, 76344 Eggenstein-Leopoldshafen, Germany

[2] Department of Energy Technology, Aalborg University, DK-9220 Aalborg, Denmark

[*] seyed.madani@kit.edu

Abstract. A precise interpretation of lithium-ion battery (LIB) heat generation is indispensable to the advancement and accomplishment of thermal management systems for different applications of LIB, including electric vehicles. The internal resistance of a lithium titanate oxide (LTO)-based LIB was determined at different state of charge (SOC) levels and current rates to understand the relationship between internal resistance and heat generation. Random and different pulse discharge current step durations were applied to consider the effect of different SOC interval levels on heat generation. The total generated heat was measured for different discharge rates and operating temperatures in a Netzsch IBC 284 calorimeter. It was seen that a 6.7% SOC decrease at high SOC levels corresponds to 0.377 Wh, 0.728 Wh, and 1.002 Wh heat generation for 26A, 52A, and 78A step discharge, both at 20 °C and 30 °C. However, a 1.85% SOC decrease at medium SOC levels corresponds already to 0.57 Wh, 0.76 Wh, and 0.62 Wh heat generation. It can be inferred that the impact of SOC level on heat generation for this cell is more prominent at a lower than at a higher SOC.

1. Introduction

Quantifying localized heat loss within a lithium-ion cell is a challenging research area due to unpredictability regarding cells' inner architecture, active materials, and thermal properties [1]. Some researchers used a calorimeter. Previous investigations employed tiny cells and samples to accomplish tests through different calorimeters [2-4]. Only a few researchers investigated commercial pouch cells [5]. Experimental and simulation investigations have been accomplished to investigate the impact of heat generation of LIB on temperature distribution. Outcomes demonstrate that the battery's local temperature is the largest at the positive and negative terminals at the discharge end, attributable to the more significant current density at these positions [6-8]. A thermal model was investigated by incorporating a different approach. The effect of coolant flow configurations on temperature distributions was studied to consider other aspects of the model. Besides, the thermal performance was evaluated for various thermal management strategies [9].

In the previous investigations of the authors [10-14], various experimental research has been accomplished on the thermal performances of LIB, and their heat generation was studied. In addition, the effect of the current rate, state of charge (SOC), and temperature on heat generation at different current rates and operational temperatures was determined.

In [10,11], the lithium-ion battery was subjected to different charge and discharge experiments under working temperatures of 30 °C, 40 °C, and 50 °C employing the experimental methodology explained in [10-14]. Sixteen charge and discharge cycles ranging from 13 A to 104 A were applied to the lithium titanate oxide (LTO) battery. Numerous experiments were accomplished at different working temperatures, including 30 °C, 40 °C, and 50 °C. A trend almost identical to the one seen at working temperatures of 30 °C was seen for the heat generation in working temperatures of 40 °C and 50 °C. The heat losses during the 8 C discharge cycle for working temperatures 50 °C, 30 °C, and 40 °C were 7.3, 8, and 8 times bigger than the 1C cycle. The outcomes demonstrate that the heat loss for discharge condition began with 6952 J, 5511 J, and 3550 J at 50 °C, 40 °C, and 30 °C during the 1C discharge cycle and reached 50852 J, 43302 J, and 28138 J during 8C discharge cycle. When a specific loading pattern was applied, the sum of total heat generations during 16 charging and discharging cycles was different under dissimilar working conditions. As a result, at a working temperature of 40 °C, the cell undergoes a 60% increase in the total heat generation compared to a working temperature of 30 °C measured under identical conditions. The cell at a working temperature of 50 °C produced 32% and 112% more heat than working temperatures of 40 °C and 30 °C, respectively. Nevertheless, the percentage growth in total heat generation due to a change in working temperature from 30 °C to 40 °C was more significant than the ratios achieved due to a change in working temperature from 40 °C to 50 °C. Accordingly, working temperatures of 30 °C were considered more efficient and better suited than those of 40 °C and 50 °C [10-14].

The efficiency throughout charging began with 93.86% for 1C and decreased to 49.19% for 8.5C. The efficiency throughout discharging began with 93.22 % for 1C and declined to 68.67% for 8C. Heat loss throughout discharging cycles showed nearly an identical behaviour as for the charging cycles, which was a growing pattern from 1C to 8C. The heat generation began with 8424 J for 1C throughout the discharge cycle and attained 40,100 J for the 8C cycle. The most effective rising rate was from the 7C cycle to the 7.5C cycle (a growth of 11.1%). The smallest increase was from 7.5C cycle to 8C cycle, which was approximately 100 J. A five-fold increase was seen in heat generation from 1C cycle to 8C cycle during discharge. The alteration between discharge and charge heat losses was approximately identical for all current rates at 50 °C and 40 °C. On the contrary, this alteration experienced various oscillations at 30 °C with the most considerable quantity in the 8C cycle [10-14].

The charge efficiency for 40 °C declined by approximately 27%, even though this percentage during discharge was approximately 35%. The heat losses during the 8C charging cycle for 50 °C, 40 °C, and 30 °C were 8.4, 12.5, and 8.5 times bigger than for the 1C cycle. The total heat generation of the cells during the 91 A discharge cycle was approximately 1100 J more than the total heat generation during the 91 A charge cycle in working temperatures of 30 °C. It was observed that notwithstanding the heat generation throughout 26 A charging cycle and discharging cycle are almost identical in working temperatures of 30 °C, approximately an additional 3500 J has to be transferred from the cell at the end of the discharge cycle, assuming its temperature should be maintained at working temperatures of 40 °C. A considerable rise could be seen in the cell's heat generation as the working temperature increased from 30 °C to 50 °C. Those mentioned above could demonstrate a considerable decrease in cell performance and unsatisfactory electrochemical efficiency at significant temperatures. In the current rates range of 13 A to 104 A, the average heat flux quantified for the working temperature of 50 °C was more significant than for the working temperatures of 30 °C and 40 °C. Heat loss following discharge and charge cycles contributes considerably to the cells' entire heat generation. The average percentages of contribution for discharge and charge cycles ranging from 13 A to 104 A were 79.3%,77.6%, and 79.3% of the entire heat generation at 30 °C, 40 °C, and 50 °C. Accordingly, the heat of mixing cannot be disregarded. It was concluded that the ratio of heat loss to electrical energy lost attributable to alteration was not on the same order of magnitude for all charge and discharge cycles [10-14].

In another work [12], numerous experiments were accomplished at a working temperature of 20 °C. Thirty-one charge and discharge cycles ranging from 13 A to 110 A were applied to the same LTO-based LIB. The thermal efficiency throughout charging was smaller than that for discharging, except for 13 A, 19.5 A, and 26 A, which was more significant. The difference between charge and discharge efficiencies raised at bigger current rates. This discrepancy was most significant at 97.5 A. It was seen that the heat loss during charging was lower than that during discharging, excluding 104 A and 97.5 A, where it was bigger. The difference between charge and discharge heat losses stayed nearly identical for all current rates. The discrepancy was most significant at 19.5 A. The heat loss for discharging and charging cycles was determined to be 34,100 J and 35,793 J for the 8C cycle and 8.5C cycle, respectively. The average quantity of heat flux at 20 °C and 110.5 A was 9.5 W. This value is 19 times bigger than at 3 A, which is 0.5 W. The efficiency throughout charging began with 97.69%, 96.92%, and 95.24% for 1C and attained 82.8%, 70.36%, and 62.1% for 8C. The heat loss throughout charging began with 2697J, 3264J, and 5744J for 1C and attained 22926J, 40839J, and 48185J for 8C [10-14].

In [13], a thermal model was derived from the computation of the heat loss from the entropic heat coefficient and the internal resistance measurements. Calorimetry data were compared with estimates from the thermal model. The experiment agreed well with the modelling results, which were gained at different current rates. The modelling was validated by comparing the exploratory consequences with the modelling discharge figures at various discharge rates and environmental temperatures. A medium heat capacity was assumed for the cell in the time-conditional and unsteady condition simulation. Experimental investigations were restricted to surface temperatures integrated with heat dissipation's thermal or analytical modelling to achieve cell temperatures and determine the total heat loss of the cell [12]. The surface fitting and voltage alterations for 50%, 60%, 70%, and 80% SOC were determined throughout the thermal process. The slope of these surfaces corresponded to the average cell voltage, which was determined as a function of the SOC. Besides, the cells' entropic heat coefficient was calculated as a function of the SOC. The investigated cell's internal resistance was determined using the hybrid pulse procedure. Heat generation and efficiency of the cell as a function of the SOC were determined. It was found that the heat generation showed an almost diminishing pattern from 10 % to 90% SOC for all current rates. The cells' potential and reversible entropic heat showed a declining pattern towards a significant SOC [10-14].

Heat generation at different current rates of charge and discharge cycles was separated into four sections corresponding to the operational temperature of the LIB. It was concluded that between 20 °C and 50 °C, heat generation for 13 A and 104 A charge was in the range of 3000-7000 J and 23000-48000 J, respectively. The difference between charge and discharge heat generations was calculated. It was shown that the variation tendency of heat generation differences at various temperatures was different. This variation reaches its minimum value at higher current rates, such as 78 A and 91 A, for 20 °C, but it reaches its maximum value at 40 °C and 78 A. It can be concluded that heat generation differences at various current rates and temperatures do not follow a regular pattern. Thermal characterization of the cell revealed that the surface temperature distributions of the cell raised with the charge and discharge rate (1C, 2C, 3C, to 10C) and operational temperature (20°C, 30°C, 40°C, 50°C). It was seen that 50°C and 30°C cases created the highest and lowest heat generation [10-14].

Notwithstanding, the heat generation decreased from 20 °C to 30 °C. The heat generation dropped from 20 °C to 40 °C except for high current rates, including 104 A, 91 A, and 78 A. This phenomenon was not observed in the literature for the investigated cell type. It may be attributable to internal cell resistance, polarization losses, phase change mechanisms, and overpotential [10-14].

As was described comprehensively in their previous investigations [10-14], the authors did not perform thermal characterizations of the LTO-based cells that focused on periodic charge and discharge pulses. In addition, to the author's knowledge, all other investigations in the literature [15-32] were limited to the heat generation measurement of lithium-ion batteries during constant current charge and discharged complete cycles. Accordingly, the present work aims to measure the heat generation rates of LTO-based lithium-ion cells for different intervals of SOC levels.

The heat generation of LIB is a challenging task because of the complex electrochemical reactions occurring and their dependence on different parameters. The overpotential represents system irreversibility and is considered a discrepancy between the cells' terminal voltage and its theoretical open circuit potential. Heat is generated that can be attributed to transport resistances and electrochemical reactions inside a LIB [22,23].

Different methods were used in the literature for thermal behaviour and heat generation determination of LIB [24-34]. Most of them were restricted to small coin types and small cylindrical batteries. These batteries are not appropriate for large-scale applications, including electric vehicles. Besides, almost all previous research employed low discharge rates, not representing the electric cars' electrical demands. The aforementioned shows the necessity of new research approaches for superior comprehension of LTO-based cells' heat generation mechanisms. LTO-based cells are characterized by significant power capability throughout discharging and charging, high safety, good thermal stability, and a long lifetime, which is generally attributable to the small operating voltage. LTO-based LIB has become a valuable new energy source for certain electric vehicles, accompanied by grid power storage systems and portable electronic devices. LTO is presently one of the leading nominees as anode material for the lithium-ion battery as a consequence of its lifetime and safety properties [35].

There are lithium titanate nanocrystals on the anode surface of LTO-based lithium-ion batteries. The aforementioned makes them charge faster than other LIBs. They have several features, including lifetime and safety properties, making them an appreciable and promising technology for different applications such as e-mobility and grid storage. LTO-based lithium-ion batteries with different cathode materials were assessed and designed for realistic applications [35,36].

Consequently, in the accessible literature, there was a lack of articles that systematically analyzed the LTO-based LIB heat generation focusing on random and periodic charge/discharge current pulses. The aforementioned is closer to the actual application of lithium-ion batteries, especially in electric vehicles. Preceding investigations have mainly concentrated on studying LIB packs' thermal performance. Notwithstanding, an isothermal battery calorimeter was rarely employed as a proper and precise instrument for quantifying heat generation inside LIB.

Many studies have been accomplished on the thermal analysis of lithium-ion cells, but only a few employed an experimental calorimetry procedure to determine LIB heat generation. In most of the previous investigations, the amount of heat generation was only calculated. In the present work, a Netzsch IBC 284 isothermal battery calorimeter was employed to precisely determine heat generation. In addition, internal resistance and open-circuit voltage were determined to assist in understanding the thermal behaviour and heat generation of LTO-based cells. The results' originality and novelty are that an isothermal battery calorimeter was used to study the LTO-based cells' thermal behaviour with a particular focus on periodic charge and discharge pulses.

Because LTO-based cells are still quite a new chemistry, their thermal characteristics and behaviour are not entirely understood. The current work might assist in establishing a thermal model of the cell and employing it unaccompanied by extensive comprehension of electrochemistry. Furthermore, the priority was on employing straightforward and uncomplicated methodologies for parameter determination.

2. Experimental

2.1. Cell description

The lithium-ion battery cells employed in this study are 13 Ah high-power lithium titanate oxide-based lithium-ion battery cells with $LiNi_xMn_yCo_zO_2$ (NMC) on the cathode side and $Li_4Ti_5O_{12}$ (LTO) on the anode side. The minimum and maximum voltage are 1.5V and 2.9V, respectively. The nominal voltage is 2.26 V. Characteristics of the LTO Li-ion battery based is demonstrated in Table 1.

Table 1. Characteristics of LTO Li-ion battery based.

Property	Value
Minimum voltage	1.5 V
Maximum charging voltage	2.9 V
Nominal voltage	2.26 V
Nominal capacity	13 Ah
Maximum discharge current	130 A
Maximum charge current	130 A
Anode	Li4Ti5O12
Cathode	Li(NixMnyCoz)O2
Calendar life	25 years

2.2. Setup of the experiments

This research employed the isothermal IBC284 calorimeter (Netzsch) to characterize the heat loss related to constant current discharge (CCD) and constant current charge (CCC) procedures. Four different current profiles were applied with random and periodic current pulses. Therefore, ten different step time durations were applied for SOC determination. The cell under investigation was inserted into the calorimeter chamber, their terminals were connected to a Maccor battery cycler, and four thermocouples T1-T4 were attached to the cell, as illustrated in Figure 1.

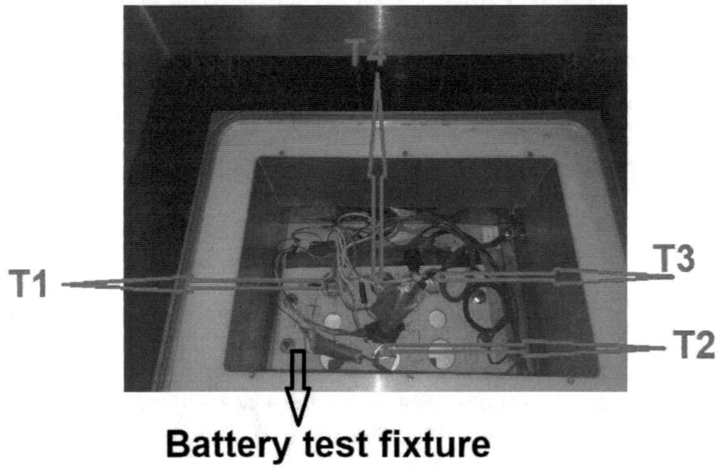

Figure 1. Setup of the battery cell and thermocouples inside the calorimeter chamber.

Figure 2 shows the experimental system setup for determining heat generation and internal resistance. Figure 2(a) depicts the calorimeter system setup, while Figure 2(b) explains the connection to the Maccor cycler.

The maximum discharge and charge current was 130 A. Attributable to safety concerns, accompanied by the simple connection of the voltage cables and power, the cell was examined in the laboratory by installing it on a fixture, which prevents volume alteration. The volume expansion is because of working with high temperatures and high current rates.

This research employed calorimetry to characterize the heat loss related to different discharge and charge profiles. The cell thermal analysis was accomplished, employing Proteus Analysis Software and MATLAB. Consequently, calibration was done before each experiment to achieve higher accuracy. Besides, the thermocouples heat flux sensors were used in this research.

The isothermal battery calorimeter is a robust instrument fabricated for correctly determining heat flux produced by lithium-ion batteries while being charged or discharged under the experiential situation. The isothermal battery calorimeter unit contains one empty cabinet (650 lbs./300kg). The thermal fluid appends 250 lbs./114 kg of extra weight, and the computer links to the instrument cabinet through two USB 2.0 cables. Multiple procedures of commercially accessible temperature control were used for the experiments. Dissimilar and independent loading were exerted on the cell. Besides, comprehensive characterization tests were done for these cells.

To estimate the heat generation of the LTO-based cell with an isothermal battery calorimeter, the cell and the calorimeter chamber need to reach a specified steady-state temperature before and after the cell cycle. The calorimeter and the cell with the expected state of charge are taken to the specified temperature while the cell is in the isothermal battery calorimeter. It takes about six hours for the cell and the isothermal battery calorimeter chamber to reach thermal equilibrium. The aforementioned is mainly attributable to the considerable thermal mass of the isothermal battery calorimeter. Afterwards, a specified load is applied to the cell with the Maccor battery cycler. Afterwards, the isothermal battery calorimeter needs another six hours for the temperature to equilibrate.

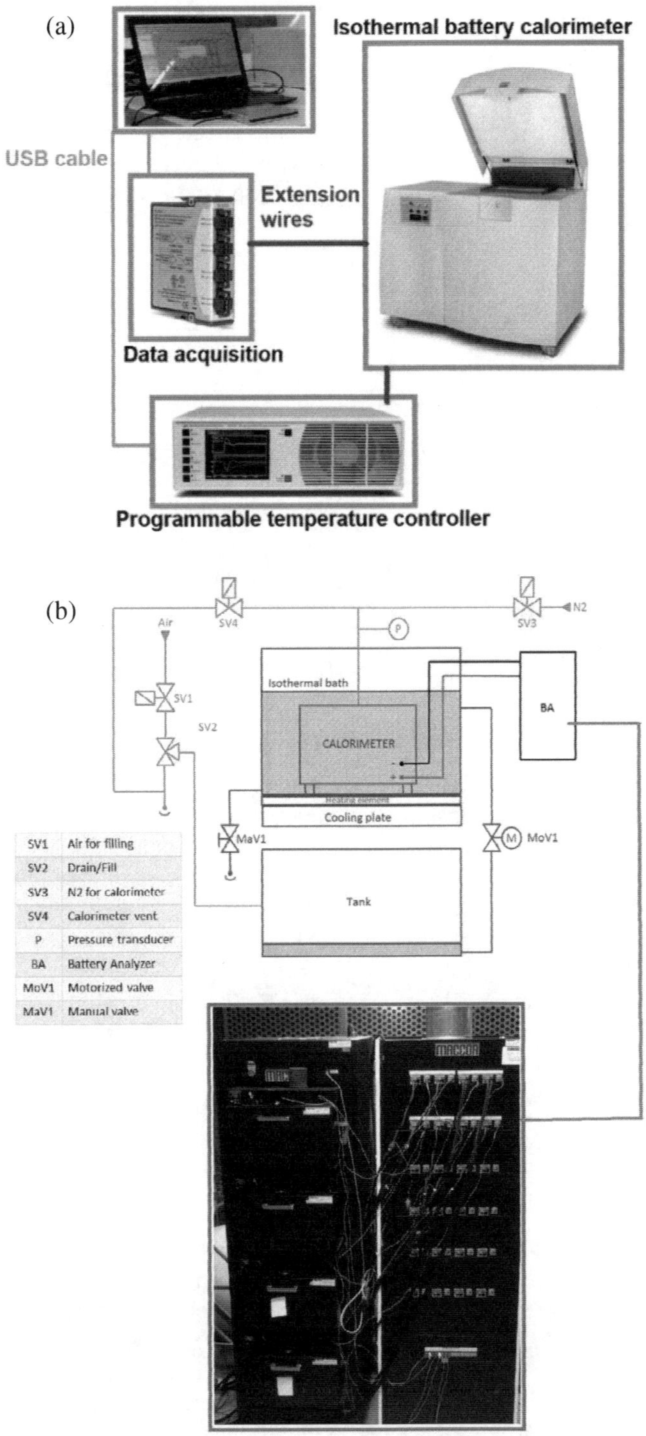

Figure 2. Experimental facilities for determination of LTO-based cell heat generation and internal resistance (a) Schematics of the experimental setup, (b) Connecting the battery cycler.

2.3. Procedures for the experiments
In the present investigation, the cell was discharged to the selected state of charge level with different current rates. Once the selected state of charge level was obtained, the cell was left to stabilize. The experiments demonstrated that this stabilization time was approximately the same for different state of charge levels. Table 2 describes the procedures for the experiments.

Table 2. The procedures for the experiments. (HG: heat generation).

Step 26 A discharge 20 °C	Step 52 A discharge 20 °C	Step 78 A discharge 20 °C	Cut off voltage	State of charge	HG: Step 26 A discharge 20 °C	HG: Step 52 A discharge 20 °C	HG: Step 78 A discharge 20 °C	HG: Step 26 A discharge 30 °C
3.25 A-CCC	3.25 A-CCC	3.25 A-CCC	2.132	76.14				0
3.25 A-CCD	3.25 A-CCD	3.25 A-CCD	2.8	100				
3.25 A-CCC	3.25 A-CCC	3.25 A-CCC	1.5	0	1.863	1.925	1.96	0.7461
3.25 A-CCD	3.25 A-CCD	3.25 A-CCD	2.8	100				
3.25 A-CCD	3.25 A-CCD	3.25 A-CCD	2.474	74.92				
26 A-PD	52 A-PD	78 A-PD	2.387	68.23	0.377	0.728	1.002	0.2249
26 A-PD	52 A-PD	78 A-PD	2.315	62.69	0.390	0.730	1.012	0.2210
26 A-PD	52 A-PD	78 A-PD	2.262	58.61	0.386	0.708	1.017	0.2290
26 A-PD	52 A-PD	78 A-PD	2.224	55.69	0.401	0.726	1.005	0.2310
26 A-PD	52 A-PD	78 A-PD	2.198	53.69	0.400	0.776	1.064	0.2290
26 A-PD	52 A-PD	78 A-PD	2.176	52.00	0.413	0.749	1.077	0.2280
26 A-PD	52 A-PD	78 A-PD	2.153	50.23	0.429	0.785	1.090	0.2390
26 A-PD	52 A-PD	78 A-PD	2.123	47.92	0.470	0.851	1.193	0.2510
26 A-PD	52 A-PD	78 A-PD	2.099	46.07	0.570	0.760	0.620	0.2249

The initial state of charge of the cell was 76.14 % for all experiments at 20 °C. Afterwards, the cell was applied to constant current discharge (CCD) and constant current charge (CCC) procedures. The battery cell was then charged until it reached the upper voltage limit, corresponding to a 100 % SOC. Discharge from 74.92 % to 46.07 % SOC demonstrated a quasi-linear voltage decrease concerning SOC.

With the intention of understanding the internal resistance impact on heat generation, as one of the principal sources of irreversible heat generation, the internal resistance was determined at different SOC levels using hybrid pulse cycle experiments. The cells' internal resistance changes under a dissimilar SOC, causing the discrepancy in heat generation under the identical working current. The complete procedure for the determination of the internal resistance is described in Figure 3. Various loading patterns were applied to the cell to get the internal resistance.

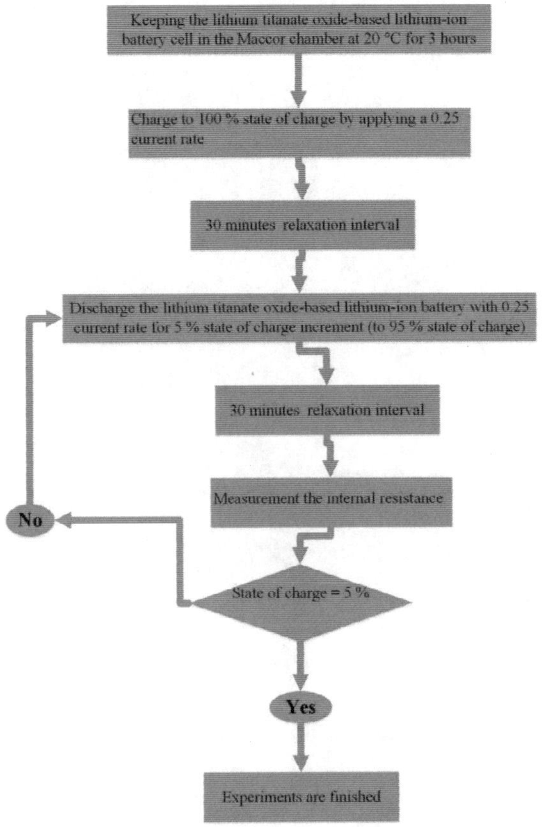

Figure 3. The procedure for measuring the internal resistance.

Heat loss could be estimated by the sum of the quantitative determination of the irreversible and reversible heat sources for LIB. Reversible heat generation of LIB usually is a linear function of current rates, notwithstanding irreversible heat generation of the LIB is a function of current squared [1,43]. Lithium-ion battery internal resistance is one of the principal sources of irreversible heat generation. Besides, it is dependent on the temperature. Open-circuit voltage (OCV) is one of the primary sources of reversible heat generation. Therefore, the determination of internal resistance and open-circuit voltage assists in understanding the thermal behaviour and heat generation of LIB. Figure 4 describes the procedure for measuring the open-circuit voltage.

Results and discussion

Discharge from 74.92% to 46.07% SOC demonstrated a quasi-linear voltage decrease concerning SOC. Figure 5 demonstrates the dependence of the cell's internal resistance on the SOC during discharging. The internal resistance of the LTO-based cell was almost constant, with around 3 ohms at a SOC >40% and increased up to a factor of three at lower SOC levels. It should be noted that 1C corresponds to 13Ah.

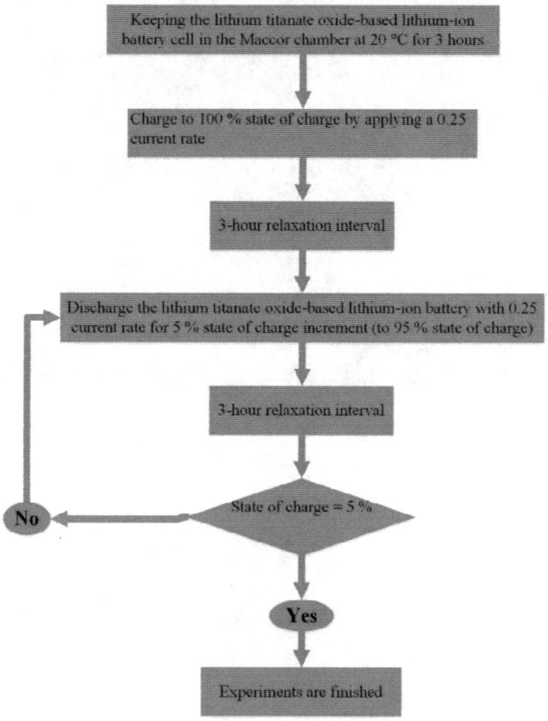

Figure 4. The procedure for measuring the open-circuit voltage.

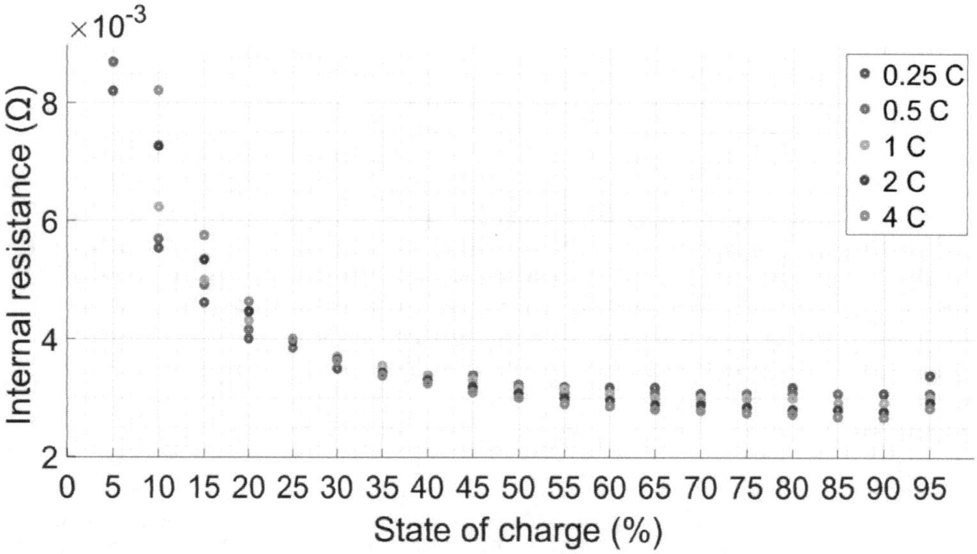

Figure 5. Dependence of the internal resistance (Ω) on the state of charge during different discharging pulses.

Consistent with most of the LTO-based cell parameters, the OCV depends entirely on the working states, including the SOC. The primary purpose of performing OCV measurements was to obtain the correlation between SOC and OCV for discharging conditions and a wide range of SOC levels. Figure 6 shows the dependence of the OCV on the SOC during different discharging pulses.

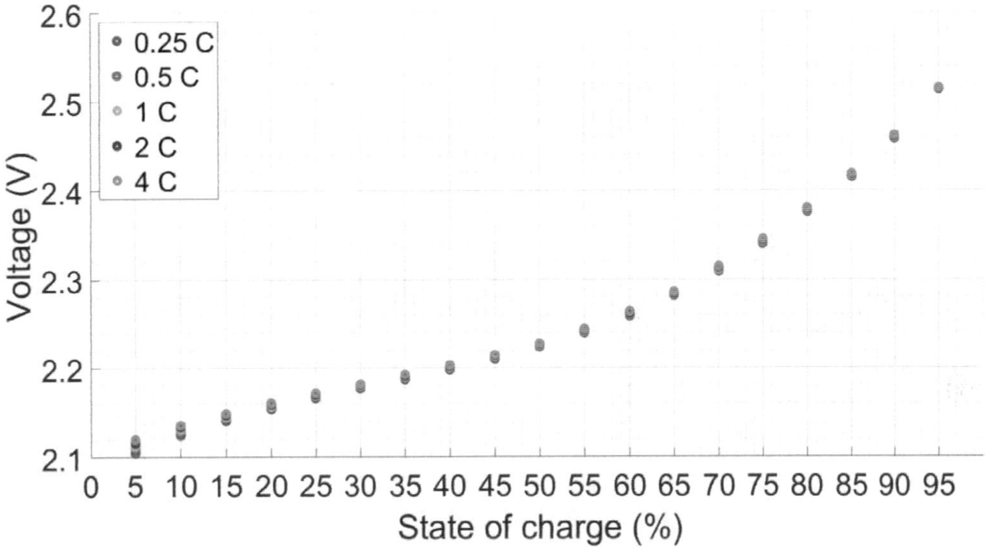

Figure 6. Dependence of the open-circuit voltage (V) on the state of charge during different discharging pulses.

The heat generation curves of the cell that have been determined using the isothermal calorimeter are illustrated in Figure 7(a) and the derived total generated heat is shown in Figure 7(b). Four different current profiles were applied with random and periodic current pulses. Therefore, ten different step time durations were considered for the SOC. Different SOC levels were selected so that cell demonstrated a quasi-linear voltage decrease as a function of SOC. The time-related amalgamation of the heat generation of the cell at different discharge steps, as measured using the isothermal battery calorimeter, corresponds to the total generated heat within a period interval. An overall increasing pattern was seen for the heat generation during discharging, and an overall decreasing pattern was seen for the heat generation during the rest period for each discharge pulse. When the cell was discharged from 74.92% to 68.23% SOC at 20 °C, the heat generation was 0.377 Wh, 0.728 Wh, and 1.002 Wh for 26 A, 52 A, and 78 A step discharge rates, respectively. However, a 1.85% SOC decrease at medium SOC levels corresponds already to 0.57 Wh, 0.76 Wh, and 0.62 Wh heat generation. It can be observed from Fig. 7(b) that the total generated heat curves for the 26 A, 52 A, and 78 A overlap each other until SOC 74.92%,. As the stored energy is discharged more, the three curves start to deviate from each other. In the SOC range from 74.92% to 46.07%, the average heat generation rate during 78 A step discharge cycles was more significant by 0.25 Wh and 0.85 Wh than that observed for the 26 A and 52 A step discharge, respectively. A decrease of 2 % in SOC level causes an increase of 0.4 Wh, 0.776 Wh, 1.064 Wh, and 0.229 Wh in heat generation for Step 26 A, 52 A, 78 A discharge at 20 °C and step 26 A discharge at 30 °C, correspondingly.

It can be concluded from these experimental data that the SOC level change at different SOC points has a different influence on the heat generation rate. The rise in irreversible heat generation at more considerable current rates might be attributable to elevated overpotential. Overpotential resistance decreases with an increasing state of charge level towards the end of discharge.

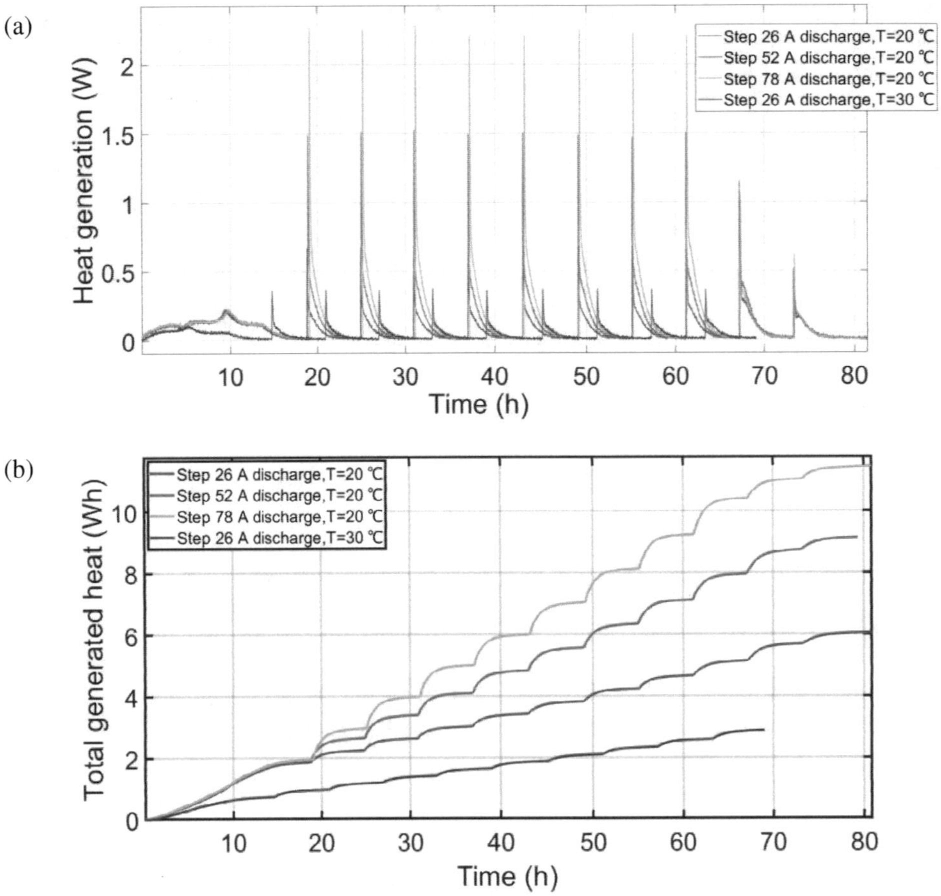

Figure 7. Isothermal calorimeter measurements on the LTO-based cell: (a) Heat generation curves for different current rates (b) Evolution of total generated heat as derived from (a).

The quantity of total generated heat decreases significantly at 30 °C in comparison with the working temperature of 20 °C. It can be inferred that the operating temperature of the LTO-based cell is the most crucial factor affecting heat generation. Notwithstanding, SOC level change could be significant in some SOC points; nonetheless, the impact of irreversible heat and internal resistance is not insignificant.

Open-circuit voltage demonstrated an increasing pattern towards a higher charge state; conversely, internal resistance showed approximately an overall decreasing pattern towards a higher state of charge. Considering the changing pattern of heat generation at different state of charge levels and total generated heat, it can be concluded that internal resistance has a greater influence on heat generation of the LTO-based cell in comparison with open-circuit voltage.

The calorimeter experiments demonstrated that the heat generation rate changed non-monotonically at different SOC levels and increased significantly towards the beginning of discharge. The aforementioned may be explained by the different contributions of the heat of mixing, heat of phase change, and irreversible and reversible heat at different state of charge levels. An overall increasing pattern was observed for the total generated heat of the LTO-based cell throughout the discharge process. Notwithstanding, in the literature, various lithium-ion batteries showed different behaviour in their heat generation. In ref. [28] heat generation of prismatic LIB was measured, and a decrease in heat generation rate at a depth of discharge of almost 30 % was observed. In addition, an observable secondary plateau in LIB heat generation was recognized for elevated battery working temperatures at a depth of discharge of around 80 %. In ref. [5] heat generation experiments were accomplished in a high-power prismatic cell. The results indicate that entropic heat is endothermic throughout discharge between 60–40% state of charge and exothermic until 80% state of charge. As mentioned above, the change becomes exothermic at an equal and less than 20% state of charge.

Figure 8 shows the surface temperature evolution of the cell. It can be seen that for all the selected locations on the cell's surface, the temperature increases during discharging conditions, which shows that the electrochemical reaction is exothermic throughout the discharge pulses. However, the temperature decreases throughout the rest period. The maximum increase in temperature of the cell for different discharge pulses is proportional to the amount of heat generation.

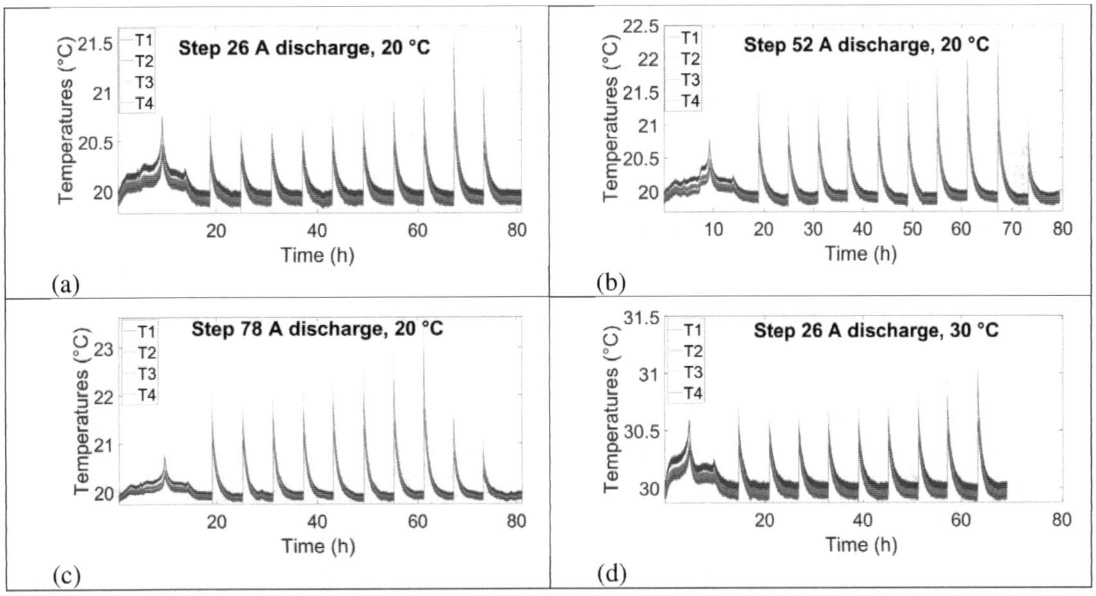

Figure 8. Surface temperature evolution of the LTO-based cell. (a) Step 26 A discharge, 20 °C, (b) Step 52 A discharge, 20 °C, (c) Step 78 A discharge, 20 °C, (d) Step 26 A discharge, 30 °C.

The chemical reactions cause the heat to be stored within the cell. Because the heat dissipation from the cell to the chamber is not large enough, the heat generation increases primarily in the discharge process and is reduced while the overpotential decreases. This phenomenon results in a fast increase in the cell's temperature at the beginning of the discharge process and a subsequent further increase.

Discharge and charge at various current rates exhibited a relatively dissimilar orientation in the temperature profile. During charge and discharge, this turns out to be negligible, as shown by an insignificant variation in the slope of the temperature profile. At different current rates, the most extensive heated parts were located at the neighbouring regions of the tabs, and the temperatures were relatively more considerable in the neighbourhood of the negative tab.

The quantity of irreversible heat, including resistance heat, may turn out to be more significant at a higher current rate. Throughout the 78 A step discharge, the heat generation was remarkably higher than that of the 26 A and 52 A step discharge throughout all the discharge process. The produced heat during each step discharge cycle was not proportional to the amount of SOC decrease. Nevertheless, it was seen that this relationship might change according to the quantity of step discharge current. The maximum heat generation for 26 A step discharge cycles was from 47.92% SOC to 46.07% SOC, while for both 52 A and 78 A step discharge cycles, it was from 50.23% SOC to 47.92% SOC. The minimum heat generation for 26 A step discharge cycles was from 74.92% to 68.23% SOC, while for 52 A, it was from 62.69% to 58.61% SOC and for 78 A step discharge cycles, it was from 47.92% to 46.07% SOC. At 30°C, discharging from 74.92% to 68.23% SOC, the heat generation was 0.22 Wh for 26 A step discharge rates. The maximum heat generation for 26 A-step discharge cycles was from 50.23% to 47.92% SOC. These changes could be associated with irreversible and reversible heat source parts. These parameters are more significant at the lowest and highest SOC levels. Besides, this might also be related to reversible entropic variations or diffusion limitations as well as the heat of mixing and heat of phase changes.

The lithium titanate oxide battery's heat generation rate and surface temperature distribution depend on different factors, such as the charge or discharge profile, the initial SOC, and the initial temperature. It may also depend on other parameters which were not investigated in this research, such as the LIB construction, design, and chemistry. The pouch cell's temperature distribution turned out to be heterogeneous. It could be inferred that the zone of the largest non-uniformity and temperature is the position where the heat generation is the largest. For the investigated LTO-based pouch cell, the most significant source of heat generation seems to be the tab to the current collector boundary.

The heat generation rate increases faster at the end of discharge than at its beginning. This finding might be due to the reduced terminal voltage while the cell is at a low SOC. In addition, the lithium-ion concentration gradient at the end of discharge might be more significant than at the beginning. The temperature gradient over the cell surface is lower at the beginning of the charge and discharge processes than at their ends.

The evolution of heat loss as a subordinate of the current charge/discharge cycles was determined. This heat flux quantification could be employed for judging the heat transmission of the inner portion of the lithium-ion battery cell. The results of this investigation can be used as supporting benchmark data for correlation to simulation outcomes, accompanied by lithium-ion battery heat generation characteristics to assist in the design and analysis of an efficient thermal management system.

When the operating temperature is 50 °C, the heat generation amount is approximately the same as at 20 °C at the lower current rates and increases comparatively towards significant current rates. The heat generation level increases with the increase of charge and discharge current rates. It is very appealing that there is a heat generation boundary for the investigated cell with individual compounds and materials as a function of operating temperature. Besides, it varies for different current rates.

3. Conclusions

A method using an isothermal calorimeter was introduced to improve the understanding of the thermal phenomena that occur within a LTO-based pouch cell at different operating temperatures and under different operating conditions. Interestingly, the majority of the parameters exhibited an overall increasing or decreasing behaviour towards significant current rates. The isothermal battery calorimeter was employed to study the operational temperature influence on the cells' heat dissipation and efficiency. It was concluded that the current rates were directly proportional to the heat dissipation rate. Using this experimental methodology will assist researchers, cell developers, and producers in deriving essential parameters, including heat generation and lifetime analysis at different current rates in particular operating cases of charge and discharge cycles. It was observed that the discharge and charge current rates substantially impact the thermal behaviour. This study connected the charge and discharge cycles with various current rates to efficiency, peak temperature, and heat generation. An almost linear relationship was revealed between these parameters and the current rate. Constant current discharge pulses at dissimilar current rates were applied. The hybrid pulse power characterization experiment was used to determine the internal resistance at different state of charge levels. The results showed that the reliance of various parameters on the cells' reversible and irreversible heat exhibits substantial changes at different state of charge levels and working conditions.

Nevertheless, their contribution is dissimilar. It was concluded that the heat generation is not negligible even at low rates of pulse discharge current step duration. The heat released during step discharge cycles was not equal in any of the cell's state of charge levels investigated. Total generated heat demonstrated a significant dependence on the SOC level, pulse discharge current step duration, ambient temperature, and discharge rate. Nevertheless, these influential parameters may have a different impact on the heat of mixing, the heat of phase change, and irreversible and reversible heat at different SOC levels. The impact of SOC level on heat generation of the cell turned out to be larger at a lower SOC compared to a higher SOC. It can be concluded that at a high SOC level, the significance of the internal resistance affecting the heat generation is lower than that of the level of SOC change, whereas, at a small SOC level, the working temperature is most critical, followed by internal resistance and SOC level.

The total generated heat for a step current discharge at 78 A increased more than for other currents. A decrease of 6.69 % in SOC level caused an increase of 0.377 Wh, 0.728 Wh, 1.002 Wh, and 0.2249 Wh in heat generation for Step 26 A, 52 A, 78 A discharge at 20 °C and stepped 26 A discharge at 30 °C respectively. Such values and findings can be used at first to assess, examine, and develop LTO-based modules and battery packs. Second, they help to elaborate a much more accurate thermal model of the cell. Third, they are valuable to tailoring thermal management systems by optimizing cooling capacity and sketching cooling routes via the battery packs. Accordingly, this can cause safer operation and a longer lifetime of LIB. This research can contribute to future studies about thermal behaviour and management of LIB.

References

[1] Manikandan B, Yap C, and Balaya P 2017 Towards Understanding Heat Generation Characteristics of Li-Ion Batteries by Calorimetry, Impedance, and Potentiometry Studies, *J. Electrochem. Soc.* **164** no. 12, A2794–A2800

[2] Guldbæk L, Karlsen and Villadsen J 1987 Isothermal reaction calorimeters - I. A literature review, *Chem. Eng. Sci.* **42** no. 5, 1153–1164

[3] Untereker D F 1978 The Use of a Microcalorimeter for Analysis of Load - Dependent Processes Occurring in a Primary Battery, *J. Electrochem. Soc.* **125** no.12, 1907–1912

[4] Tudron F B Dynamic Microcalorimetry: Thermal Effects of Miniature Alkaline Cells Under Load., *J. Geophys. Res.* **79** no. 2, 437–439

[5] Abdul-Quadir Y et al. 2014 Heat generation in high power prismatic Li-ion battery cell with LiMnNiCoO2 cathode material, *Int. J. Energy Res.* **38** no. 11, 1424–1437

[6] Awarke A, Jaeger M, Oezdemir O, and Pischinger S 2012 Thermal analysis of a li-ion battery module under realistic ev operating conditions, *International Journal for Energy Research*, 1099-114

[7] Jeon D H and Baek S M 2011 Thermal modeling of cylindrical lithium ion battery during discharge cycle, *Energy Conversion and Management* **52** 2973-2981.

[8] Kim U S, Shin C B and Kim C S 2008 Effect of electrode configuration on the thermal behavior of a lithium polymer battery *Journal of Power Sources*, 180, 909-916

[9] Shen J, Wang Y, Yu G and Li H, 2020 Thermal Management of Prismatic Lithium-Ion Battery with Minichannel Cold Plate, *J. Energy Eng.* **146** no. 1.

[10] Madani S S, Schaltz E and Kær S K 2018 Study of Temperature Impacts on a Lithium-Ion Battery Thermal Behaviour by Employing Isothermal Calorimeter *ECS Trans.* **87** no.1

[11] Madani S S, Schaltz E and Kær S K 2018 Investigation of the Effect of State-of-Charge and C-Rates on the Heat Loss and Efficiency of a Lithium-Ion Battery *ECS Trans.* **87** no.1, 51–58

[12] Madani S S, Schaltz E and Kær S K 2018 Heat loss measurement of lithium titanate oxide batteries under fast charging conditions by employing isothermal calorimeter *Batteries* **4** no.4, 59

[13] Madani S S, Schaltz E and Kær S K 2018 Thermal Modelling of a Lithium Titanate Oxide Battery *ECS Trans* **87** no.1, 315–326
[14] Madani S S 2020 Experimental Study of the Heat Generation of a Lithium-Ion Battery *ECS Trans.* **99** no.1, 419
[15] Hall F, Touzri F, Wußler S, Buqa H, and Bessler W G 2018 Experimental investigation of the thermal and cycling behavior of a lithium titanate-based lithium-ion pouch cell, *J. Energy Storage*, 17, 109–117
[16] Wu W, Xiao X, and Huang X 2012 The effect of battery design parameters on heat generation and utilization in a Li-ion cell, *Electrochem. Acta*, 83, 227–240.
[17] Chen S C, Wan C C, and Wang Y Y 2005 Thermal analysis of lithium-ion batteries, *J. Power Sources*, **140** no. 1, 111–124
[18] Jeon D H and Baek S M 2011 Thermal modeling of cylindrical lithium ion battery during discharge cycle. *Energy Conversion and Management*, 52, 2973-2981
[19] Kim U S, Shin C B and Kim C S 2008 Effect of electrode configuration on the thermal behavior of a lithium polymer battery *Journal of Power Sources*, 180, 909-916
[20] Yang H and Prakash J 2004 Determination of the Reversible and Irreversible Heats of a LiNi0.8Co0.15Alsub0.05O2/Natural Graphite Cell Using Electrochemical-Calorimetric Technique, *J. Electrochem. Soc.* **151** no. 8, A1222
[21] Kobayashi Y et al. 1999 Electrochemical and calorimetric approach to spinel lithium manganese oxide, *J. Power Sources*, 81–82, 463–466.
[22] Kantharaj R and Marconnet A M 2019 Heat Generation and Thermal Transport in Lithium-Ion Batteries: A Scale-Bridging Perspective, *Nanoscale Microscale Thermophys. Eng.*, **23** no. 2, 128–156
[23] Ma S et al. 2018 Temperature effect and thermal impact in lithium-ion batteries: A review, *Prog. Nat. Sci. Mater. Int.* **28** no. 6, 653–666
[24] Bandhauer T M, Garimella S, and Fuller T F 2011 A Critical Review of Thermal Issues in Lithium-Ion Batteries, *J. Electrochem. Soc.* **158** no. 3
[25] Onda K, Kameyama H, Hanamoto T, and Ito K 2003 Experimental Study on Heat Generation Behavior of Small Lithium-Ion Secondary Batteries, *J. Electrochem. Soc.* **150** no. 3, A285
[26] Bang H, Yang H, Sun Y K, and Prakash J 2005 In Situ Studies of Li[sub x]Mn[sub 2]O[sub 4] and Li[sub x]Al[sub 0.17]Mn[sub 1.83]O[sub 3.97]S[sub 0.03] Cathode by IMC, *J. Electrochem. Soc.* **152** no. 2, A421
[27] Zhang J et al. 2014 Comparison and validation of methods for estimating heat generation rate of large-format lithium-ion batteries, *J. Therm. Anal. Calorim.* **117** no. 1, 447–461
[28] Chen K, Unsworth G, and Li X 2014 Measurements of heat generation in prismatic Li-ion batteries, *J. Power Sources* 261, 28–37
[29] Bandhauer T M, Garimella S and Fuller T F 2014 Temperature-dependent electrochemical heat generation in a commercial lithium-ion battery, *J. Power Sources* 247, 618–628.
[30] Schuster E, Ziebert C, Melcher A, Rohde M, and Seifert H J 2015 Thermal behavior and electrochemical heat generation in a commercial 40 Ah lithium ion pouch cell *J. Power Sources* 286, 580–589
[31] Lin C, Xu S, Li Z, Li B, Chang G and Liu J 2015 Thermal analysis of large-capacity LiFePO4 power batteries for electric vehicles *J. Power Sources* 294, 633–642
[32] Drake S J et al. 2015 Heat generation rate measurement in a Li-ion cell at large C-rates through temperature and heat flux measurements *J. Power Sources* 285, 266–273
[33] Madani S S, Schaltz E and Kær S K 2018 Review of parameter determination for thermal modeling of lithium ion batteries *Batteries*. **4** no.2, 20
[34] Madani S S, Swierczynski M J and Kaer S K 2017 A review of thermal management and safety for lithium ion batteries 2017 *12th International Conference on Ecological Vehicles and Renewable Energies*, EVER 2017, 1-20

[35] Ohzuku T, Ueda A and Yamamoto N 1995 Zero-Strain Insertion Material of Li[Li1/3Ti5/3]O4 for Rechargeable Lithium Cells *Journal of The Electrochemical* **142** no.5, 1431–1435

[36] Yi T F, Jiang L J, Shu J, Yue C B, Zhu R S and Qiao H B 2010 Recent development and application of Li4Ti5O12 as anode material of lithium ion battery *Journal of Physics and Chemistry of Solids* **71**,1236 – 1242

Graphene-enhanced sulfur cathode with high interface stability in Li-S batteries

Yu Polishchuk[1], S Dubinevych[2], V Zinin[2] and E Shembel[1,3]

[1] Ukrainian State University of Chemical Technology, Dnipro, 49005, Ukraine
[2] LLC "MaxAh" K, Kiev, 01001, Ukraine
[3] Enerize Corporation, Coral Springs, FL, 33067, USA

yu.v.polishchuk@gmail.com

Abstract. The physicochemical properties of graphene and graphene materials obtained by the method of controlled detonation gas synthesis are presented. The fundamental possibility of controlling the graphene and graphene materials physicochemical properties by changing the synthesis conditions is shown. Dynamics of changes in the impedance spectra of Li-S and Li-S-Li batteries with graphene in non-aqueous liquid electrolyte 0.7 M LiIm, 0.25 M $LiNO_3$, DME:DOL (2:1) were studied. The results of electrochemical testing of experimental samples and prototypes of elements of the lithium-sulfur system by the methods of cyclic voltammetry and galvanostatic cycling are presented. The effect of interface stabilization was expressed in a decrease in impedance spectra, an increase in discharge capacity, and more stable long-term cycling with high parameters. The investigation showed a positive effect of graphene materials in the active mass of the multilayered S-based electrode with special design.

1. Introduction

Li-S battery has been one promising alternative electrochemical energy storage system with high energy density. Along with their obvious advantages, lithium-sulfur batteries have a number of significant disadvantages that prevent their successful commercialization: low electronic conductivity of sulfur; active material dissolution, and the well-known shuttling effect [1]. Substances dissolved in the electrolyte diffuse from the cathode side, thereby causing an irreversible loss of capacitance. In addition, polysulfides form a secondary redox cycle between the anode and cathode, resulting in a high degree of self-discharge.

A strategy with a positive effect is the use of carbon materials. On the one side, they act as an effective electrically conductive additive. On the other side, porous carbon materials with a highly porous surface make it possible to effectively immobilize sulfur or its compounds, including polysulfides [2].

Carbon materials with high electrical conductivity and variable structures are extensively used in lithium non-aqua batteries. Relatively new forms of carbon are fullerenes and graphenes. Graphene is a potent conductor of electrical and thermal energy, extremely lightweight, chemically inert, and flexible with a large surface area.

Recently, significant interest in graphene materials has been caused by the prospect of its use in biology and medicine [3-5]. However, the available information on the toxicity and safety of graphene and materials based on it is insufficient and contradictory. As a relatively new material, graphene raises understandable doubts due to its possible impact on human health. Due to its special structure, graphene, for example, can penetrate the lungs and pose a threat comparable to dust or even asbestos fibers [6]. It is noted that the degree of toxicity of graphene and its derivatives may depend on its physicochemical properties such as size, surface charge, shape, number of layers, surface functional groups and particle state [3, 7]. Some studies [8] provide evidence of the inertness of graphene and the does not interaction with living organisms when it is used for scientific and practical biomedical purposes. At the same time, most researchers insist on the need for a long-term accumulation of data, additional studies of the safety of using graphene. The analysis given in [9] shows a comparison of obtaining graphene methods from the point of view of the environmentally sustainable production of such materials. Some production methods have a significantly lower environmental impact. That is why special attention should be paid to the choice of a method for obtaining graphene materials, as well as ensuring its environmental safety.

Graphene can be obtained by various methods, which are traditionally divided into mechanical and chemical. Also, recently, other methods of obtaining graphene are being developed, for example, radio-frequency plasma-enhanced chemical vapor deposition [10], high pressure–high temperature (HPHT) growth process [11], method for high-performance production of graphene by detonation of carbon-containing material [12] and others.

In recent years, graphene has begun to play an important role in power sources application due to its ultra-high electrical conductivity, large surface area, and flexible two-dimensional nanostructure [13]. As a rule, graphene is used for multilayer electrode materials: as flexible self-supporting graphene / S paper [14]; as graphene / S / C layered nanocomposite [15]; as surface modification sulfur by graphene [16], etc. Earlier, we showed the possibility of increasing the electrical parameters of a Li-S battery due to the use of a special layered design of S-based cathode [17]; due to the use of graphene materials [18].

2. Experiment

In this investigation, we used graphene materials obtained by a unique technology developed by LLC "MaxAh" K company. The positive electrodes were fabricated by slurry coating. The aluminum foil was used as the current collector. Disk electrodes based on sulfur had the following parameters: d – 1.6 cm, working surface – 2 cm^2. Cathode mass had the following composition:75% - natural sulfur; 20% - graphene materials produced by LLC "MaxAh" K or graphite/carbon black mix and 5% - polyvinylidene fluoride PVdF (6020/1001, Solef) as a binder. The drying of the cathode was at 60°C for 6 h under vacuum and 2 h at 120°C without vacuum. The multilayer cathode was produced using carbon materials by the technology described in [17].

The negative electrodes were made of lithium foil. The microporous polypropylene Celgard® 2300 with a thickness of 25 μm was used as a separator. Electrolyte composition was 0.7M bis(triuoromethanesulfone) imide (LiIm), 0.25M $LiNO_3$, dimethoxyethane (DME) : dioxolane (DOL) (2:1 by volume). All procedures were carried out in an Ar-filled glove box.

The following measurements were used: XRD spectroscopy with monochromatic Co-Kα radiation. Identification of compounds (phases) was carried out by comparing the interplanar distances (d, A) and relative intensities of the experimental curve with the data of the PCPDFWIN electronic base. The spectra of the Stokes scattering component were recorded, excitation at a wavelength of 532 nm, power density - 0.25 mW/μm^2, spectral resolution ≈ 2 cm^{-1}. Impedance spectra of Li-S batteries were obtained using VoltaLab PGZ 301 analytical radiometer. The frequency range was from 100 MHz to 100 kHz. Cyclic voltammetry was performed in the potential range from 1.5 to 2.9 V at a scanning rate of 0.5 mV/s. Cycling was carried out in a three-electrode cell, where the sulfur-based electrode is the working electrode, the auxiliary and reference electrodes are metallic lithium. Galvanostatic cycling was performed in the range of potentials from 1.8 to 2.9 V at discharge current - 100 μA/cm^2, and charge current – 50 μA/cm^2. The measurements were conducted at a temperature of 25 °C.

3. Results and discussion

3.1. Characteristics of the graphene powders

We used powders of graphene and graphene materials obtained by the method of controlled detonation gas synthesis. The method of obtaining graphene was based on the method of Prof. Sorensen and Dr. A. Nepal [12]. The method was developed in terms of additional reagents, catalysts, and methods for extracting finished graphene. Then graphene was subjected to thermal and mechanical activation for 5, 10, 15 and 20 minutes. A vortex layer device was used for activation. Processing of the obtained graphene powder was carried out in an air atmosphere. The activation can also take place in an inert gas atmosphere. The carbon material obtained by this method is graphene with a special "pompom" morphology, shown in figure 1.

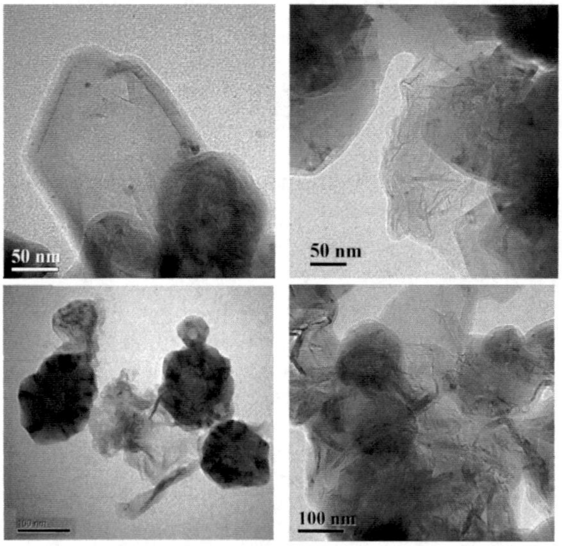

Figure 1. TEM image of "pompom" graphene.

The characteristics of the original and processed graphene powders are shown in figures 2 and 3. According to X-ray diffraction analysis, all materials contain graphite. The amount of crystalline carbon in the original sample is 60%. After processing at 20 minute, the amount of carbon with crystalline phase decreases to 32%. The amount of carbon crystalline phase was estimated by area on the XRD diagram. For example, natural graphite, which was also used in the work, has a clearly expressed crystalline structure and the absence of an amorphous phase, figure 2 (b). The average size of crystallites L for natural graphite is 270 Å.

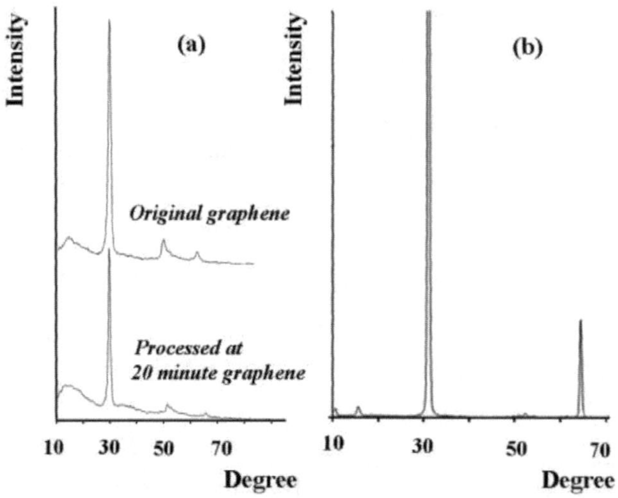

Figure 2. XRD spectra of graphene (a) and natural graphite (b) samples.

An increase of the thermomechanical activation time also leads to an increase of Raman peaks intensity, as shown in figure 3. The processed graphene sample is characterized by the presence of some graphene oxide, as evidenced by an intense peak at 1350 cm^{-1} on the Raman spectrum. The average size of crystallites L of the original sample is 267 Å, and, for the processed at 20 minutes graphene sample, L is 168 Å. The Scherrer equation (also known as the Debye-Scherrer equation) was used to determine the size of the crystallites. The calculation took into account the position of the peaks (2θ/2), the physical broadening of the line in the diffraction pattern, and the wavelength of X-ray radiation [19].

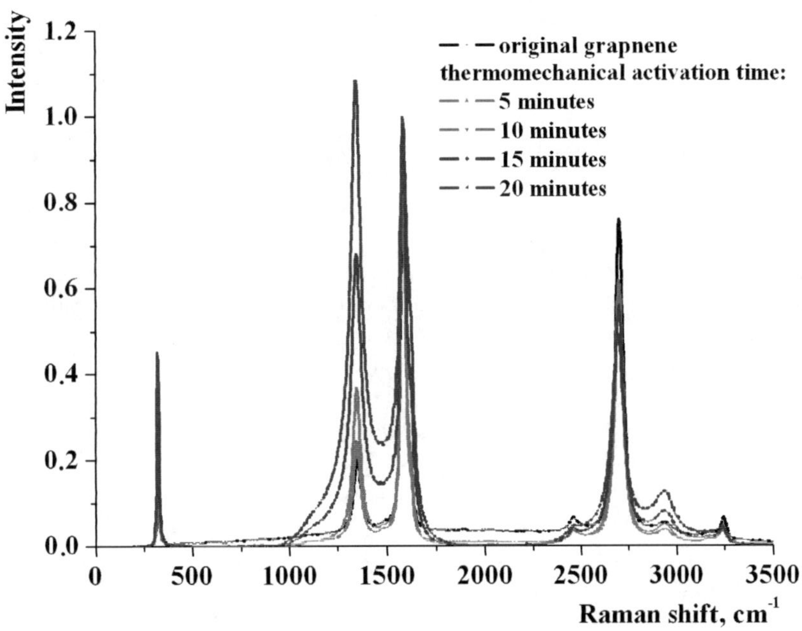

Figure 3. Raman spectra of the initial and processed graphene samples.

The calculation of the graphene layers number based on the change in the position of the G-band [20, 21] and the ratio of the intensities of the 2D and G bands was performed according to the methodology presented in [22]. For a more accurate determination of the maximum position the characteristic spectral region was approximated by peaks with the shape of the Lorentz contour. In terms of mass and position of the G-band the amount of double-layer graphene is about 40%, 3-6 layer - up to 30%, 7 layers and more - up to 20%, monolayer graphene - up to 10%. The number of layers calculated based on the ratio of the intensities of the 2D and G-bands in both types of graphene samples is equal to two.

3.2. Results of impedance spectroscopy
The replacement of carbon materials in the cathode mass with graphene does not lead to a significant decrease in the complex resistance of the Li-S system at first day, figure 4. We attribute this to the predominance of passive film formation on lithium. However, the resistance of S-based electrodes with graphene in a Li-S-Li cell is several orders of magnitude lower than that of cathodes with a mixture of graphite and carbon black, as shown in figure 5. This trend has been observed for several days of storage.

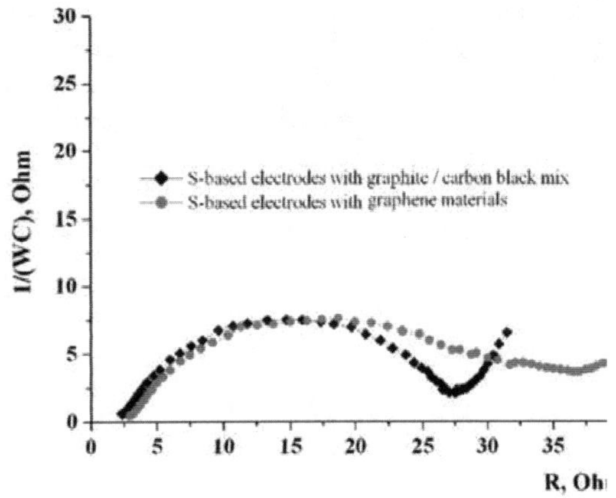

Figure 4. Impedance spectra of Li-S cells.

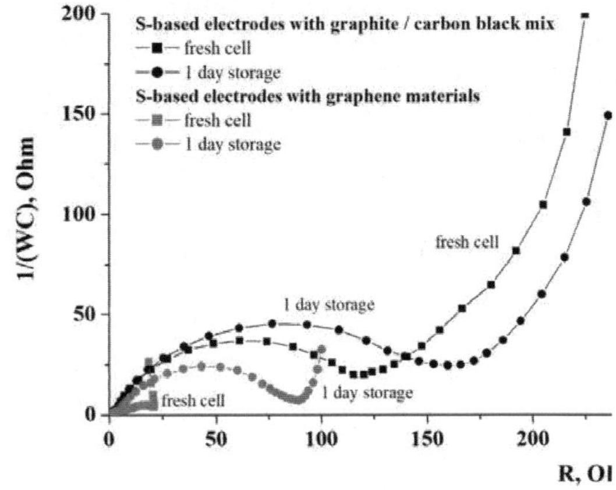

Figure 5. Impedance spectra of S-based electrodes in Li-S-Li cells.

3.3. Results of cycling volt-amperometry

The effect of graphene is more significant during voltammetry cycling. The first cathode curve of the current-voltage characteristic of the S-based electrodes with graphite/carbon black mix shows two clear peaks at potentials of 2.2 and 1.9 V, as shown at figure 6 (a).

In this range of potentials, the formation of various compounds is possible, for example, Li_2S and Li_2S_2. During cycling, there is a small shift of the potential of the first cathode peak to 2.3 V, which may correspond to Li_2S_4 formation.

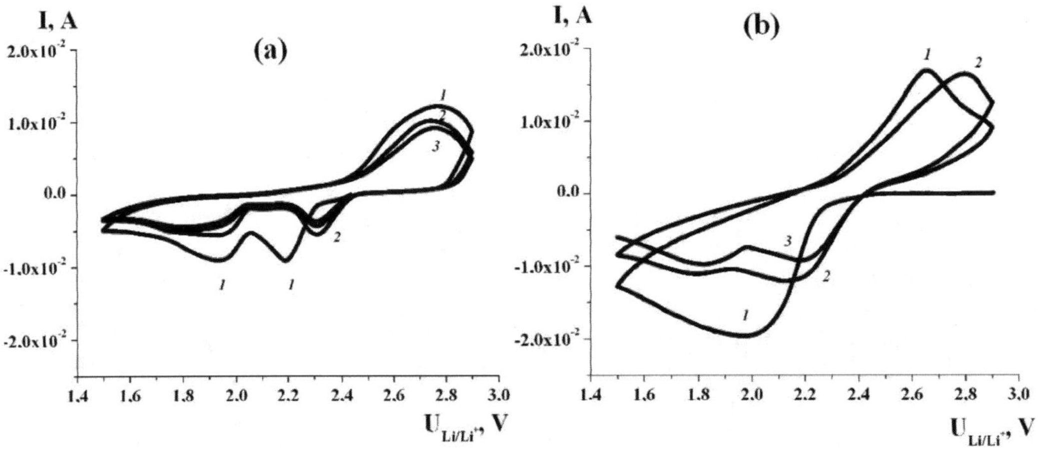

Figure 6. Change of Li-S cell volt-ampere curves. The slurry cathode contains 75% sulfur, 20% graphite/carbon black mix and 5% binder (a); 75% sulfur, 20% graphene and 5% binder (b). The numbers on the curve are the numbers of the cycles.

The anode curve has a clear peak at a potential of 2.6-2.7 V for both types of carbon materials in cathode mass. It should be noted that the use of graphene in S-based electrodes leads to a two fold increase in the current value in the cathode and anode peaks, figure 6 (b).

3.4. Results of galvanostatic cycling
Figure 7 shows the dynamics of specific capacity changes in Li-S cell. Acetylene black was used as the carbon layer. A decrease in the rate of formation of polysulfides and decreasing of shuttle effect due to the use of graphene materials as an electrode additive confirm the results of galvanostatic cycling.

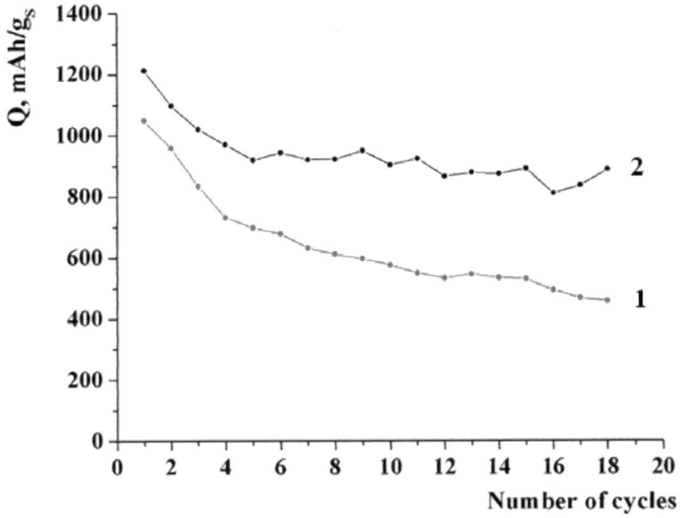

Figure 7. Dynamics of specific capacity changes in Li-S cell. The slurry cathode contains 75% sulfur, 20% graphite/carbon black mix and 5% binder (1); 75% sulfur, 20% graphene and 5% binder (2). Electrode area is 2 cm^2.

Also, figure 8 shows the initial galvanostatic cycling results of prototypes Li-S elements with prismatic design, working area of the electrode is 12 cm^2.

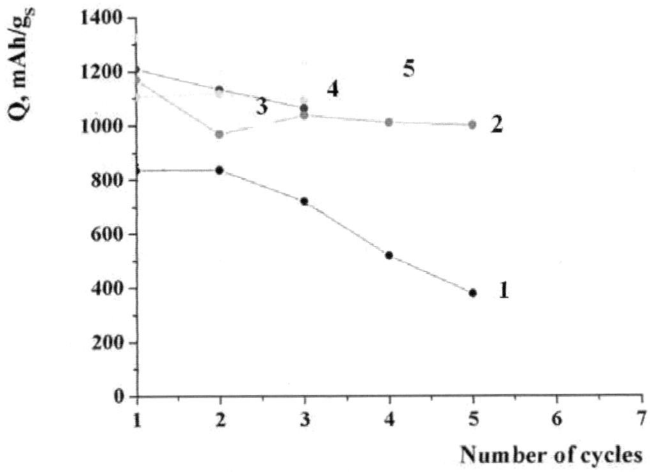

Figure 8. Dynamics of specific capacity changes in Li-S cell. The slurry cathode contains 75% sulfur, 20% graphite/carbon black mix and 5% binder (1); 75% sulfur, 20% graphene and 5% binder (2-5). Electrode area is 12 cm^2.

In this case, graphene is used not only as an electrode additive, but also as a material for the carbon layer. Such a combination of graphene in a multilayer cathode leads to a significant stabilization of the capacity for at least 5 cycles. The average value of the specific discharge capacity is ~1200 mAh/g_S.

4. Conclusion
A complex of investigations allows us to conclude that the unique structural features and physical parameters of graphene materials and the use of a special electrode design make it possible to reduce

the resistance of sulfur-based electrodes as well as significantly stabilize the discharge capacity of lithium-sulfur batteries during cycling.

Acknowledgments
Financial support under contract number 42/216724 "Development of innovative high-energy power sources based on graphene materials" and NATO SPS 985148 project "Development of new cathodes for stable and safer lithium-sulfur batteries" is gratefully acknowledged.

References
[1] Manthiram A, Fu Y Z, Chung S H, Zu C X and Su Y S 2014 *Chemical Reviews* **114** pp 11751–87
[2] Wang H L, Yang Y, Liang Y Y, Robinson J T, Li Y G, Jackson A, Cui Y and Dai H J 2011 *Nano lett.* **11** pp 2644–47
[3] Yao J, Wang H, Chen M and Yang M 2019 *Microchim Acta* **186** 395
[4] Priyadarsini S, Mohanty S, Mukherjee S, Basu S and Mishra M 2018 *J Nanostruct Chem* **8** pp 123– 137
[5] Kostarelos, K 2016 *Nat Rev Mater* **1** 16084
[6] Bradley D 2012 *Materials Today* **15**(6) p 230
[7] Ding X, Liu H and Fan Y 2015 *Adv. Healthcare Mater.* **4** pp 1451–1468
[8] Savicheva A, Tapilskaya N, Spasibova E, Gzgzyan A, Kogan I *et al.* 2020 *Gynecological Endocrinology* **36**(51) pp 48-52
[9] Beloin-Saint-Pierre D and Hischier 2021 *R Int J Life Cycle Assess* **26** pp 327–343
[10] Wang J J, Zhu M Y, Outlaw R A, Zhao X, Manos D M and Holloway B C 2004 *Appl. Phys. Lett.* **85** p 1265
[11] Parvizi F, Teweldebrhan D, Ghosh S, Calizo I, Balandin A A. Zhu H and Abbaschian R 2008 *Micro & Nano Lett.* **3**(1) pp 29–34
[12] Sorensen *et al* US Patent 2016 *Process for high-yield production of graphene via detonation of carbon-containing material* 9,440,857
[13] Xu C H, Xu B H, Gu Y, Xiong Z G, Sun J and Zhao X S 2013 *Energy & Environmental Science* **6** p 1388
[14] Jin J, Wen Z, Ma G, Lu Y, Cui Y, Wu M, Liang X and Wu X 2013 *RSC Advances* **3** pp 2558-60
[15] Jin K, Zhou X and Liu Z 2015 *Nanomaterials* **5** (3) pp 1481–92
[16] Yu M, Li R, Wu M and Shi G 2015 *Energy Storage Materials* **1** pp 51–73
[17] Polishchuk Yu, Shembel E and Strakova-Fedorkova A 2020 *ECS Transactions* **99** pp 133–142
[18] Dubinevych S, Zinin V, Redko V, Blyuss B, Shembel E, Zaderey N, Polishchuk Yu, Markevych A, Strakova-Fedorkova A and Tkach M2021 *ECS Transactions* **105** pp 247–260
[19] Suryanarayana C and Grant Norton M 1998 *X-Ray Diffraction: A Practical Approach* (NY: Springer New York) p 273
[20] Malard L M, Pimenta M A, Dresselhaus G and Dresselhaus M S 2009 *Physics reports* **473**(5-6) pp 51–87
[21] Wang H, Wang Y, Cao X, Feng M and Lan G 2009 *Journal of Raman Spectroscopy* **40** (12) pp 1791–96
[22] Reina A, Jia X, Ho J, Nezich D, Son H, Bulovic V, Dresselhaus M D and Kong J 2009 *Nano lett.* **9** (1) pp 30–35

Nanostructured innovative carbon-based materials modify electrodes and dramatically improve the efficiency of thin, flexible lithium batteries

Y Pustovalov[1], E Shembel[2,3], D Kaszuba[1], V Redko[2], T Pastushkin[2], N Zaderey[3], Yu Polishchuk[3], A Markevich[3], I Sagirov[1]
[1] INT Sp. z o.o.,Ul Pilsudskiego 17/4, 34-074, Rzeszów, Poland
[2] Enerize Corporation, Coral Springs, FL, 33067, USA
[3] Scientific Research Laboratory of Chemical Power Sources of Ukrainian State University of Chemical Technology, Dnipro, 49005, Ukraine

juripudt@gmail.com

Abstract. Developed nanostructured innovative carbon materials (nano graphite) have a unique combination of properties. These include: high electronic conductivity, magnetic properties, high level adhesion and cohesion properties, low density. Using these properties of nano graphite allow fabricating the electrode for batteries without binder, and effectively modifying electrodes, which affect the mechanism of redox electrochemical processes. As a results, the energy, power and stability of the parameters of thin flexible lithium batteries increase.

1. Introduction

Carbon nanomaterials are at the forefront of the emerging technologies of the millennium and represent a vast area of required knowledge for the design of tomorrow's devices. INT Sp. z o.o. Poland develops and produces innovative nanostructured carbon materials (nano graphite) on the basis of many years of research by collaborators for various applications. The goal is to using the nano graphite to increase the energy, power and stability of parameters of Li-ion, Mg-ion, Li-S, and other batteries, including thin flexible battery [1], change the properties of polymer materials [2], and in hydrogen energy for catalysts of fuel cells [3].

Thin flexible batteries provide special interest. The market of this batteries from US$98 million in 2020 will grow to US$296 million by 2025. This means that the market will grow by an average of 24.7% per year from 2020 to 2025. [4,5]. Our publication is devoted to the project of INT Sp. z o.o., the purpose of which is to develop a technology for manufacturing thin flexible high-energy lithium-ion current sources based on nano graphite. In the battery being developed, the anode will be based on thin-layer nano graphite plates without binder, the cathode modified by nano graphite without binder, the electrode current lead will be a thin-layer of nano graphite plate also without binder. In this article we present as example results of investigation and using the nano graphite CNS-300, which produced by INT Sp. z o.o.

2. Structure of the powder and plate from of innovative nanostructured carbon materials in compare with a natural Zavalievskiy graphite.

Parameters the structure of the nano graphite affect the micro- and macro structural characteristics of the modified electrodes, increase the efficiency of current distribution over the volume of the electrodes, and increase the energy and power of the lithium batteries. In this article as example we present results of investigation and application the nano graphite CNS-300.

In figure 1 we illustrate the AFM structural of plates, which are fabricated from the nano graphite powders that were synthesized at various parameters. These can be used as anode for Li-ion batteries or as current plates lead of electrodes. The surface structure of these plates determines the impedance of the electrodes that are based on these plates.

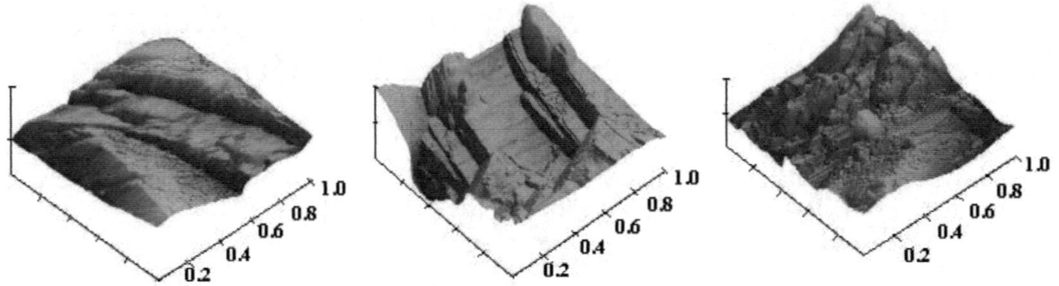

Figure 1. AFM structural characteristics of plates from the nano graphite, which aresynthesized under various parameters.

In figure 2 we illustrate the SEM structural of samples of powders: of nano graphite CNS-300 (fig.2 (b) and (c)), which was synthesized by INT Sp. z o.o., and the natural Zavalievskiy graphite (fig. 2 (a)), which is widely used in lithium batteries. The SEM structure of these powders predict the characteristics of the electrodes, which are based on these powders.

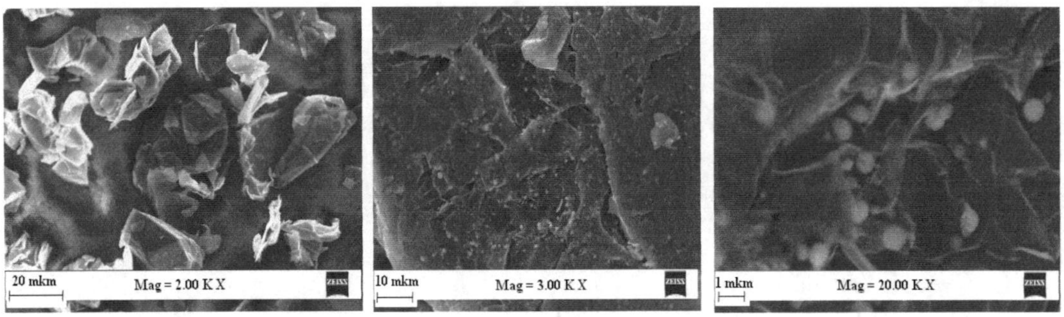

(a) **(b)** **(c)**

Figure 2. SEM structural characteristics of: (a) natural Zavalievskiy graphite, which is widely used in lithium batteries; (b), (c): Synthesized nano graphites, CNS-300, which are synthesized by company INT Sp. z o.o..

In Section 3, we compare the conductivities of nano graphite CNS-300 powder and natural Zavalievskiy graphite powders. The results show that the conductivity of nano graphite CNS-300 is significantly higher than that of natural Zavalievskiy graphite.

In Section 4, we compare the electrochemical performance of sulfur-based electrodes that are modified with nano graphite CNS-300 and sulfur-based electrodes that include natural Zavalievskiy graphite. The results show the following. Sulfur-based electrodes modified with CNS-300 nanographites have low impedance characteristics and high current-voltage characteristics. Sulfur-based electrodes, which include natural Zavalievskiy graphite, have high impedance characteristics and low current-voltage characteristics. These results confirm the effective increase in the energy and power of electrodes modified with nano graphites.

3. Electronic conductivity of nano graphite powder and natural Zavalievskiy graphite powder. Function of magnetic field frequency and powder density

For the CNS-300 synthesized nano graphite powder and natural Zavalievskiy graphite powder the conductivity of over a wide range of magnetic field frequencies was measured by an innovative non-contact, non-destructive electromagnetic method developed by Enerize Corporation [7 - 9]. This device is illustrated on figure 3.

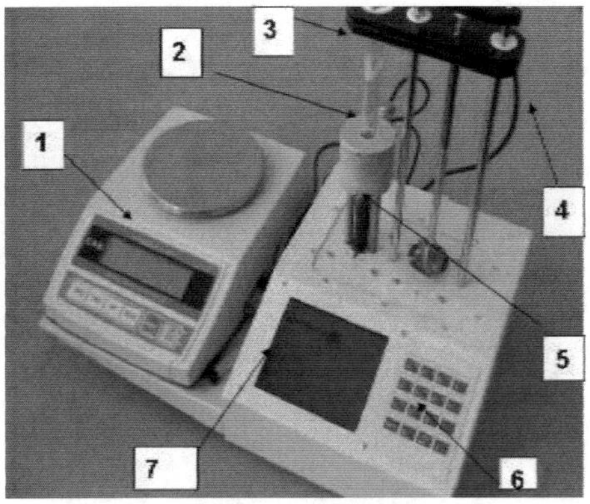

Figure 3. Device for non-destructive non-contact electromagnetic test of the conductivity of powdered materials. Description of the front side of the device: 1 -electronic scales; 2 -removable resonator chamber; 3 -glass containers; 4 -optical-mechanical device for measuring volume of powder; 5- radial vibro seal; 6 -keyboard for control and data entry; 7 -graphic display.

The data presented in figures 4, 5, and 6 illustrate the dependence of the conductivity of graphite powders on their density and magnetic field frequency. Two types of graphite powders are presented:

- CNS-300, synthesized nano graphite powder, and
- Natural Zavalievskiy graphite powder, which is widely used in Li-ion batteries. The data, which presented, illustrate the high electronic conductivity and low density of the synthesized innovative nano graphite compared to the natural Zavalievskiy graphite, which is widely used in lithium batteries.

An important factor is that the conductivity of the nano graphite depends on the frequency of the magnetic field. These parameters affect the mechanisms of the conduction and the mechanisms of electrochemical redox reactions of modified anodes and cathodes of Li batteries. As a result the energy and power of Li batteries increase. The low density of the nanostructured carbon powders ensures its uniform distribution in the volume of the electrode mass.

Li-ion batteries, which anodes and cathode are modified with nano graphite powders, have low impedance parameters, high reversibility of discharge and charging processes, stability of characteristics during numerous charge-discharge cycles, and high energy in a wide range of currents and temperatures.

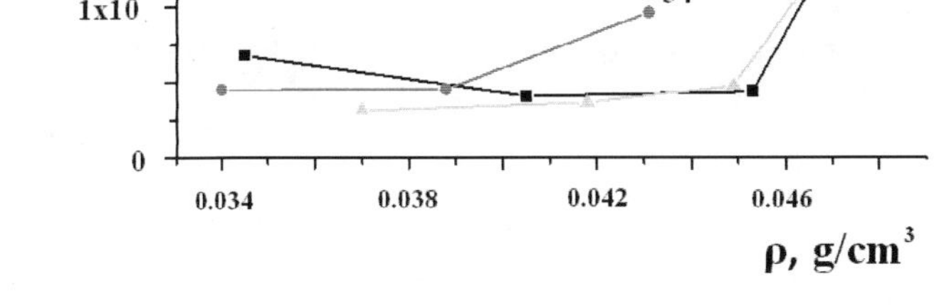

Figure 4. Conductivity of nano graphite powder CNS-300 as a function of powder density and magnetic field frequencies (MHz), at which the measurements were made,
The numbers on curves are magnetic field frequencies (MHz).

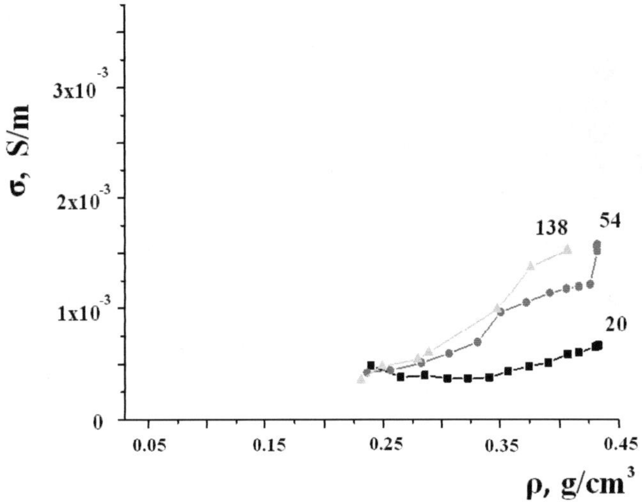

Figure 5. Conductivity of natural Zavalievskiygraphitepowder as a function of powder density and magnetic field frequencies (MHz), at which the measurements were made,
The numbers on curves are magnetic field frequencies (MHz).

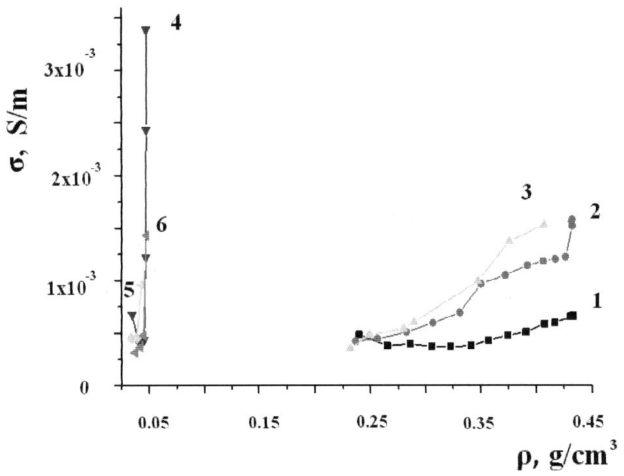

Figure 6. Comparison conductivity of two samples of powders: curves 4, 5, 6 – nano graphite CNS-300; curves 1, 2, 3– natural Zavalievskiy graphite.
Presented the conductivity as a function of powder density and magnetic field frequencies (MHz), at which the measurements were made.
The frequency values of the magnetic field are as follows: Curves 1, 4 - 20 MHz. Curves 2, 5 - 54 MHz. Curves 3, 6 - 138 MHz.

4. Modification of the electrodes based on the sulfurusing nano graphite affects the electrochemical properties of the electrode

The figures 7 and 8 show photographs of electrodes that do not contain a binder. This because these electrodes are used the nano graphite CNS-300, which has the strong adhesive properties and organize strong mechanical stable the electrodes without binder.

Figure 7 presents a photo of electrodes, which include only nano graphire CNS-300 and could be used as anode for Li-ion batteries.

Figure 8 present photos of electrodes, which includes the sulfur and nano graphite CNS-300, and could be used as cathode for Li-S batter.

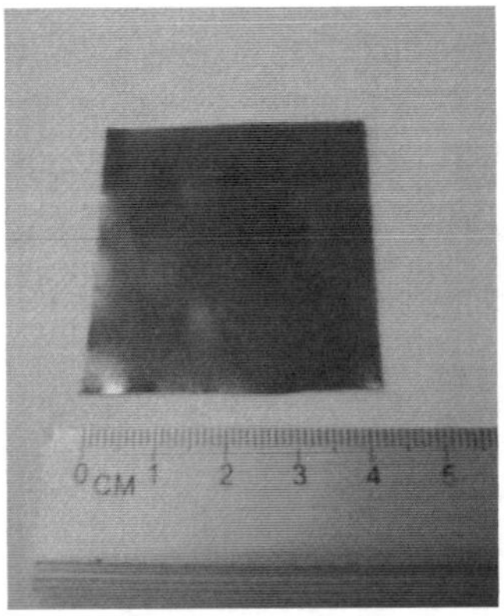

Figure 7. Photograph of electrode manufactured from nano graphite CNS-300 and without a binder. This electrode is used as anode in Li-ion battery.

Figure 8. Photograph of electrode manufactured from the sulfur and nano graphite CNS-300. These electrodes also without binder. This electrode is used as cathode of Li-S batteries.

Impedance of electrodes is an important factor, which influences the power of batteries. If the value of impedance of each electrode of the batteries decreases, the total resistance of the battery decreases. This means that the battery power increases, the current source can operate at high currents without reducing the operating parameters.

We test the impedance of electrodes in cells where the sulphur-based electrode is the working electrode, and the auxiliary electrode is metallic lithium. All measurements were carried out in an Ar-filled glove box.

In figures 9 and 10 results of the impedances test of two types of electrodes, based on sulfur are compared:

Type 1, figures 9: sulfur 79%, nano graphite CNS-300 - 21%, without binder.

Type 2, figures 10: sulfur 75%, natural Zavalievskiy graphite 10%, carbon: 10%, binder 5%.

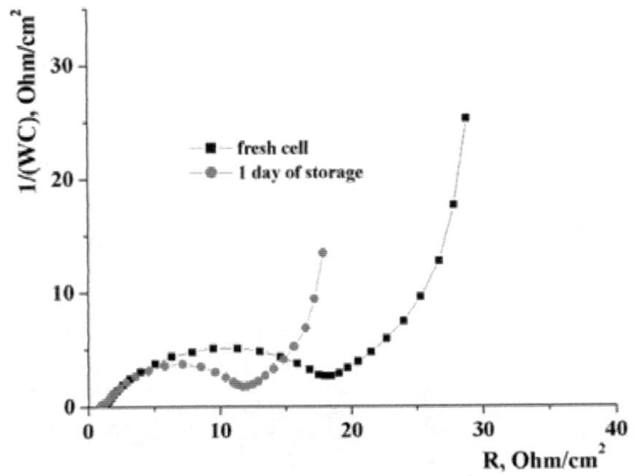

Figure 9. Impedance characteristics of the electrode, which includes:
sulfur 79% and
nano graphite CNS-300 - 21%.
Electrolyte: 0.7M bis(triuoromethanesulfone) imide (LiIm), 0.25M $LiNO_3$, DME:DOL (2:1).

Figure 9 shows the dynamics of change in the resistance of a sulphur - based cathode that includes the nano graphite and does not contain a binder. The complex resistance, which is defined as the diameter of a semicircle, of a freshly assembled element is 18.60 Ohm/cm^2. After a day, it decreases to a value of 11.58 Ohm/cm^2.

Figure 10 shows the dynamics of change in the resistance of an cell with a sulfur-based cathode that contains binder. The hodographs of cells with a sulfur-based cathode and binder, have a fundamentally different character. The impedance spectra have two semicircles on the figure 10. The complex resistance of the system is much higher than in the case of electrodes containing nano-carbon and without binder. For a freshly assembled cell, the complex resistance is 35.44 Ohm/cm^2. Within a day, it rises to a value of 99.85 Ohm/cm^2, which is approximately 9 times higher than the current value for electrodes without a binder.

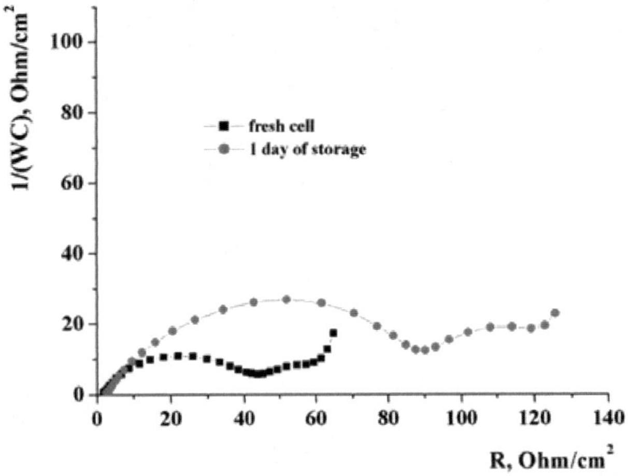

Figure 10. Impedance characteristics of the electrode, which includes: sulfur 75%, Natural Zavalievskiy graphite: 10%, carbon: 10%, binder: 5%. Electrolyte: 0.7M bis(triuoromethanesulfone)imide (LiIm), 0.25M $LiNO_3$, DME:DOL (2:1).

Figures 11 and 12 below show the results of current-voltage cycling in a three-electrode cell, where the working electrode is based on sulfur, the auxiliary electrode is metallic lithium, and the reference electrode is metallic lithium.

Figure 11 shows a cyclic voltammogram of a sulphur -based electrode, which include the nano graphite CNS-300, and do not containing binders. The current-voltage characteristic of such electrodes has one peak on the cathode branch initially a potential of about 1.7 V. During the next cycling, the cathode peak is stable neat at a potential of about 2.0 V. We can confirm that the cyclic voltammogram of this electrode is stable.

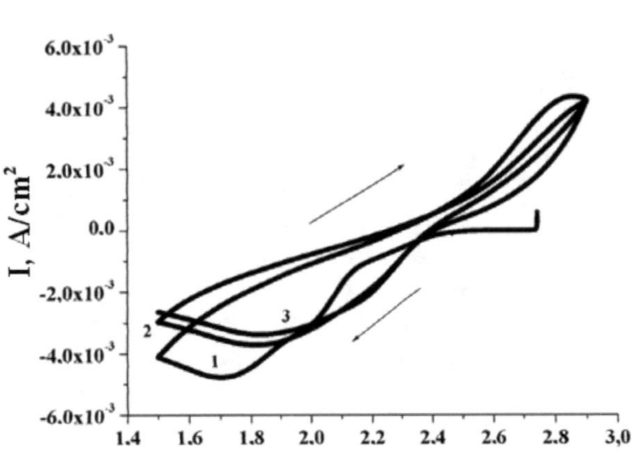

Figure 11. The current-voltage characteristic of the S cathode in cell Li-Li-S battery. The potential range of 1.5 - 2.9 V. The potentialscan rate is 0.5 mV/ sec. Cathode includes: sulfur 79%, and nano graphite CNS-300 21%. Electrolyte: 0.7M bis(triuoromethanesulfone) imide (LiIm), 0.25M $LiNO_3$, DME:DOL (2:1). The numbers of the curves show the numbers of cycles.

The character of the curves indicates a multistage mechanism of sulfur reduction and possible reactions involving the products of sulfur reduction - polysulfides.

$16Li^+ + S_8 + 16e \leftrightarrow 8Li_2S$, E = 2.2 B vs. Li/Li^+, (1)

$Li_2S_4 + 2Li^+ + 2e \rightarrow 2Li_2S_2$, E = 1.9 ~ 2.1V vs. Li/Li^+, (2)

$Li_2S_4 + 6Li^+ + 6e \rightarrow 4Li_2S$, E = 1.9 ~ 2.1V vs. Li/Li^+. (3)

The absence of a cathodic peak at a potential of 2.3 V indicates the impossibility of the formation of Li_2S_4 by the reaction:

$4Li^+ + S_8 + 4e \rightarrow 2Li_2S_4$, E = 2.3V vs. Li/Li^+. (4)

Figure 12 shows a cyclic voltammogram of a sulfur-based electrode, which contains natural Zavalievskiy graphite, carbon, and binder. The current-voltage characteristic of such electrodes has one first peak on the cathode branch at a potential of about 2.1 V. During the next cycling, the cathode peak almost completely degenerates. The anode peak also degenerates during cycling.

The presented results show that the sulfur-based electrode modified with nano graphite and containing no binder is stably cycled in the voltammogram regime.

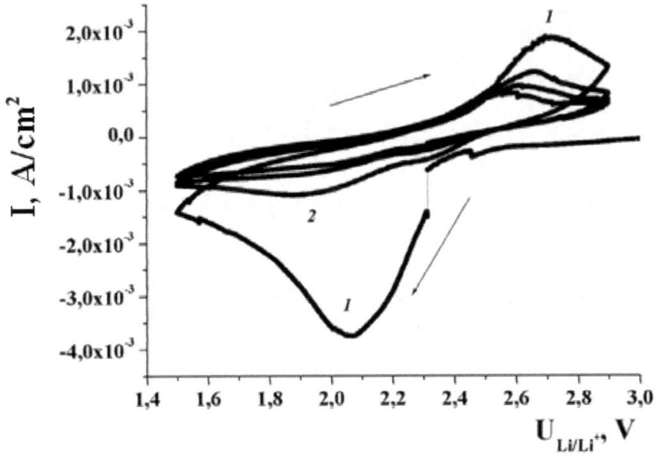

Figure 12. The current-voltage characteristic of the sulphur cathode in the Li-Li-S battery. The potential range of 1.5 - 2.9 V. The potential scanrate is 0.5 mV/ sec. Cathode includes: sulfur 75%, natural Zavalievskiy graphite: 10%, carbon 10%, binder 5%. Electrolyte: 0.7M bis(triuoromethanesulfone)imide (LiIm), 0.25M $LiNO_3$, DME:DOL (2:1). The numbers of the curves show the numbers of cycles.

5. Conclusion

1. Designed and manufactured nanostructured carbon powders materials have a high level of conductivity.

2. Using the innovative nanostructured carbon compositions for modification of the sulfur-based cathodes and not using the binder allow dramatically decrease the impedance of Li-S battery and increase the energy, power, and stability of the electrochemical process.

3. Reducing the density of nanostructured carbon powders provides a uniform distribution of this carbon material over the volume of the electrode mass and increases its energy.

4. High adhesion and cohesion of the nanostructured carbon powders to the particles of the electrode material optimizing the macro and micro the porous structure of the electrode mass, optimizing the distribution of the electrochemical reaction in the porous system of the electrode, and improves the performance of the electrodes and the energy and power of current source as a whole.

5. High adhesion and cohesion between the particles of the innovative nanostructured carbon powders to the particles of the electrode material makes it possible to exclude the binder from the composition of the electrode material, reduces the internal resistance of the current source and significantly increase the energy and of the current source.

Acknowledgments
Innovative technologies and materials, which are developing and implementing for high energy batteries and are presented in this article, are as follows

1. Company INT Sp. z o.o., Poland, develops and produces of unique nanostructured carbon-based materials for modify the electrodes and the solid electrolytes for develop and produce high energy flexible lithium batteries, including Li-ion, Mg-ion, and Li-S batteries, etc;
2. Enerize Corporation develops the non-destructive non-contact testing methods & devices for use during development and production of batteries.
3. Ukrainian State University of Chemical Technology (USUCT) investigates electrochemical properties of electrodes, which are modified by nanostructured carbon-based materials for Li-batteries including Li-ion and Li-S.
4. USUCT, Ukraine; Pavol Jozef Safarik University, Slovak Republic; and Enerize Corporation, which is an industrial partner of NATO project, together are realizing the NATO SPS 985148 project "Development of New Cathodes for Stable and Safer Lithium-Sulfur Batteries" / Science for Peace and Security Programme with the following progress of the Li-S batteries using the modification the Li-S batteries with the unique nanostructured carbon-based materials, which is developed and produced by the Company INT Sp. z o.o., Poland.

References

[1] Gómez I J, Sulleiro M V, Mantione D and Alegret N J 2021 *Carbon Nanomaterials Embedded in Conductive Polymers: A State of the Art Polymers* **13** p 745
[2] Peera S G, Koutavarapu R and Akula S J 2021 C*arbon Nanofibers as Potential Catalyst Support for Fuel Cell Cathodes: A Review Energy Fuels***35**, 15 pp 11761–99
[3] https://thebatteries.tech/product/
[4] https://www.marketsandmarkets.com/Market-Reports/flexible-battery-market-190884508.html
[5] Pustovalov Y, Shembel E, Redko V, Pastushkin T, Zaderey N, Markevich A, Vishnyakov L and Sagirov I J 2019 Innovating Carbon Materials Open New Possibilities for Increasing Performance of Li-ion Batteries *RS Global J. Word Sience* **2** pp 4-8
[6] Redko V, Shembel E, Pastushkin T, etc. *Method and Device for Rapid Non - Destructive Quality Control of Powdered Materials.* US Patent No. 8,102,181
[7] Redko V, Shembel E. etc. *Method and Apparatus for Measuring Conductivity of Powder Materials Using Eddy Currents.* US Patent No. 7,288,941
[8] Pustovalov Y, Shembel E, Vishnyakov Letc. *Carbon material for electrodes of lithium-ion power sources and method of production thereof.* US Patent application No. 10/898631
[9] Redko V, Shembel E etc. *Methods and Systems for Non-Destructive Determination of Fluorination of carbon Powders.* US Patent application No. 10/898631

Appraisal the effective diffusion coefficients of Li-ions by PITT and network thermodynamic methods in thin-layer Fe-sulfidic electrodes of Li-accumulator

R D Apostolova[1], O V Markevych[1] and E M Shembel[1,2]
[1]Ukrainian State University of Chemical Technology, Dnipro, Ukraine, 49005
[2]Enerize Corporation, Coral Springs, FL, 33067, USA

e-mail: apostolova.rd@gmail.com

Abstract. In this work, we explored the possibilities of network thermodynamics to determine the most important kinetic parameter of the diffusion process, the diffusion coefficient of lithium ions (D_{Li}) in iron sulfide. Iron sulfides have been electrolytically synthesized in thin aluminum-based layers for implementation in miniature lithium batteries. A comparison is made of the results obtained by the method of network thermodynamics and the method of potentiostatic pulse titration PITT. The theoretical aspects of both methods are presented. The task was reduced to obtaining curves I (current) - time (t), their analysis and calculation of coefficient values (D_{Li}), using the theoretical foundations of diffusion and the methods used. At the early stages of applying the method of network thermodynamics to the study of diffusion in lithium current sources, it was not clear why the method is suitable for determining D_{Li} not in the entire working range of potentials. A wide range of studies using the PITT method helps to answer questions related to the application of the network thermodynamic method. When using both methods, it is important to establish the potential range with diffusion kinetics and to reveal the accompanying electrode processes. Thus, it was found that both methods used are unable to provide reliable results in the Fe_xS_y electrode potential range of 2.8–1.8 V, since the diffusion nature of the electrode process is not a priority in this potential range.

The efficient conversion of Li-battery systems is associated with the solving the problems of heterogeneous kinetics of multistage electrode processes with phase transitions, the transformations at the interface, diffusion, and other reasons. The most important parameter of diffusion kinetics in a Li-battery is the diffusion coefficient of Li-ions (D_{Li}). A number of methods are proposed with their characteristic boundary conditions, assumptions, and shortcomings to determine D_{Li}. Different methods were used by us to determine D_{Li} in V_2O_5 electrodes [1].
In this study, we compared the values of the effective diffusion coefficient of Li-ions obtained by the method of pulsed potentiostatic titration (PITT) and the method of network thermodynamics for the further optimization of electrochemically synthesized iron sulfides in Li-battery.

1. Technology of synthesis and the structure of thin-layer Fe-sulfide electrodes.

1.1 Electrochemical synthesis of Fe-sulfide electrodes.
Fe-sulfidic electrodes with an active mass of approximately 1.5–1.8 mg/cm^2 of aluminium base were obtained by the methods [2–4].

Electrolyte composition, mole/L:
– $FeSO_4 \cdot 7H_2O$ – (7.0-7.5) $\times 10^{-2}$;
– $Na_2S_2O_3 \cdot 5H_2O$ – (4.0-4.5) $\times 10^{-2}$;
– H_3BO_3 – 2.0. pH (4.0-4.5);
Electrolysis parameters:
– $i_{cathode}$= 2.0–2.5 mA/cm^2;
– $T_{electrolyte}$ = 20 – 22°C.

1.2. Structure of Fe-sulfides.
Samples of Fe-sulfides in the form of compact precipitates on an aluminium base were studied. X-ray diffractograms were recorded on a DRON-2.0 device in the interval 2θ=10-80 degrees using a LiF single crystal as a monochromator (CoKα radiation). Operating mode of the installation: 35 kV, 20 mA.

According to the X-ray phase analysis, Fe_3S_4 phase and FeS phase admixture are present in the studied sediments of Fe-sulfidic material. On the aluminium base, the predominant structure is the cubic greigite Fe_3S_4 with detected interplanar reflections of d 3.82; d 1.88; d 1.73.

1.3. Electrochemical investigations of Fe-sulfidic electrodes.
The conversion of the active electrode material in the electrochemical reaction with lithium took place in a three-electrode cell with electrolyte EC (ethylene carbonate, Merck), DMC (dimethyl carbonate, Merck), 1 M $LiClO_4$ (Iodobrome), lithium counter electrode and Li/Li^+-reference electrode in the potential range 2.8–1.1 V. The discharge-charge curves of the chemical current source prototype were obtained in galvanostatic mode on a computer-controlled test bench. In the scheme of research, the degree of discharge (degree of lithiation) of the electrode was first determined. Then the potential balanced at a certain degree was shifted by 5–10 mV in the cathode direction and the change of current (I) was recorded for 3600 s. The dependence I (current) – t (time) was received. Recording of (I–t) – curves were performed using an analytical radiometer VoltaLab PJZ 301. Subsequently, the obtained chronoamperometric (I–t) responses were analyzed using theoretical aspects of PITT and network thermodynamics.

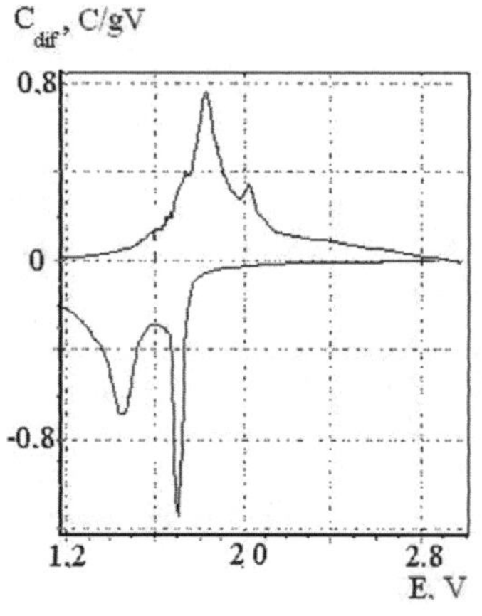

Figure 1. Differential intercalation /deintercalation capacitance of the Fe_xS_y-electrode obtained from galvanostatic discharge-charge curves after a long cycle. I_{disch} = 5 μA/cm^2.

In Fe_xS_y samples, the electrode intercalation process takes place in the range of 2.8–1.1 V in two stages (figure 1). Preliminary analysis of the process was limited by us to the first stage of 2.8 – 1.6 V (figure 2 a, b). The second stage is in the range of 1.6 –1.2 V (figure 1).

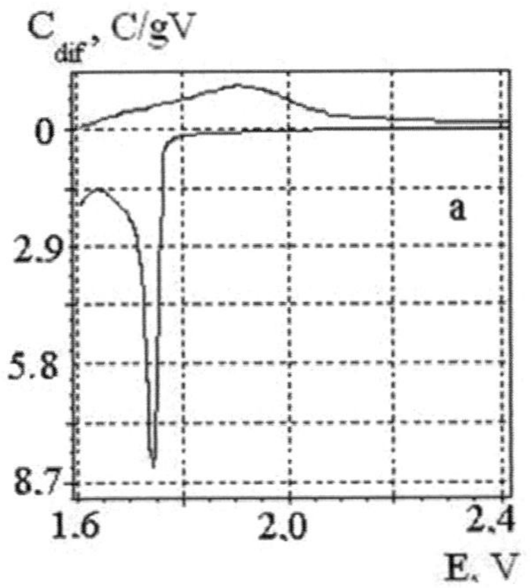

Figure 2a. Differential capacitance of Fe_xS_y–electrode vs. E (V), obtained from galvanostatic curves.

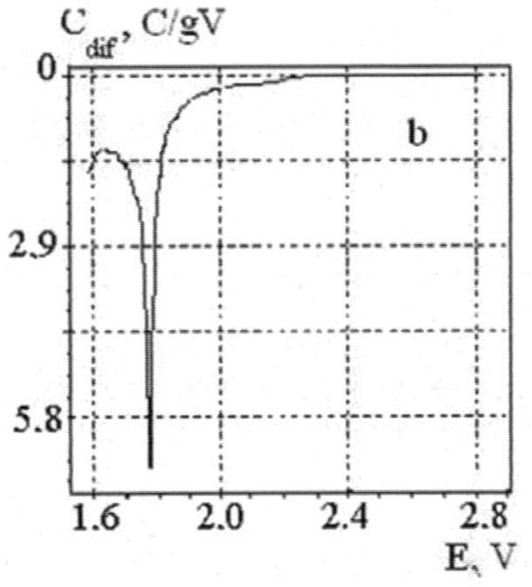

Figure 2b. Differential capacitance of Fe_xS_y–electrode vs. E (V) obtained from PITT responses.

2. Theoretical aspects and the results of the investigations.

2.1. Theoretical aspects of PITT method and the obtained results.

Using the approximation to the response of the electrode to the low-amplitude electrical signal, we can determine the diffusion time constant for the lithiation process of the electrode material [5] (1):

$$\tau = \left(Q_m \Delta x / \pi^{1/2} It^{1/2}\right)^2 = \left\{ Q_m \left[\frac{\Delta x}{\Delta E} \Big/ \left(\pi^{1/2} It^{1/2} / \Delta E\right) \right] \right\}^2 = \left(C_{dif} / \pi^{1/2} It^{1/2} / \Delta E\right)^2, t \langle\langle \tau \qquad (1)$$

where Q_m – the charge passing through the electrode after the imposition of an electrical signal, c/g; X – the degree of lithiation of the electrode material; $It^{1/2}$ – Cottrell slope, determined from the linear section of the curve I (current) – t (time), $A \cdot s^{-1/2}/cm^2$; C_{dif} – differential intercalation/deintercalation capacity, c/g V.

Effective coefficient (D_{Li}) can be determined as (2):

$$D = L^2/\tau, \tau \langle\langle t \qquad (2)$$

where L is the characteristic diffusion length.

The dependence of the differential intercalation/deintercalation capacitance (C_{dif}) on the potential (E) for equilibrium systems can be calculated from the corresponding PITT– studies or galvanostatic discharge-charge curves.

Based on the experimental PITT research, the effective chemical diffusion coefficient of Li^+ (D_{Li}) of the order of 10^{-11} cm^2/s was determined in the Fe_xS_y-electrodes. The calculations took L = (3.0–4.5) $\times 10^{-4}$ cm, based on the size of the aggregate particles, which are formed in turn from submicron units determined by electron microscopy of electrolytically obtained Fe sulfides [6]. If we take L = (0.30–0.45) $\times 10^{-5}$ cm into account the size of the spherical submicron particle, which consists of micro aggregates of the electrode material, the value of D_{Li} is obtained 2 orders of magnitude lower (10^{-13} cm^2 /s). However, in both calculations, only the conditional effective value of the chemical diffusion coefficient Li^+ was obtained in the electrode material with a porosity of 18% [3], the pores of which are filled with electrolyte. The value of D_{Li} (if L = 3×10^{-5} cm) varies with the potential in the

1.40–1.56 V potential interval of the second stage of recovery of Fe_xS_y-electrode as a result of curve analysis in figure 3. The order of D_{Li} values is set to (4.3×10^{-13}; 3.0×10^{-13}; 1.0×10^{-13}; 4.5×10^{-13}) in the E-range of (1.42; 1.44 ; 1.50; 1.56 V, respectively) when $L=3\times10^{-5}$ cm.

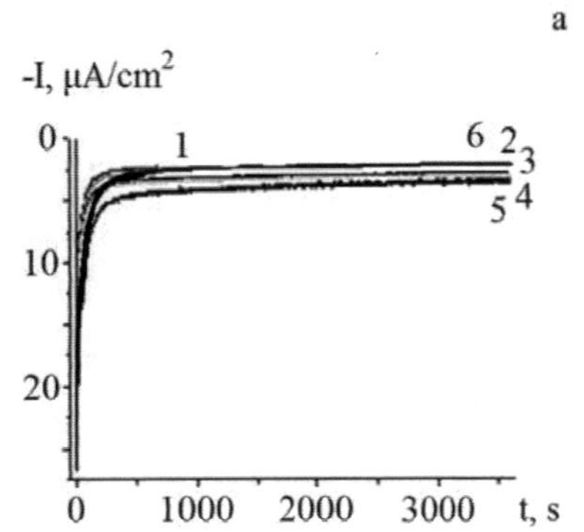

Figure 3a. PITT response of Fe_xS_y – electrode depending on the E, V in I – t coordinates. E, B: 1) 1.419, 2) 1.449, 3) 1.479, 4) 1.554, 5) 1.499, 6) 1.514.

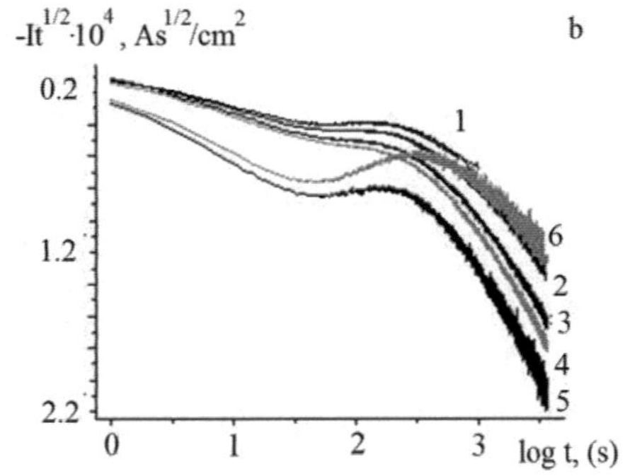

Figure 3b. PITT response of Fe_xS_y – electrode depending on the E, V in $It^{1/2}$ – log t coordinates. E, B: 1) 1.419, 2) 1.449, 3) 1.479, 4) 1.554, 5) 1.499, 6) 1.514.

In the potential range of 2.8–1.8 V, the Cottrell slope is not observed (figure 4).

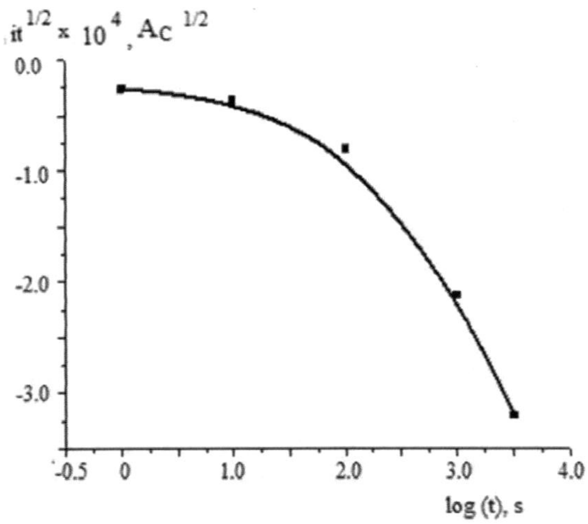

Figure 4. PITT response of Fe_xS_y–electrode at E=1.85 V.

Usually, the Cottrell slope in the $It^{1/2}$–log t coordinates appears as maxima, since the diffusion process is accompanied by other electrode processes. In figure 4, the absence of maxima indicates the non-ruling role of the diffusion process in Fe_xS_y.

2.2. *Theoretical aspects of network thermodynamical method and the obtained results.*

D. Oster, A. Perelson, A. Katchalsky gave birth to thermodynamics, which A. Katchalsky called "network" and discussed its relevance and potential for modern biophysics [7–9].

Network thermodynamics is considered as a method that provides an opportunity to provide a system with an electrical circuit and allow to apply the technique of network analysis to problems of thermodynamics.

Network thermodynamics began and initially actively developed in the fields of biology and in relation to human behaviour during the 60-70s of the last century until 2009-2010 years. And then there is a lull in the research of network thermodynamics in these areas, which the leading researcher D. Mikulecky explains to the complexity of network thermodynamic theory and the lack of interest of scientists in this connection. He predicted the remoteness of the successful finish of the network in thermodynamics [10].

Electrochemists also did not make a significant contribution to the development of network thermodynamics. A well-known study in electrochemistry in which a simple network thermodynamic technique was applied to a network model of membrane polarization of styrene particles and in giant water bodies using the spice program [11].

The possibility of using network thermodynamics in electrochemistry is evidenced by works [12, 13]. The method of network thermodynamics was used to determine the effective coefficient of diffusion of lithium ions in V_2O_5 oxide in an aprotic medium [1], as well as in MoO_3 oxide [14].

The method of the network thermodynamics is based on the mathematical model of diffusion of the Li^+ in electrode material using the diffusion equation for a one-dimensional system (3):

$$\frac{\partial c}{\partial t} = D \cdot \frac{\partial^2 c}{\partial x^2} \tag{3}$$

Equation (3), according to principles based on the network thermodynamics, may be transformed to the equation (4):

$$Z(t) = A1 \cdot \left[A0 \cdot t - 2 \cdot \sum_{k=1} \frac{1}{k^2} \exp\left(k^2 \cdot A0 \cdot t\right) \right] + A2 \tag{4}$$

where

$$A0 = \frac{\pi^2 \cdot D_{Li^+}}{\delta^2} \tag{5}$$

$$A1 = \frac{RT \cdot \delta}{(nF)^2 \cdot c \cdot_{Li^+} \cdot \pi^2} \tag{6}$$

$$A2 = R_f + A1 \cdot \frac{\pi^2}{3} \tag{7}$$

k – quantity of the exponents in the equation (4); t – relaxation time, s; δ – thickness of the electrochemical active layer, cm; D_{Li} – diffusion coefficient of the Li^+ in the electrode structure, (cm^2/s); R – gas constant, 8.31 J/mole·K; T – temperature, K; n – number of electrons in the electrochemical process; R_f – resistance of the electrode, $\Omega \cdot cm^2$.

Equation (4) was used for the approximation of the relaxation curves.

The description of research methods:

1. Method of current relaxation and transformation of this curve to relaxation of impedance during the time.

According to [13], the system impedance is described by the following

$$(p)=\frac{RT}{c(nF)^2\sqrt{D_{Li^+}p}}\coth\left(\delta\sqrt{p/D_{Li^+}}\right)+R_f \quad (8)$$

where c is the concentration of Li^+ in the film; p is the Laplace transform parameter (c^{-1}); δ is the film thickness of the electrochemically active material, cm; n is the number of electrons participating in the electrochemical process; F is the Faraday number; R_f is the electrode resistance, Ohm·cm^2.

2. Approximation of the relaxation curves according to the equation obtained on the basis of principles of the network thermodynamics. The goal of this approximation is determination of the variable factors (A0, A1 and A2) in the equation (4).

3. Calculation of the D_{Li} coefficient into Fe_xS_y electrode according to the equation (5).

The discharge curve of synthesized Fe_xS_y sulfide in the range of E=2.8-1.1 V is carried out in two stages: 1 – with a maximum near 1.68–1.71 V, 2 – in the range of 1.60–1.1V. Figure 5 shows an example of close agreement between the experimental and approximated data used to determine D_{Li} at E = 1.71 V. Inset: experimental results and the error values.

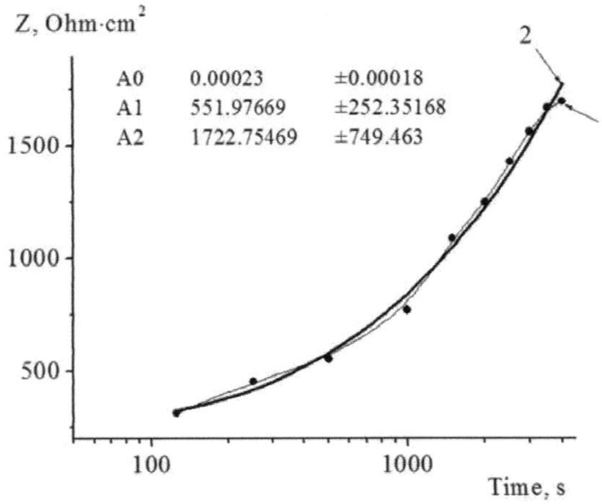

Figure 5. Experimental (1) and approximated (2) curves at E=1.71 V for experimental A0=0.00023.

The resulting D_{Li} value is equal to 2.3×10^{-11} cm^2/s when $L=3.0\times10^{-4}$ cm and to 2.3×10^{-13} cm^2/s when $L = 3.0\times10^{-5}$ cm at E=1.71 V.

However, in the potential range of 2.8–1.8 V, the experimental and approximated curves do not coincide. Figure 6 shows a significant divergence of these curves at a potential of 1.85 V.

In this case, the error value exceeds the value of parameter A0. This is explained by the non-prevailing role of the diffusion component in the Fe_xS_y sulfide electrode process and the significant contribution of the ohmic resistance and other components of the total electrode process to it.

The infringement of the diffusion component with varying degrees is also inherent in Fe-sulfide in the potential range of 1.6-1.1 V, for example, at a potential of 1.514 V (figure 7).

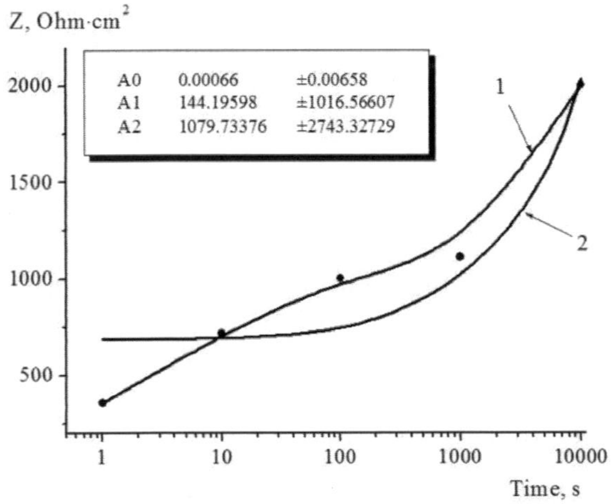

Figure 6. Experimental (1) and approximated (2) curves at E=1.85V for A0=0.00066.

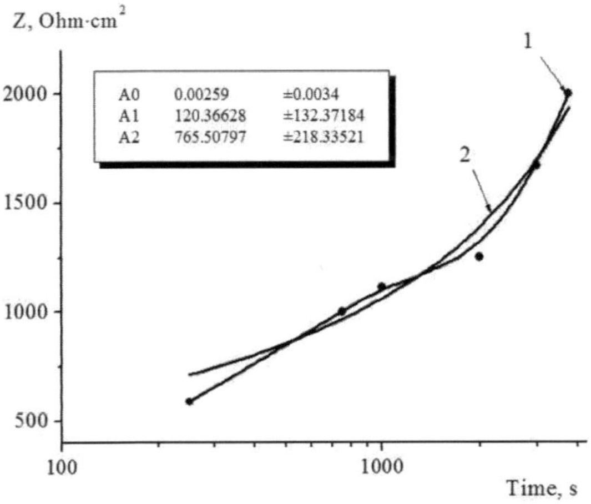

Figure 7. Experimental (1) and approximated (2) curves at E=1.514 V for experimental A0=0.00259.

The resulting D_{Li} value is equal to 2.3×10^{-11} cm^2/s when L = 3.0 $\times10^{-4}$ cm and to 2.3×10^{-13} cm^2/s when L = 3.0 $\times10^{-5}$ cm at E=1.71 V.

In the range of the discharge curve of 2.80–1.85 V, it is not possible to determine D_{Li}, neither by the PITT method, nor by the method of network thermodynamics, since the original semiconductor iron sulfide has a high ohmic resistance. As Fe_xS_y lithiates, it decreases.

The value of D_{Li}, corrected for the voltage drop across the high resistance of Fe_xS_y, and the slow charge transfer in [15], exceeds that of D_{Li} in this study using the PITT method by an order of magnitude. Obviously, in the future, it is necessary to improve the method of network thermodynamics to determine the correct D_{Li}, taking into account non-diffusion electrode processes.

Conclusions

In accordance with the task, the possibility of determining the effective diffusion coefficient of lithium ions in the synthesized iron-sulfide electrode material for a lithium battery in a wide range of potentials was evaluated using the well-known, often used PITT method and the approbated method of network thermodynamics. Both methods are based on the diffusion model of the electrode process. Known studies of iron sulfides, including PITT studies, help to explain some of our results.

1. Experimental data indicate that reliable information about the diffusion coefficient D_{Li} of synthesized Fe-sulfides can be obtained in a limited range of the electrode potential. The semiconductor nature of the initial iron sulfide makes a significant contribution to the electrode process, leveling the role of the diffusion component. This is clearly shown in the range corresponding to the values of the open circuit potential (OCP) 2.8 –1.8 V vs Li^+/Li-electrode.

2. As the sulfide is lithiated, its conductivity increases and the role of diffusion kinetics increases. Both methods for determining the D_{Li} coefficient show solidarity in its assessment in the range (OCP) 2.8 –1.8 V vs Li^+/Li-electrode.

3. At the maxima of the electrochemical activity of iron sulfide, the diffusion model can more reliably reflect Li^+ transfer process in the structure of iron sulphide.

4. Side processes associated with electrolyte decomposition, reduction and oxidation of iron sulfide during lithiation/delithiation veil the diffusion component, leading to a decrease in the D_{Li} coefficient. This was clearly shown by us in our studies of iron sulfides using the PITT method.

5. In the future, correction of the mathematical model in determining the rate of Li^+ mass transfer from diffusion to migration, influence of phase transition factors in iron sulfide or side processes factors is required when applying the method of network thermodynamics, as was done in experiments using the PITT method.

References

[1] Apostolova R, Tysyachny V, Markevich A, Ksenjek O, Shembel E, Markovsky B and Aurbach D. 2014 *Voprosi khimii i khimicheskoy tekhnologii* **2** 4–8

[2] Apostolova R and Shembel E. 2020 *Electrochemically synthesized iron-sulfide compounds for miniature lithium batteries* Monograph (Dnipro: LIRA LTD) 212

[3] Apostolova R., Zaderey N, Samojlov S and Shembel E. 2008 *Voprosi khimii i khimicheskoy technologii* **1** 123–129 (In Russian)

[4] Shembel E, Apostolova R, Nagirnij V, Baskevich A and Lytvin P. 2004 *Electrokhimiya* **40** 821–829 In Russian

[5] Wen C, Boukamp B and Huggins R.1979 *J. Electrochem. Soc.* **126** 2258–2266

[6] Apostolova R and Shembel. 2019 *Thin-layer electrolytically synthesized sulfides and oxides of transition metals for lithium current sources* Monograph (Dnipro; LIRA) 264 In Ukrainian

[7] Oster G, Alan A, Perelson S and Katchalsky A. 1973 *Quarterly Review of Biophysics* **6** 1–134. Published online by Cambridge University Press: 17 March 2009

[8] Oster G, Perelson A and Katchalsky A. 1971 *Nature (Lond.)* **234** 393–399

[9] Apostolova R, Tysyachny V, Markevich A, Ksenjek O, Shembel E, Markovsky B and Aurbach D. 2014 *J. Voprosi khimii i khimicheskoy tekhnologii* **2** 4–8

[10] Mikuleck C. 2011 *Axiomathes* **21** 455-471

[11] Horno J, González-Fernández C, Hayas A and González-Caballero F. 1989 *Biophys J* **55** 527–535

[12] Ksenzhek O and Petrova S. 1996 *Electrokhimija* **32** 1477–1485 (In Russian)

[13] Ksenzhek O and Petrova S. 1998 Advances in mathematical modeling and simulation of electrochemical process and oxygen depolarized cathodes Ed. by Van Zee and T. Fuller. “Elecrtochem. Soc. Inc.” 132–139

[14] Markevich A, Ksenzhek O, and Shembel E. 2005 *Voprosi khimii i khimicheskoy tekhnologii* **3** 164–168

[15] Apostolova R, Kolomoyets O and Shembel E. 2012 *Proc. XII Int. Conf. Fundamental problems of energy conversion in lithium electrochemical systems* (Krasnodar Russia October 1–6) 59–62

Melanin as biological organic polymer with semiconductor properties is unique effective modifier for MnO_2 cathode and increases the energy of Li-MnO_2 battery

I Maksyuta[1], E Shembel[1,2], V Kyrychenko[3], V Redko[2], T Pastushkin[2] and N Zaderey[1]
[1] Scientific Research Laboratory of Chemical Power Sources of Ukrainian State University of Chemical Technology, Dnipro, 49005, Ukraine
[2] Enerize Corporation, Coral Springs, FL, 33067, USA
[3] Sunoil-Agro, Pavlograd, 51400, Ukraine

irmax777@gmail.com

Abstract. This article presents results on the modification of a composite mass based on MnO_2 with an organic polymer – melanin. The main goal was to evaluate the possibility of increasing the efficiency of electrode processes during cycling and their stabilization. MnO_2, synthesized chemically on the basis of Ukrainian manganese ores, was used for modification. This manganese dioxide has a high density (about 2.64 g/cm^3), a high specific surface area (28 m^2/g), and electronic conductivity also. The final product of the synthesis is γ-MnO_2 powder with 5% admixture of γ-Mn_2O_3.

1. Introduction

Improvement and modernization of existing Li-ion are aimed at increasing their energy intensity, reliability and safety. The specific characteristics of reversible lithium electrochemical systems are constantly improving due to regular research aimed at finding new and improving existing electrode materials.

The leading position among the compounds for the positive electrode is occupied by manganese oxides. This material has a number of advantages: high capacity, high positive potential, low cost and low toxicity. Li-MnO_2 systems are widely used in the market as primary current sources. They are also successfully used as reusable cycling current sources. The advantages of lithium-manganese dioxide cells include the possibility of obtaining a sufficiently high specific energy. The disadvantage of the cells is their low specific power due to weak discharge currents, and their limited performance at low temperatures (<–20°C). Preliminary heat treatment of the active cathode mass improves the discharge characteristics and increases the safety of the battery. A promising direction for improving the energy characteristics of Li-ion is the modification of electrode materials [1].

The main goal of this work was to evaluate the possibility of stabilizing and increasing the efficiency of electrode processes during cycling of a composite electrode based on MnO_2 by modifying it with an organic polymer semiconductor, melanin. A positive effect from the addition of melanin was previously obtained by us for a composite electrode based on $LiMn_2O_4$ [2]. Melanin, which is used for modification of the cathode based on MnO_2, is produced on the basis of innovative technologies developed by Sunoil-Agro Company. Principal common structure of Melanin is

$C_{18}H_{10}N_2O_4$; density is from 1.6 to 1.8 g/cm^3; molecular weight is 318 g/mol. The innovative technologies for synthesizing and producing the MnO_2 by a chemical method is based on Ukrainian manganese ores, and are developed by Enerize Corporation company USA [3].

2. Experimental methodology

Modification of MnO_2 was carried out in a water-alcohol solution. A certain amount of this solution was added with continuous stirring to a mixture of dry components MnO_2 and melanin until a paste-like mass was formed, followed by heat treatment at temperatures of 250°C.

The starting material for the modification was MnO_2 synthesized chemically from the Ukrainian manganese ores. This manganese dioxide has a high density, about 2.64 g/cm^3, a high specific surface area of 28 m^2/g, and also electronic conductivity.

The compositions of the active mass of the composite electrode were as following:

1) MnO_2 - 85%, graphite (EUSM) - 5%, acetylene black -5%, fluoroplastic binder brand 4FD 5%;

2) Mixture (MnO_2 - 64% and melanin - 16%) 80%, graphite (EUSM) -7.5%, acetylene black (7.5%) and fluoroplastic binder brand 4FD - 5%.

The evaluation of the effect of melanin on the characteristics of MnO_2 during multiple cycling was carried out in prototypes of lithium batteries of a disk design, size 2325. The discharge current was 0.1 mA/cm^2, the charge current was 0.05 mA/cm^2. Electrolyte: 1 M $LiClO_4$ in a mixture of EC and DMC (1:1), (Merck). The methods of continuous galvanostatic cycling and cyclic voltammetry with potential sweep were used. The auxiliary electrode was metallic lithium.

The assembly of lithium battery models and the preparation of electrolytes were carried out in a dry sealed glove box in an argon atmosphere. The control of the phase composition of MnO_2 was carried out by X-ray phase analysis using a DRON-3 diffractometer in monochromatic Cu-Kα radiation. The electrical conductivity of the initial MnO_2 powder was determined by a non-contact non-destructive electromagnetic method on a laboratory bench developed by Enerize Corporation.

The elements were tested at room temperature from +25°C and to +60°C. The charge-discharge curves in the galvanostatic mode in the potential range of 2.0 ÷3.6 V were recorded on an automated stand.

3. Results and Discussion

In figure 1 the results of investigation of the conductivity of two types of MnO_2, which are synthesized by chemical and by electrochemical methods, are presented. The measurement of the conductivity of these powders is realized using an innovative non-contact non-destructive electromagnetic method, which was developed by Enerize Corporation [4, 5]. Dependence of the conductivity of the powder on its density are presented in these figures. Powder compaction is carried out by vibration.

Results, presented on figure 1, confirmed that the conductivity of the powder MnO_2, which is the chemically synthesized powder of MnO_2 from Ukrainian manganese ores, is higher than the conductivity of MnO_2 powder, which is from the market. Investigation, which are presented below, have been provided with powder of MnO_2 that were chemically synthesized on the basis of Ukrainian manganese ores.

The electrochemical properties of cathodes based on MnO_2 and MnO_2 with the addition of melanin were analyzed in terms of the following parameters: average discharge voltage, specific primary and reversible capacitance, and rate of discharge capacitance decrease per cycle during discharge/charge testing.

The discharge-charge curves of the composite electrode based on melanin-modified MnO_2 are shown in figure 2. The nature of the curves of the unmodified MnO_2 is similar to the curves of the modified one.

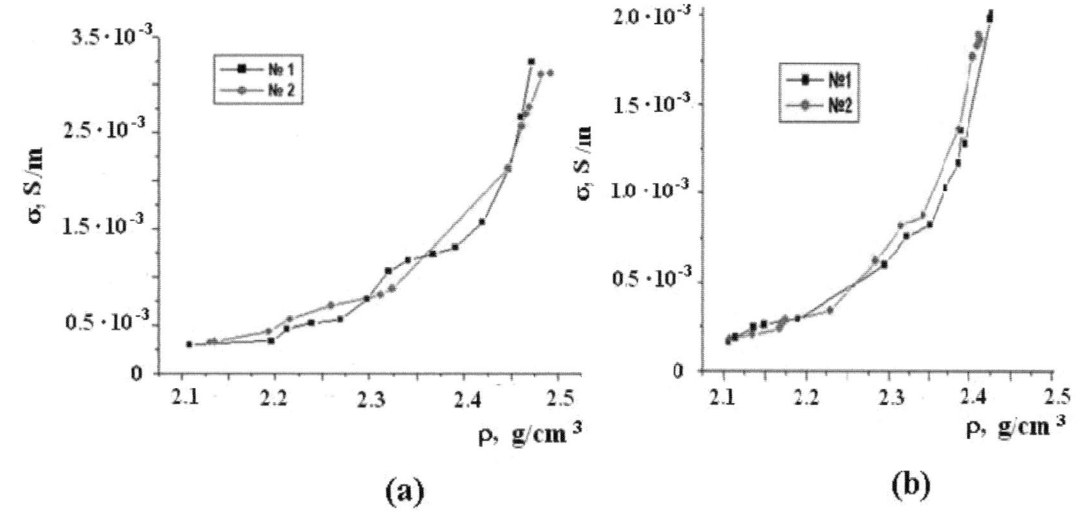

Figure 1. Comparison of the conductivities of powders of MnO_2, which is synthesized by various methods: (a) chemically synthesized MnO_2 by Enerize Corporation on the basis of Ukrainian manganese ores and (b) electrochemically synthesized MnO_2 from the market.

The process of intercalation of lithium ions into the crystal lattice of manganese dioxide on the discharge curve of the first cycle (fig. 2, curve 1) takes place almost at a constant voltage of 2.9-2.8 V.

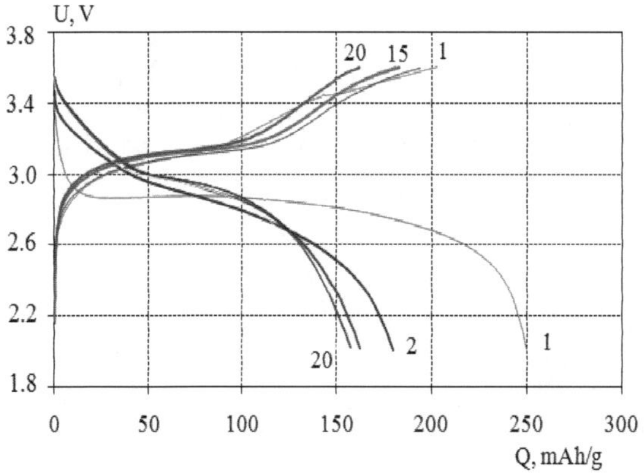

Figure 2. Discharge - charge curves of electrode based on MnO_2 that modified with melanin. Current of discharge 0,1mA/cm^2, current of charge 0,05 cm^2. Electrolyte: 1M $LiClO_4$, DMC, EK. The numbers near the curves are cycle numbers.

The profile of the discharge curve of the second and subsequent cycles differs from the profile of the first cycle. The discharge voltage in the second and subsequent cycles is higher than in the first, the discharge capacity is reduced. This is due to the fact that the phase composition of manganese dioxide during charging in the second cycle is described by the formula $LixMnO_2$, which differs from the initial MnO_2. On the charging curves of the composite electrodes, there are two horizontally inclined voltage sections near 3.0-3.15 V and 3.4-3.50 V. This indicates that the process of de- intercalation of lithium ions from $LixMnO_2$ during charging proceeds in two stages.

A comparative analysis of the test results showed that the modification with melanin contributes to a significant increase in the discharge capacity in the first cycle. Comparison of the discharge-charging characteristics of the studied composite masses shows (fig. 3) that the discharge capacity of the modified MnO_2– electrodes on the 1st cycle reaches 258 mAh/g, for an unmodified electrode this value is 220 mAh/g. The utilization rate of modified MnO_2 in the first cycle is 84%, which is 14% higher than for unmodified samples.

Modification with MnO_2 melanin makes it possible to increase the stability of its cycling. The degree of reduction in the discharge capacity of the modified cathode is in the range of 1.3 - 1.5 mAh/g per cycle. For unmodified MnO_2, this figure is 3.3 mAh/g per cycle, which is 2 times higher than that of the modified one (fig. 4).

The data presented in figure 4 show that the introduction of melanin into the MnO_2 composite mass makes it possible to increase not only the primary but also the reversible discharge capacity. The value of the reversible discharge capacity of the modified electrodes at a current density of 0.05 mA/cm^2 for the first 10 cycles varies from 188 to 167 mAh/g.

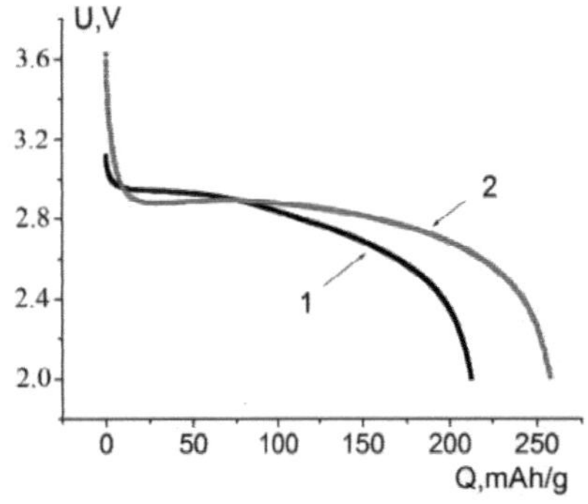

Figure 3. Comparison of curves of the разряда. Current of discharge 0,1mA/cm^2. 1 - electrode based on the initial $LiMnO_2$; 2 - electrode based on modified with melanin MnO_2.

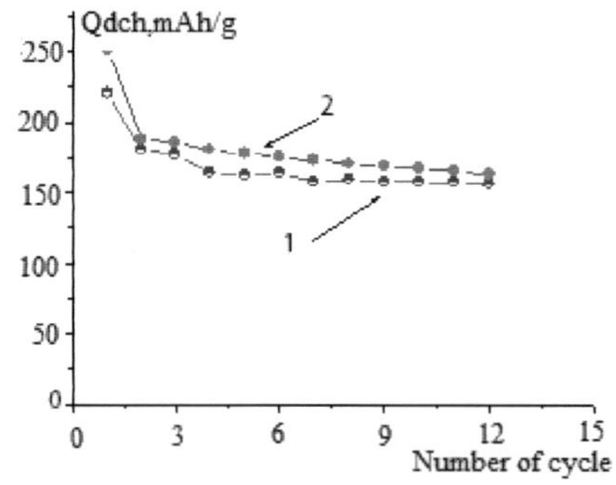

Figure 4. Discharge capacity of MnO_2 electrodes during cycling. 1 - electrode based on the initial MnO_2; 2 - electrode based on modified with melanin MnO_2.

For unmodified electrodes this value is equal to 180-157 mA h/g and has a strong tendency to decrease during cycling. A comparative analysis of the results shows that the modification of MnO_2 with melanin makes it possible to increase the stability of its cycling. Thus, the degree of reduction in the discharge capacity of the modified cathode is in the range of 1.5 - 2.3 mAh/g per cycle. For unmodified MnO_2, this indicator is from 2.5 to 5.6 mAh/g per cycle, which is 2 times higher than that of the modified one (fig. 4).

The positive effect of the modification of MnO_2 with the addition of melanin was confirmed by cyclic voltammetry. Comparison of I-U - curves shows that in the first cycle the current value at the maximum of the cathode region at a voltage of 2.4 V for an unmodified cathode is 4.56 mA (fig. 5(a), curve 1), and for MnO_2 with the addition of A melanin - 7.56 mA (fig. 5 (b), curve 1).

The magnitude of the current at the maximum of the current-voltage curve depends on a number of factors, one of which is the mass of the active substance. The mass of MnO_2 in the unmodified cathode is 0.0407 g. This is 1.5 times higher than that for which is equal to 0.0279 g. Therefore, the cathodic peak should be higher for the unmodified electrode. The data obtained clearly show the positive effect of modifying the composite mass based on MnO_2 with the addition of melanin. The introduction of melanin into the composition of the cathode mass makes it possible to increase the current peak at the maximum by 1.7 times in the first discharge cycle.

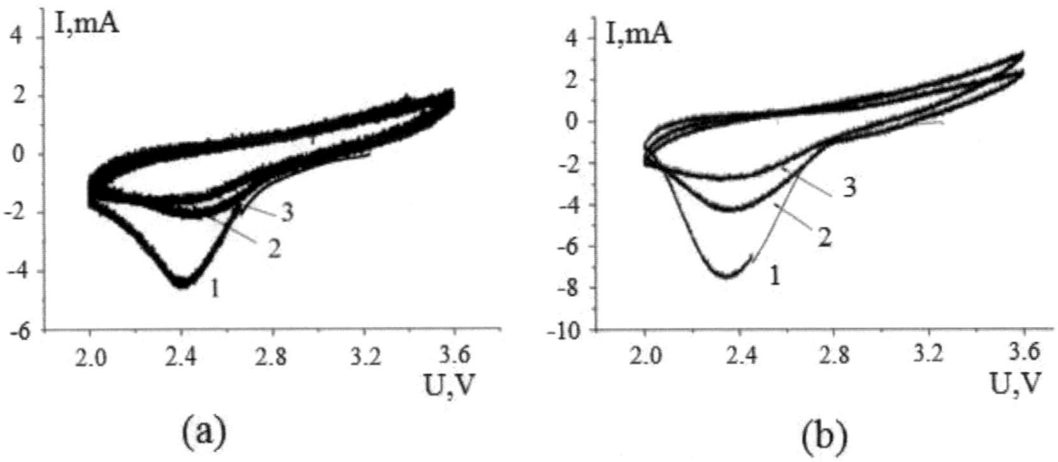

Figure 5. Cycling I-U – curves of Li – MnO_2: (a) electrode based on the initial MnO_2; (b) electrode based on modified with melanin MnO_2. Temperature 25°C. Potential scan rate 0.5 mV/s. Electrolyte: 1M $LiClO_4$, DMC, EK (1:1). The numbers near the curves are cycle numbers.

The efficiency of the reversible process of intercalation/deintercalation of lithium ions into the structure of an electrode modified with melanin and unmodified with MnO_2 at a temperature of +60°C was studied by cyclic voltammetry.

Cycling of composite unmodified MnO_2–electrodes at a temperature of +60 °C is characterized by an almost constant value of the discharge peak for 1, 2 and 3 cycles, the value of which for these cycles is 14.5 mA, figure 6 (b), curves 1, 2, 3. This value is much higher than the peak currents obtained for the system under study at room temperature (25°C). At room temperature, the maximum current for 1 cycle is - 4.58 mA; on cycle 2 - 2.12 mA; 3 - 1.72 mA, figure 5 (a).

The next feature of the current-voltage cycling of the composite MnO_2 electrode at +60°C, compared to room temperature, is a significant increase in the peak current in the anode cycle, which corresponds to the process of deintercalation of lithium ions from the MnO_2 crystal lattice. The peak values of the anode currents for cycles 2 and 3 are close to each other (118 mA and 125 mA) and significantly exceed the peak values of the discharge currents, fig. 6 (a), curves 2, 3.

The effect of temperature on the characteristics of the MnO_2 electrode, which modified with melanin, is shown in Fig. 7. Cycling at a temperature of + 60°C increases the rate of deintercalation of lithium ions in the crystal lattice of MnO_2. This conclusion is confirmed by the appearance of well-defined anodic peaks, compared with the data at room temperature, both for modified with melanin (compare fig. 5 (b) and 7) and for unmodified MnO_2 (compare fig. 5 (b) and fig. 7 (a)).

A comparative analysis of the obtained data shows that the discharge at a temperature of + 60°C for an unmodified electrode in the first cycle shows a maximum of 14.5 mA (fig. 6 (b)). For the modified one, this value is 10.75 mA (fig. 7 (b)). At first glance, the results for the unmodified electrode are higher than those for the modified one. However, one must take into account the fact that the mass of MnO_2 in the unmodified cathode is 1.5 times higher than that for the modified one. Therefore, the difference in the cathode peaks is due to the difference in the mass of the active substance of the studied elements.

The electrochemical process of intercalation/deintercalation of lithium ions into the structure of manganese dioxide in a non-aqueous electrolyte includes a number of stages, such as: transport of charged particles in the volume of the electrolyte and in the pore space of the electrode; charge transfer through the surface film/composite interface; charge migration in a solid electrolyte surface film; diffusion of lithium ions in the active material of the electrode; charge transfer across the composite/current collector interface.

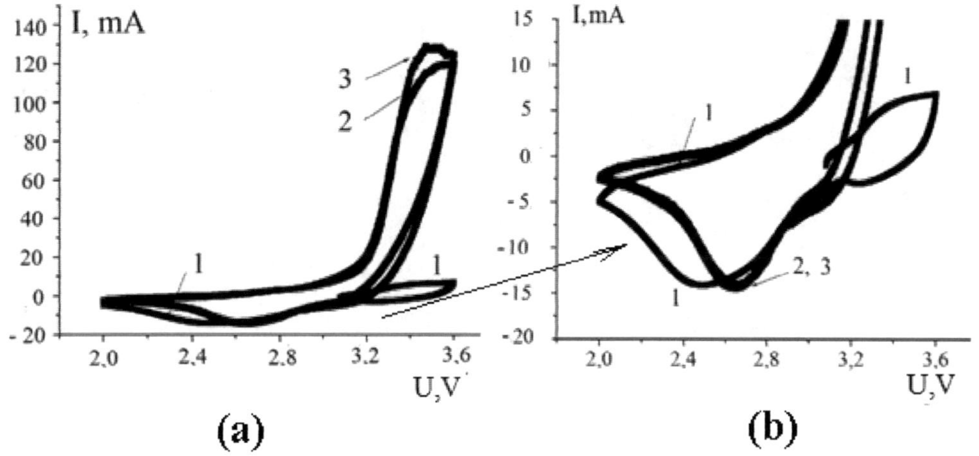

Figure 6. Cycling I-U–curves of non modified MnO_2: (a) I-U – curves for 1-3 cycles; (b) cathode area I-U-curves. Temperature 60°C. Potential scan rate 0.5 mV/s. Electrolyte: 1M $LiClO_4$, DMC, EK (1:1). The numbers near the curves are cycle numbers.

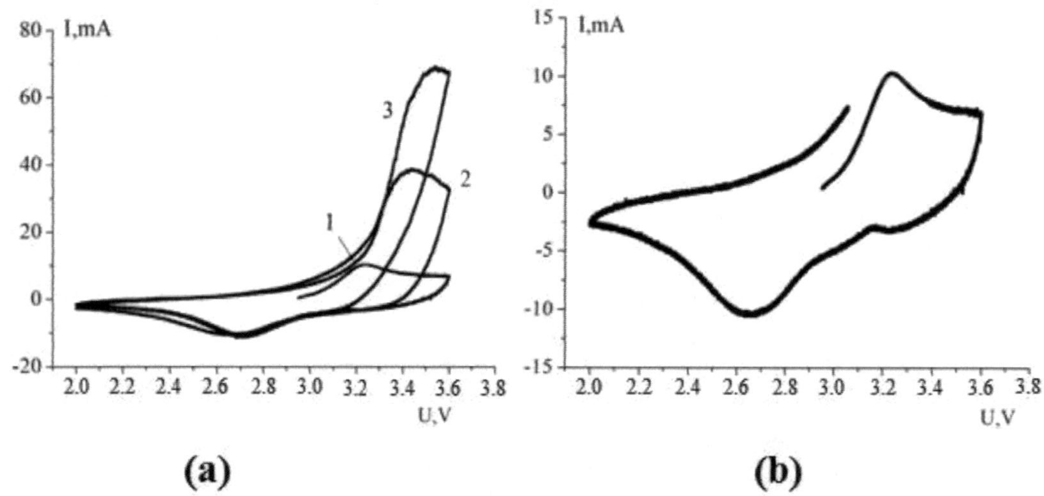

Figure 7. Cycling I-U – curves of modified with melanin MnO_2: (a) I-U – curves for 1-3 cycles; (b) I-U-curve 1st cycle. Temperature 60°C. Potential scan rate 0.5 mV/s. Electrolyte: 1M $LiClO_4$, DMC, EK (1:1). The numbers near the curves are cycle numbers.

The data obtained at this stage of research do not allow us to establish which of the listed stages have a decisive effect on the electrochemical efficiency of manganese dioxide. It can be assumed that the increase in the discharge characteristics of the composite MnO_2 with increasing temperature is due, to a greater extent, to the transfer of charges through the interface of the SEI film/MnO_2 composite, the resistance of which decreases with increasing temperature.

It is important that the introduction of A melanin into the composite mass makes it possible to increase the utilization factor of MnO_2 in the first discharge cycle by 15%, as well as to increase the reversible discharge capacity by 8% and to increase its stability during cycling by 2 times.

An increase in the discharge capacity of a cathode based on MnO_2 modified with melanin can possibly be explained by the effect of melanin as a semiconductor on the properties of the solid electrolyte film (SEI/MnO_2) at the electrode/electrolyte interface. This also confirms the positive effect of melanin on the electrochemical characteristics of the MnO_2 cathode.

References

[1] Johnson Ch 2007 Development and utility of manganese oxides as cathodes in lithium batteries *J. of Power Sources* **165** pp 559–65

[2] Shembel E, Kyrychenko V, Maksyuta I, Redko V, and Zaderey N 2020 Melanin as semiconductor with polymer structure is effective modifier for electrodes of high energy Li - ion batteries *ECS Transactions* **99** (1) pp 47-56

[3] Shembel E, Pisniy V, Globa N, etc. *Manganese Dioxide for Lithium Battery Cathode*. US Patent Application 10/583,886

[4] Redko V, Shembel E, Pastushkin T, etc. *Method and Device for Rapid Non - Destructive Quality Control of Powdered Materials.* US Patent No. 8,102,181

[5] Redko V, Shembel E. etc. *Method and Apparatus for Measuring Conductivity of Powder Materials Using Eddy Currents*. US Patent No. 7,288,941

Different carbon processes for lithium-sulfur batteries

N Hoffmann[1], M G Ortiz[1,2], J E Thomas[3], A Visintin[1]

[1]Instituto de Investigaciones Fisicoquímicas Teóricas y Aplicadas (INIFTA), Facultad de Ciencias Exactas (UNLP), CONICET, Diag. 113 y 64, La Plata, Argentina

[2]Centro de Investigación y Desarrollo en Ciencia y Tecnología de Materiales (CITEMA), Universidad Tecnológica Nacional - CICPBA, 60 y 124, 1923, Berisso, Argentina

[3]YPF Tecnología S.A., Av. del Petróleo Argentino S/N (entre 129 y 143), B1923 Berisso, Buenos Aires, Argentina

nhoffmann@inifta.unlp.edu.ar

Abstract. In this work, different carbon modification processes have been employed and the results have been compared with the unmodified Super-P carbon. The active materials were used in respective cathodes materials for Li/S batteries. The electrochemical performance was studied by galvanostatic charge-discharge cycling, cyclic voltammetry, electrochemical impedance spectroscopy and rate capability. CV measurements showed the modified carbons started out charging and discharging at set voltages, with their peaks broadening as part of the cycling-induced decay. It was found that the carbon processed with nitric acid delivered the highest capacity retention after 100 cycles at C/10. The results indicate that the modification of carbon with nitric acid could be promote the decrease of the shuttle effect.

1. Introduction

The progress on lithium-ion batteries is close to reaching its physical limits, and thus different chemical systems must be used to achieve higher energy densities. Among the chemistries currently under research, the system based on the reaction between lithium and sulfur is closest to adoption, with several prototypes demonstrating its viability. Lithium-sulfur batteries are expected to reach a capacity of 600 Wh/kg, over twice the 270 Wh/kg that lithium-ion batteries deliver. However, the complex chemistry of the system requires a strategy to circumvent its disadvantages [1, 2]. Such a development is of key technological and social importance, as the recently proliferating electric vehicles can strongly gain from such batteries: it has been long recognized that a range of 500 km on a single charge will be a game-changer for the industry once it is economically achievable at scale. A denser battery allows both for more energy and less weight, both of which make it more feasible to achieve such a driving range [1].

Improvements in battery technology also drive the substitution of fossil fuels in consumer applications, just like with cars. Even though the mitigation of greenhouse gas emissions through grid power is generally negligible unless the grid mix does not depend in fossil fuel-based power generation [3], local improvements in air quality are possible [4], and utilities can realize gains in efficiency that are impossible in consumer-scale devices, such as the use of co-generated gas and heat power [5]. Such small improvements in efficiency can become more significant through greater improvements in battery technology, as they expand the possible applications of grid-based power.

Firstly, the sulfur chemistry inside a battery requires the generation of several polysulfide varieties, molecules comprising eight or fewer atoms of sulfur and two lithium atoms. These molecules span several compositions, changing as the cell charges, discharges, and even at rest; their polarities change with each composition, and many of them are soluble in the electrolyte [6]. Their dissolution allows them to deposit more efficiently on the cathode, but it also allows them to reach the anode, which results in capacity loss and electrochemical blockage of the lithium surface. This issue, the "polysulfide shuttle", is perhaps the most difficult challenge for the creation of efficient and long-lasting lithium-sulfur batteries, inspiring numerous strategies to anchor polysulfide molecules in the cathode while still allowing for the transport of lithium ions across the electrolyte [7].

The use of carbon as the support material for sulfur in the cathode still allows for numerous different strategies, such as using a different source for carbon to vary the final structure [8, 9]. As pure carbon structures are nonpolar, they have very weak interactions with the highly polar polysulfides. Furthermore, the carbon atoms in a pure carbon structure are electrically neutral. Thus, additions of heteroatoms with other oxidation states and electronegativity, or even functional groups, are desirable in order to mitigate the shuttle effect inherent to the system [10]. Certain specific structures can even go so far as to completely alter the chemical reactions happening inside the cell [11]. The chemical structure of the carbon support used can be altered too, to introduce moieties to it. These moieties can be single atoms of nitrogen, oxygen [12, 13], or even sulfur, as well as more complex moieties.

Due to their polarity, these more complex groups interact more strongly with the polysulfide molecules, which are polar and soluble in the electrolyte: they are attracted to other polar elements, and thus remain attracted to the cathode over the anode throughout their life. Prevention of the polysulfide shuttle to the anode limits deposition of insulant lithium sulfide on the anode and allows for a higher availability of sulfide species in the cathode, where it provides capacity to the battery [13]. Polar materials have been successfully added to both the cathode and the separator to increase the availability of sulfur in the cathode, improving its capacity retention, and reducing the loss of capacity in the anode [14, 15, 16, 17].

In this work, the strategy selected to improve the capacity loss of the cathode is the use of chemical processes, described in the extant literature, that modify the surface of the activated carbon powders that are, in turn, used as sulfur hosts in the active material of the cathode for the lithium-sulfur cells. The selected processes can introduce a variety of moieties based on nitrogen [18] added through mechanochemistry with ammonia, a common and inexpensive reagent. The functions added to the carbon through this process are highly polar, and thus should attract polysulfides to the cathode, where they are available for the reaction. The second process also adds nitrogen-oxygen moieties [19] through simple boiling in nitric acid under reflux: the nitric anion possesses a highly polar nitrogen-oxygen link. The third process adds oxygen-based moieties through boiling with hydrogen peroxide [20].

The selected processes are simple and require inexpensive reagents, which allow for simple scaling of the creation of polar moieties on the surface of the material for enhancement of polysulfide retention.

2. Experimental

2.1. Carbon Modification

2.1.1. Nitric Acid Modification of Super-P. 1 g of Super-P and graphite carbon black (TIMCAL) was placed in a heating balloon with 100 ml of boiling, concentrated nitric acid (65%) under reflux and stirring for six hours.

After filtering to remove the nitric acid, the material was repeatedly washed in distilled water until neutral pH readings were obtained, and then dried overnight at 80°C.

*2.1.2. Ammonia Modification of Super-P. 1.*2 g of Super-P and graphite carbon black (TIMCAL) were mixed with 0.5 ml of ammonium hydroxide (Aldrich, 25%) and placed in a planetary ball mill (Pulverisette-7) with 10mm diameter steel balls. The material was ground down at 800 rpm for 30 minutes.

After the grinding was complete, the obtained carbon was washed in distilled water until neutral pH readings were obtained, and then dried overnight at 80°C.

2.1.3. Hydrogen Peroxide Modification of Super-P. 2 g of Super-P and graphite carbon black (TIMCAL) were mixed with 50 ml of concentrated hydrogen peroxide (30%, Alfa Aesar) and placed in a heating balloon. The material was heated to a boil, under reflux and with stirring, for six hours.

After filtering to remove the peroxide, the material was repeatedly washed in distilled water until neutral pH readings were obtained, and then dried overnight at 80°C.

2.2. Material Preparation

2.2.1. Active Material Preparation. Each material, as well as normal Super-P and graphite carbon black (TIMCAL) as a control, was mixed with powdered sulphur (Parafarm). The combination had a final ratio of 1:3 carbon-to-sulfur by mass. The combination was introduced in a ceramic vessel and heated in a tubular oven (Estigia) at 155°C for 12 hours under a nitrogen (Linde N2, 4.8) atmosphere, a process known as "melt-diffusion". The obtained materials were then ground at 300 rpm for 10 minutes in a planetary mill (Pulverisette 7) with 3 mm diameter stainless steel balls.

2.2.2. Electrode Preparation. After melt diffusion and a short grinding, the materials were combined with additional Super-P and graphite carbon black (TIMCAL) and PVDF (Aldrich, Mw=534000) in the following ratio: 80 % active material, 10% PVDF, and 10% Super-P. Then, these were mixed with 1-methyl-2-pyrrolidone (Fluka). The aluminum foil was coated with this mixture of materials and kept overnight at a temperature of 60°C. After 12 hours the aluminium foil was pressed in a hydraulic press, and 15 mm circular sections were cut out for use as working electrodes.

2.2.3. Cell Assembly. Prior to usage, the electrodes were weighed after 30 minutes vacuum drying at 60°C, then they were placed inside an MBRAUN MB-10 glovebox with an internal atmosphere with under 0.5 ppm of both H_2O and O_2. The cut-out electrodes were used as the cathodes in CR2032 cells, while a small disc of lithium (Aldritch 99.9%) was used as the anode. A 19 mm disc cut out of 20 micron-thick porous polypropylene (Celgard H2010) was used as separator.

The electrolyte was prepared as a 1:1 mix of anhydrous dioxolane (Sigma-Aldrich, 99.8%) and anhydrous 1,2-dimethoxyethane (Sigma-Aldrich, 99.5%) with 0.25M lithium nitrate (Aldrich, 99.99%) and 1M anhydrous lithium bis-trifluoromethanesulphonimide (Sigma-Aldrich, 99.99%). Electrolyte so prepared was added in an E/S ratio of 15 μl/mg. The cells were then sealed in a TOB-MR-120 hydraulic press at a pressure of 100-120 kgf/cm^2.

2.3. Electrochemical Characterization

2.3.1. Cyclic Voltammetry. Two cycle voltammograms were measured on an AUTOLAB PGSTAT302N galvanostat/potentiostat at a scan rate of 0.05 mVs-1 in potential ranges of 1.75- 2.75 V.

2.3.2. Charge-Discharge Cycles. Initial cycles were made with a theoretical C-rate of 1/50; that is, the sulfur amount was calculated, and a symmetrical charge-discharge rate of 33 μA-mg^{-1}(S) was applied. After three cycles, the charge-discharge rate was recalculated such that the capacity would instead be C/10. Specifically, the current was increased by a factor equal to 1/10 of the number of hours of the last discharge duration.

The device used for charging, discharging, and measuring of the capacity was an Arbin Instruments BT-2000 battery tester device. Potential windows of charging and discharging cycles were set at 1.8 and 2.8 V.

2.3.3. Electrochemical Impedance Spectroscopy. Electrochemical impedance spectroscopy (EIS) was measured at three states of charge (SoC): at full charge, at full discharge, and half charge. Furthermore, cells were measured at two different conditions: after 100 cycles at C/10, and prior to the cycling. To place a cell with 100 cycles in its half-charged state, it is fully discharged (at C/10 rate), and then charged at C/10 rate for half of its last cycle's charge duration. A cell without cycling is instead charged and then discharged at a tentative C/50 rate (33 μA-mg^{-1}(S)) following by charging it for half the duration of its discharge.

The device employed for EIS is a Gamry Instruments INTERFACE 1000B module (potentiostat/galvanostat/ZRA). The EIS was measured at an amplitude of 10 mV over a frequency range of 0.01 Hz to 100 kHz.

3. Results

The discharge capacities at C/10 are shown in Fig. 1. Though the discharge capacity values are quite similar, there is a loss followed by a recovery for the ammonia-modified carbon. The unmodified carbon shows a higher capacity at first (over 1000 mAh/g(S)), it is overtaken at different points by all the others (600-850 mAh/g). The initial high capacity of the unmodified carbon in comparison with the modified carbons indicates a higher availability of sulfur during the first few cycles, with no subsequent recovery, which is eliminated or at least limited in all modified carbons, whether altered through mechanical chemistry or through heating with a reagent. The retention capacity of nitric acid modified Super-P is highest of all, showing the best stability over 100 cycles. Figure 1b shows a normalized version, for ease of comparison, the discharge capacity retention values are more stable for nitric acid modified Super-P, remaining at values between 82-90%; while for unmodified carbon these values change from 90% to 58%. Thus, nitric moieties in the carbon could be contributing the most stability to the cathode by limiting the effects of polysulfide shuttling, though their effects wear out during cycling.

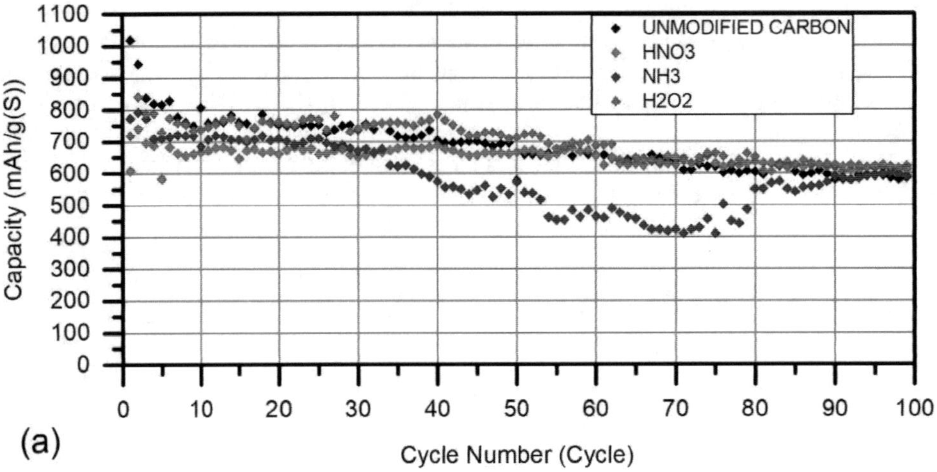

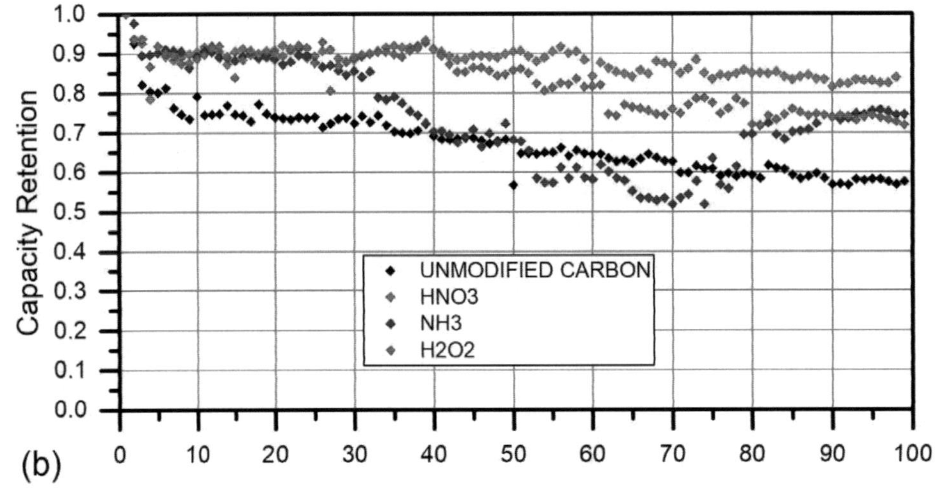

Figure 1a: Discharge capacity at C/10 rate. b: Discharge capacity retention after cycling at C/10 rate, normalized to first cycle.

The cyclic voltammetry results are shown in the following figures, Fig. 2a-b. The two cathodic peaks, located around 2.45-2.5 V and 1.8-2.05 V, correspond to the reduction of sulfur into high-order polysulfides (Li_2S_n,4<x<8) and polysulfides to solid lithium sulfides (Li_2S_2/Li_2S), respectively, while the broad anodic peak around 2.45-2.5 V could be assigned to the oxidation of polysulfides to sulfur with facile kinetics.

While all the carbons are degraded from the cycling, it is clear unmodified carbon both starts out performing worse, requiring extra voltage to complete the charge, and has a larger fraction of its discharge at lower potentials. After cycling, ammonia-modified carbon showed its reactions to be no longer discernible. It could be attributed to stronger shuttle effect during the successive charge discharge cycles in the cell.

Of the remaining two carbons, the peroxide-modified carbon preserves its single peak for charging, though it is quite broadened; however, the nitric acid-modified material discharges at a higher voltage overall.

The lower discharge peaks represent the formation of lower polysulfides after the generation of the higher ones. Both the states of unmodified carbon, and the post-cycling peroxide, show several discernible features alongside their lower polysulfide discharge (1.85V-2.1V). These features are not present in any of the modified materials prior to cycling and may be related to the extra discharge capacity shown by the early cycles of the unmodified carbon.

Furthermore, while during initial cycling the modified carbons show a single peak during charging and two peaks during discharge, the peaks appear to be more complex and broader after cycling. This is attributed to the more complex chemistry present after cycling due to the leaching of polysulfide species into the electrolyte, rather than the cathode.

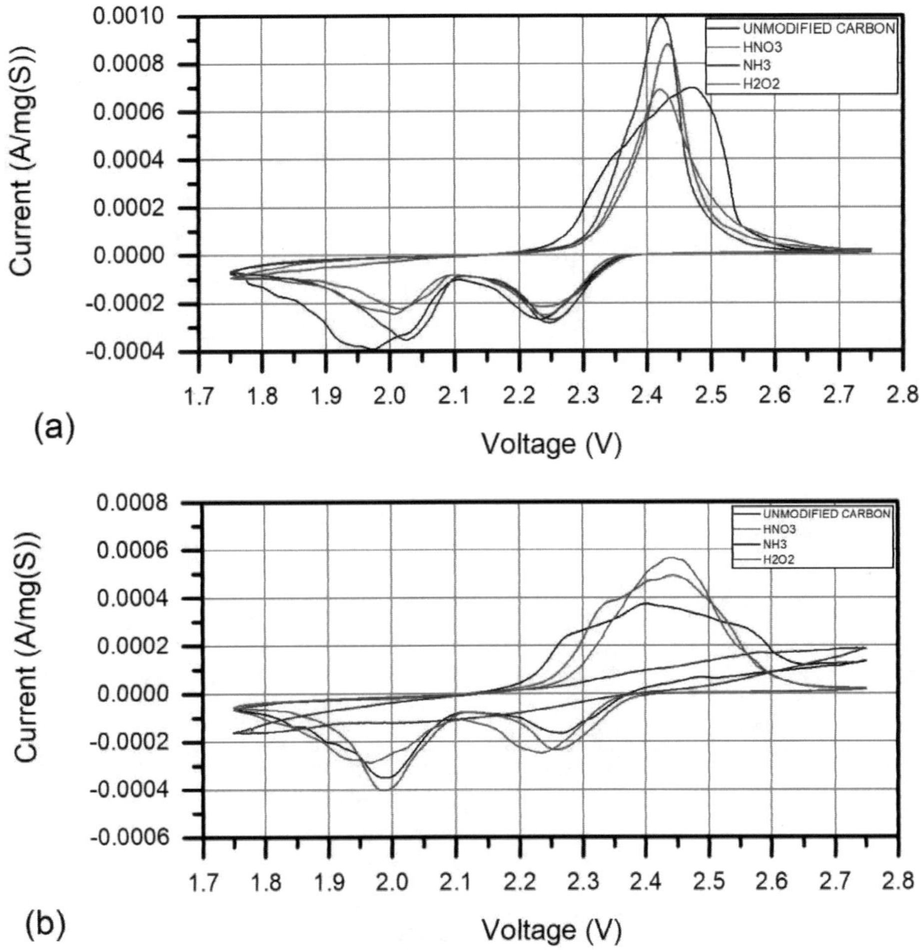

Figure 2: Cyclic Voltammetry at 25°C with a rate of 0.05 mV/s. 2a: No previous cycling. 2b: after 100 cycles at C/10.

The EIS diagrams are shown in Figures 3 for prior and after cycling, respectively. As shown, the Nyquist plots are composed of a depressed semicircle in the high-frequency region, a depressed semicircle in middle-frequency region, and a straight slopping line in the low-frequency region. These can be attributed the interface reactions, the charge transfer resistance and the features of the double layer, and the Warburg impedance, as diffusion dominates over capacitance at constant current and low frequency cycling, respectively. If the results before and after cycling are compared, as it can be seen, the impedance values increase dramatically with the cycling.

It is easily deduced that the ammonia-modified carbon shows both the highest resistance and the lowest ionic transport, thus, a drastically higher impedance at lower frequencies. At higher frequencies, the impedances are similar for unmodified carbon, peroxide-modified carbon, and nitric acid-modified carbon.

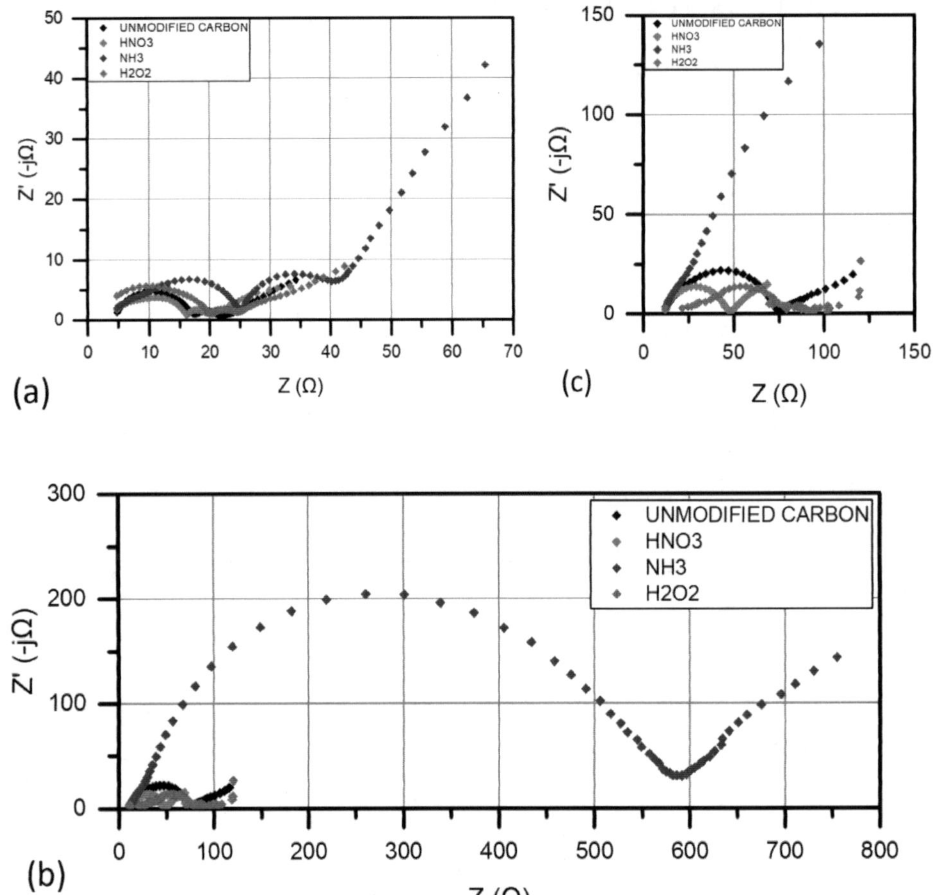

Figure 3: Nyquist plots. 3a: at full charge, with no previous cycling. 3b: At full charge after 100 cycles at C/10. 3c: Zoom of higher frequency values for figure 3b.

After 100 cycles, the ammonia-modified carbon remains the one with the highest impedance. The impedance of the peroxide-modified carbon instead is in line with the unmodified carbon, while the nitric acid modified carbon shows, in fact, lower impedance at that point. A zoom of the low resistance, low capacitance, is provided for ease of reference in Fig. 3c.

4. Conclusions

Different modifications to Super-P commercial carbon were carried out with nitric acid, ammonia and hydrogen peroxide. Modified and unmodified carbonaceous materials were studied as cathode modifiers for Li-S batteries.

It was found that the carbon processed with boiling nitric acid delivered the highest capacity for retention after 100 cycles at C/10, though most carbons had approximately similar capacities in that test. The carbon processed with ammonia yielded the lowest capacities at C/10.

The peaks observed in the modified carbons are narrower and better defined than the ones in unmodified carbon. This is taken to indicate the reactions occur uniformly in the modified carbons at first. Broadening is observed after 100 cycles, implying that there is at least an initial control of the polysulfide chemistry in the cell by the moieties added through processing.

Acknowledgements

This work was supported by the CONICET (Consejo Nacional de Investigaciones Científicas y Tecnológicas) and ANPCyT (Agencia Nacional de Promoción Científica y Tecnológica).

References

[1] Bruce P, Freunberger S, Hardwick L, Tarascon J 2011 *Nat. Mater.* **11** 19-29
[2] Zhao M, Li B, Zhang X, Huang J and Zhang Q 2020 *ACS Central Sci* **6** 1095
[3] Ellingsen L A, et al. 2016 *Environ. Res. Lett* **11** 54010
[4] Nichols B G, Kockelman K M, Reiter M 2015 *Trans. Res. D Transp. Environ.* **34** 208-218
[5] Albatayneh A, Assaf M N, Alterman D, Jaradat M 2020 *Environ. Clim. Technol.* **24** 669–680
[6] Zheng D, Wang G, Liu D, Si J, Ding T, Qu D, Yang X and Qu D 2018 *Adv. Mat. Technol.* **3** 1700233
[7] Yang X, Li X, Adair K, Zhang H and Sun X 2018 *EER* **1** 239-293,
[8] Ren J *et al* 2018 *J. Energy Chem.* **30** 121-131
[9] Yan Y, Shi M, Wei Y, Zhao C, Carnie M, Yang R and Xu Y 2017 *J. Alloys Compd.* **738** 16-24
[10] Liang J, Sun Z, Li F, Cheng H, 2016 *Energy Storage Mater.* **2** 76-106
[11] Zhao Q et al 2018 ACS *Appl. Mater. Interfaces* **10** 10882-10889
[12] Pang Q, Tang J, Huang H, Liang X, Hart C, Kam C T and Nazar L F, 2015 *Adv. Mater.* **27** 6021-6028
[13] Jiangxuan S, Terrence X, Mikhail L G, Pengyu Z, Dongping L, Ying-Bing J, Yongsheng C, Yuhua D, and Donghai W, 2014 *Adv. Funct. Mater.* **24** 1243-1250
[14] Zhou G *et al* 2015 *Nat. Commun.* **396** 542-550
[15] Yang C, Yin Y, Ye H, Jiang K, Zhang J and Guo Y 2014 *ACS Appl. Mater. Interfaces* **6** 8789-8795
[16] Zhou Z *et al* 2018 *J. Power Sources* **396** 542-550
[17] Zhang H, Lin C, Hu X, Zhu B, Yu D 2018 *ACS Appl. Mater. Interfaces* **10** 12708-12715
[18] Jin G, Mingang Z, Shijian Y, Xiaoyan Y and Shiwei W 2018 *Ionics* **24** 2219-2225
[19] Borchardt L, Leistenschneider D, Zürbes K, Schneidermann C, Grätz S, Oswald S, Wegner K, Klemmed B, Giebeler L and Eychmüller A 2018 *C* **4** 14
[20] Pongilat R, Franger S and Nallathamby K 2018 *J. Phys. Chem. C* **11** 5948-5955

Activated and carbonized metal-organic frameworks for improved cycle performance of cathode material in lithium-sulphur batteries

D Capkova[1], M Almasi[2], J Macko[1], N Kiraly[2], O Cech[3], P Cudek[3], A Strakova Fedorkova[1], V Knap[4] and T Kazda[3]

[1] Department of Physical Chemistry, Faculty of Sciences, Pavol Jozef Safarik University in Kosice, Kosice, Slovak Republic
[2] Department of Inorganic Chemistry, Faculty of Sciences, Pavol Jozef Safarik University in Kosice, Kosice, Slovak Republic
[3] Department of Electrical and Electronic Technology, Faculty of Electrical Engineering and Communication, Brno University of Technology, Brno, Czech Republic
[4] Department of Electrotechnology, Faculty of Electrical Engineering, Czech Technical University in Prague, Prague, Czech Republic

dominika.capkova@upjs.sk

Abstract. The development of new battery systems has been intensively pursued in an effort to increase energy densities. Lithium-sulphur batteries represent a group of next-generation batteries with high theoretical capacity and energy density. The electrochemical properties of lithium-sulphur batteries may be improved by the application of appropriate conductive and porous additives to sulphur in the cathode material. Recently, materials belonging to the group of metal-organic frameworks (MOFs) have been widely investigated as host materials for sulphur thanks to their unique porous structure. In this work, various types of MOFs (GaTCPP, MOF-76(Gd), MIL-101(Fe)-NH_2) were applied to the cathode material. MOFs were activated or carbonized before cathode material preparation. The structure of activated GaTCPP showed the lowest capacity fading per cycle (0.07 %) from activated MOFs during cycling at 0.5 C for 200 cycles. The carbonization process may improve the electrochemical properties of the electrode material. The best electrochemical properties showed carbonized MOF-76(Gd), and the capacity fading rate per cycle was only 0.04 % despite 200 cycles at 0.5 C.

1. Introduction

The dynamic development of modern society has created a high energy demand, especially for portable electronic devices, the automotive industry, and aerospace. The main challenge is how to store electric energy effectively and safely [1,2]. Rechargeable lithium-ion (Li-ion) batteries have become indispensable energy storage devices due to their high safety [3]. However, their specific energy density based on the intercalation process is insufficient for future applications. As an alternative, the conversion mechanism of lithium-sulphur (Li-S) batteries became the target of intensive research. Li-S batteries are considered a candidate for next-generation energy storage systems thanks to their high theoretical energy density (2600 Wh kg^{-1}), environmentally friendliness, and low sulphur cost [4-6]. However, their practical application is hampered by several issues. A volumetric expansion of sulphur during charging is up to 80 % which can cause a reduction in contact between different parts of the cathode. Another challenge is the polysulfide shuttle. The higher polysulfides (Li_2S_8, Li_2S_6, Li_2S_4) are soluble in the electrolyte and can diffuse to the lithium metal anode. Higher polysulfides can reduce to lower polysulfides (Li_2S_3, Li_2S_2) which can migrate back to the cathode. Migration process of polysulfides is known as the polysulfide shuttle [7,8]. Above mentioned shortcomings result in the loss of active material, decline in specific capacity, and decrease in Coulombic efficiency [9].

Various strategies have been proposed to alleviate the issues mentioned above. The reduction of shortcomings may be achieved by the application of conductive material with a large surface area and desirable pore size distribution. As cathode materials have been investigated porous carbon materials [6], carbon nanotubes [10], or graphene [11]. Nevertheless, the non-polar nature of carbon materials provides a weak affinity for polysulfides, they can only operate as a physical barrier against the diffusion of polysulfides. Therefore, polar materials such as metal-organic frameworks (MOFs) are needed to arrange chemical interactions to entrap polysulfides [12]. MOFs are formed by metal ions or clusters that are bridged by organic linkers to build a robust framework with a large surface area and high porosity [13, 14]. One of the defining attributes of MOFs is their local structure, which is topologically heterogeneous and symmetric. Their controllable surface properties and a variable pore size make them an eternal inspiration for materials chemistry scientists [15]. The porosity, void space enlarging ability, polysulfide bonding, and the tolerance of the volumetric extensibility of the MOF structure may suppress the drawbacks of Li-S batteries and form a stable support for sulphur in cathode material [16]. In 2011 Tarascon et al [17] published MOF's first-time application in cathode material of Li-S battery, where sulphur was impregnated in MIL-100(Cr).

Herein, various MOF materials were tested as support for sulphur in Li-S batteries. Two MOF materials (GaTCPP and MOF-76(Gd)) were used in an activated form, and in an effort to enhance the electrochemical properties two MOFs were carbonized (MOF-76(Gd) and MIL-101(Fe)-NH_2). The structure of activated GaTCPP showed the lowest capacity fading per cycle (0.07 %), representing capacity retention of 86 % despite 200 cycles at 0.5 C. It was proven that the carbonization process of MOF enhanced its properties, and the electrode material showed highly stable cycle performance. Overall, the best electrochemical properties, the highest discharge capacities, and the most stable cycle performance showed carbonized MOF-76(Gd). The capacity retention of the electrode with carbonized MOF-76(Gd) was 93 % which presents a capacity fading rate per cycle of only 0.04 % after 200 cycles at 0.5 C.

2. Experimental

2.1 Preparation of electrode materials and cell assembly

Two activated and two carbonized MOF materials were used as a host for sulphur in cathode materials for Li-S batteries. The synthesis of microporous material GaTCPP with exact chemical formula $\{[Ga_2(H_2TCPP)(OH)_2]\cdot 5DMF\cdot 2H_2O\}_n$ (H_2TCPP = (4,4′,4′′,4′′′-(5,10,15,20-porphyrintetrayl)-tetrabenzoate, DMF = *N,N′*-dimethylformamide) is described in [18], MOF-76(Gd) with the chemical formula $\{[Gd(BTC)(H_2O)]\cdot DMF\}_n$ (BTC = benzene-1,3,5-tricarboxylate) in [4, 13], and MIL-101(Fe)-NH_2 with chemical composition $\{[Fe_3(O)(BDC\text{-}NH_2)_3(H_2O)_2Cl]\cdot G\}_n$ (where BDC-NH_2 = 2-aminoterephthaliate, G = guest molecules) in [9]. MOFs GaTCPP and MOF-76(Gd) were applied in an activated state (AC), and MOF-76(Gd) and MIL-101(Fe)-NH_2 were carbonized (C) in a tubular furnace at 700 °C in a nitrogen atmosphere.

The electrode materials containing various MOFs were prepared by wet milling in a planetary ball mill. Sulphur, selected MOF material, and carbon Super P were placed into the zirconium oxide grinding jar. The electrode material components and ethanol were ground at 500 rpm for 30 min using 10 zirconia spheres with a diameter of 10 mm. Subsequently, the prepared mixture was dried in an oven at 60 °C for 24 h. To prepare the electrode slurry, polyvinylidene fluoride (PVDF) was dissolved in *N*-methyl-2-pyrrolidone (NMP), and then the composite of sulphur, MOF, and carbon Super P was added. PVDF forms connections between particles and the current collector. The resulting mass ratio of the electrode components was 60:15:15:10 for sulphur, MOF, Super P, and PVDF. The electrode slurry was homogenized on a magnetic stirrer for 24 h and then uniformly dispersed on the carbon-coated aluminium current collector using a coating bar with 200 μm thickness of the wet film. The coated electrode slurry was dried at 60 °C for 24 h. The electrodes were cut with a diameter of 18 mm, pressed (a pressure of 350 kg cm^{-2}), and dried for 24 h.

2.2 Methods and characterization
The powder X-ray diffraction (PXRD) patterns were obtained on Rigaku MiniFlex 600 powder diffractometer using CuKα radiation (λ = 1.5406 Å) in 2 theta range of 5 - 60°.

The electrode materials' surface characterization and elemental mapping were performed by scanning electron microscope TESCAN VEGA 3 equipped with an EDS analyser.

All electrochemical measurements were performed using a BioLogic VMP3 potentiostat. Cyclic voltammetry (CV) was measured in a potential window from 1.8 V to 3.0 V (*vs* Li/Li^+), and the scan rate was set to 0.1 mV s^{-1}. Galvanostatic cycling was carried out within the potential range of 1.8-2.8 V (*vs* Li/Li^+).

3. Results and discussion

3.1 Morphological and structural analyses
The crystal structure of all applied MOF materials in the present study was investigated by PXRD analysis (see figure 1). The typical patterns for MOF materials are present in all activated MOFs, and the main diffraction peaks are visible below 20° of 2 theta. PXRD patterns of carbonized MOF materials show a characteristic amorphous character without the presence of diffraction peaks at low angles. In the case of MIL-101(Fe)-NH_2(C), two diffraction peaks appear due to the presence of Fe_2O_3 as the final thermal decomposition product during carbonization.

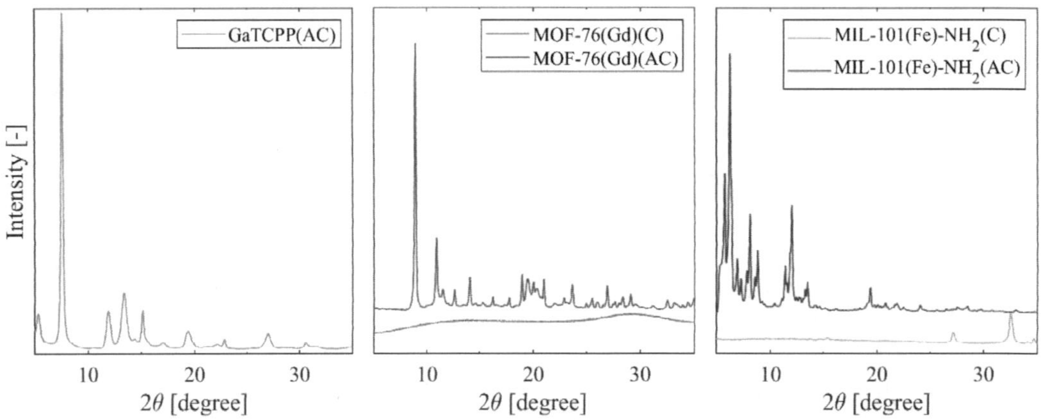

Figure 1. PXRD patterns of MOF materials: GaTCPP(AC) (green line), MOF-76(Gd)(AC) (blue line), MOF-76(Gd)(C) (orange line), MIL-101(Fe)-NH_2(AC) (purple line), and MIL-101(Fe)-NH_2(C) (yellow line).

The structure of the electrodes containing all MOF materials was investigated by SEM analysis. All prepared electrodes were without any cracks or holes and porous enough for the electrolyte penetration (see figure 2). The elemental composition of the electrodes was analysed by EDX analysis, and the elemental mapping of the distribution of carbon and sulphur despite the electrode surface is depicted in figure 2. Both elements were distributed uniformly on the surface of all analysed electrodes, indicating effective milling during cathode material preparation.

3.2 Electrochemical characterization
To explore the electrochemical behaviour of all synthesized MOFs in cathode materials for Li-S batteries, the cells were assembled with two activated MOFs (S@C/GaTCPP(AC), S@C/MOF-76(Gd)(AC)), and two carbonized MOFs (S@C/MOF-76(Gd)(C), S@C/MIL(C)).

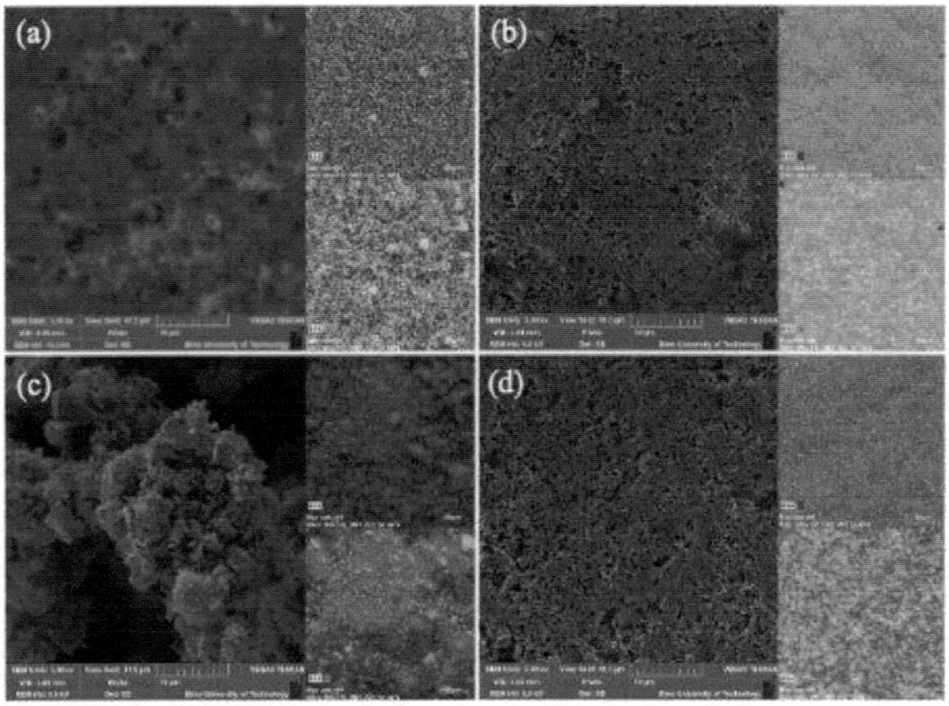

Figure 2. SEM figures and distribution of carbon (red color) and sulphur (green color) on the electrode surface of (a) S@C/GaTCPP(AC), (b) S@C/MOF-76(Gd)(AC), (c) S@C/MOF-76(Gd)(C), and (d) S@C/MIL(C) electrodes.

3.2.1 Activated metal-organic frameworks

The CV curves of activated MOF materials are illustrated in figure 3 (a). Both MOFs, GaTCPP(AC) and MOF-76(Gd)(AC), show two cathodic and anodic peaks. The first cathodic peak located around 2.36 V may be described by the transformation of sulphur to higher-order polysulphides. The second cathodic peak observed around 2.0 V is attributed to the reduction of higher-order polysulphides to lower-order polysulphides. The high stability of both electrode materials is indicated by the overlapping cyclic voltammograms for the three cycles. However, the current density is higher for the S@C/GaTCPP(AC) (green colour) than S@C/MOF-76(Gd)(AC) (blue colour), indicating improved sulphur utilization and chemical interaction between sulphur and conductive support of the cathode.

Afterwards, multi-current galvanostatic cycling was carried out from 0.2 C to 2 C for both electrodes with activated MOFs. The initial discharge capacities at 0.2 C were comparable for both cathodes (667 mAh g^{-1} for S@C/GaTCPP(AC) and 684 mAh g^{-1} for S@C/MOF-76(Gd)(AC)). However, the specific capacity of the S@C/MOF-76(Gd)(AC) increases with the number of cycles more significantly than the S@C/GaTCPP(AC) electrode. The capacity drop at cycling at 2 C was more notable for the S@C/MOF-76(Gd)(AC) than S@C/GaTCPP(AC) electrode resulting in higher discharge capacity for the electrode with GaTCPP MOF material. The discharge curves, despite cycling at 0.2 C are shown in figure 3 (c) for the S@C/GaTCPP(AC) electrode and in figure 3 (d) for the S@C/MOF-76(Gd)(AC) cathode. High- and low-voltage plateaus are clearly visible for both electrodes, although the low-voltage plateau is extended for the S@C/MOF-76(Gd)(AC) cathode with increasing number of cycles. Coulombic efficiency, despite multi-current cycling, reached the value of 93 % for the S@C/GaTCPP(AC) electrode and 91 % for the S@C/MOF-76(Gd)(AC) electrode.

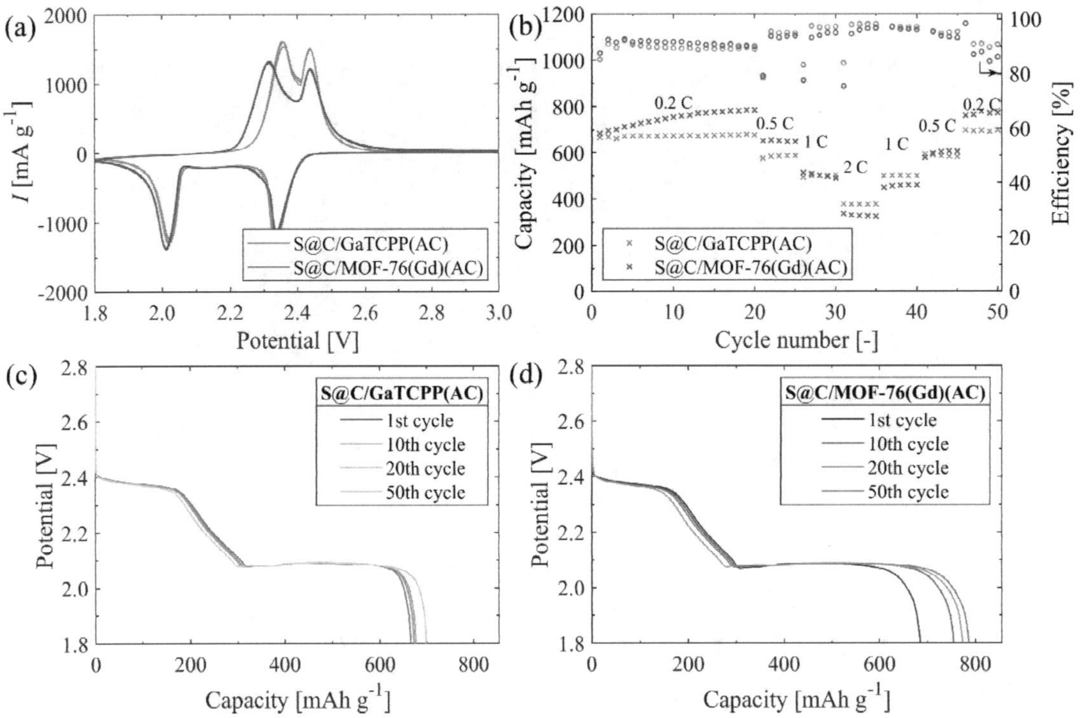

Figure 3. (a) Cyclic voltammograms of the Li-S cell with S@C/GaTCPP(AC) and S@C/MOF-76(Gd)(AC) cathodes at a scan rate of 0.1 mV s^{-1}. (b) Cycling performance of the Li-S cell with S@C/GaTCPP(AC) and S@C/MOF-76(Gd)(AC) electrodes at various C-rates. The discharge profiles of the Li-S cell despite cycling at 0.2 C of the (c) S@C/GaTCPP(AC) and (d) S@C/MOF-76(Gd)(AC) electrodes.

A further electrochemical investigation of the S@C/MOF-76(Gd)(AC) and S@C/GaTCPP(AC) electrodes was realized by long-term galvanostatic cycling at 0.5 C for 200 cycles (see figure 4). The initial discharge capacity was higher for the S@C/MOF-76(Gd)(AC) electrode (621 mAh g^{-1}) than for the S@C/GaTCPP(AC) electrode (601 mAh g^{-1}). However, after 200 cycles the capacity retention for the S@C/GaTCPP(AC) electrode reached a value of 86 % compared to 81 % for the S@C/MOF-76(Gd)(AC) electrode, which represents a capacity fading rate per cycle of 0.07 % for S@C/GaTCPP(AC) and 0.10 % for S@C/MOF-76(Gd)(AC). The potential of high and low voltage plateaus is stable for the S@C/GaTCPP(AC) electrode, but for the S@C/MOF-76(Gd)(AC) electrode is gently decreasing with the number of cycles. Moreover, the shape of the discharge curves of the Li-S cell with S@C/MOF-76(Gd)(AC) cathode changes with cell degradation, indicating attenuated electrode stability. Coulombic efficiency during cycling at 0.5 C for 200 cycles reached the value of 95 % for the S@C/GaTCPP(AC) electrode and 93 % for the S@C/MOF-76(Gd)(AC) electrode.

3.2.2 Carbonized metal-organic frameworks

Another group of studied MOF materials was carbonized MOFs, which are represented by MOF-76(Gd)(C) and MIL-101(Fe)-NH_2(C). The electrochemical behaviour of these carbonized MOFs as an additive in cathode material in Li-S batteries is depicted in figure 5. Cyclic voltammograms show two reduction and oxidation peaks for the S@C/MOF-76(Gd)(C) electrode (orange colour). However, the S@C/MIL(C) electrode (yellow colour) exhibits two reduction peaks but only one broad oxidation peak, indicating slower reaction kinetics than the S@C/MOF-76(Gd)(C) electrode. Moreover, the current density was higher for the S@C/MOF-76(Gd)(C) than for the S@C/MIL(C) electrode. Both cyclic voltammograms were stable with number of cycles without any significant decrease in current density indicating stable cycle performance of the electrodes with MOF materials.

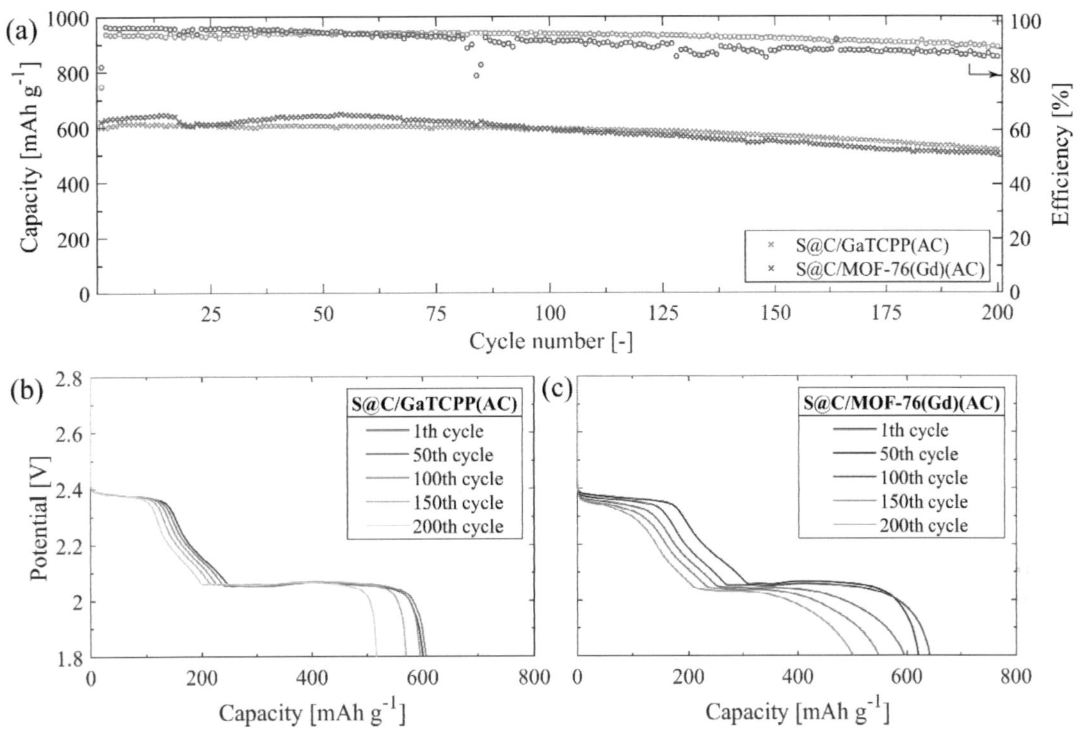

Figure 4. (a) Long-term galvanostatic cycling of the Li-S cell at 0.5 C of the S@C/GaTCPP(AC) and S@C/MOF-76(Gd)(AC) cathodes. The discharge curves during cycling at 0.5 C for the (b) S@C/GaTCPP(AC) and (c) S@C/MOF-76(Gd)(AC) electrodes.

Galvanostatic cycling was carried out at various C-rates, from 0.2 C to 2 C, for the MOF-76(Gd)(C) and MIL-101(Fe)-NH_2(C) electrodes (see figure 5 (b)). The discharge capacities for carbonized MOFs were higher than for the activated MOFs described in section *3.2.1*. The initial discharge capacity at 0.2 C for the S@C/MIL(C) electrode was 775 mAh g^{-1}, and for the S@C/MOF-76(Gd)(C) cathode was 763 mAh g^{-1}. The discharge capacity at 0.2 C was highly stable for the S@C/MOF-76(Gd)(C) electrode, while the capacity of the S@C/MIL(C) cathode decreased. Moreover, the capacity at 2 C decreased more significantly for the S@C/MIL(C) cathode (312 mAh g^{-1}) than for the S@C/MOF-76(Gd)(C) electrode (476 mAh g^{-1}). Coulombic efficiency was slightly higher for the S@C/MIL(C) electrode (97 %) than for the S@C/MOF-76(Gd)(C) cathode (94 %). Both plateaus, despite cycling at 0.2 C, are shortened for the S@C/MIL(C) electrode (see figure 5 (c)), while discharge curves for the S@C/MOF-76(Gd)(C) electrode are highly stable with increasing cycle number (see figure 5 (d)).

To further demonstrate the positive effect of the carbonization process on the electrochemical properties of MOF materials, long-cycle stability of Li-S battery cells was carried out at 0.5 C for 200 cycles (see figure 6 (a)). The specific capacity for the S@C/MIL(C) cathode was 705 mAh g^{-1} and retained 476 mAh g^{-1} in 200th cycle, representing 68 % of the initial discharge capacity. The capacity fading rate per cycle reached the value of 0.16 %. The S@C/MOF-76(Gd)(C) electrode showed the capacity of 658 mAh g^{-1} and maintained 610 mAh g^{-1} after 200 cycles, representing impressive cycling stability with a high capacity retention of 93 % and a low fading rate per cycle (0.04 %). Coulombic efficiency during galvanostatic cycling despite 200 cycles at 0.5 C was 95 % for the S@C/MIL(C) electrode and 94 % for the S@C/MOF-76(Gd)(C) electrode. The discharge curves at 0.5 C of S@C/MIL(C) and the S@C/MOF-76(Gd)(C) in the 1st, 50th, 100th, 150th, and 200th cycles are depicted in figures 6 (b) and (c), respectively. The capacity of the S@C/MIL(C) electrode continuously decreases with a cycle number. However, the specific capacity of the S@C/MOF-76(Gd)(C) electrode increases up to the 100th cycle and then gently decreases.

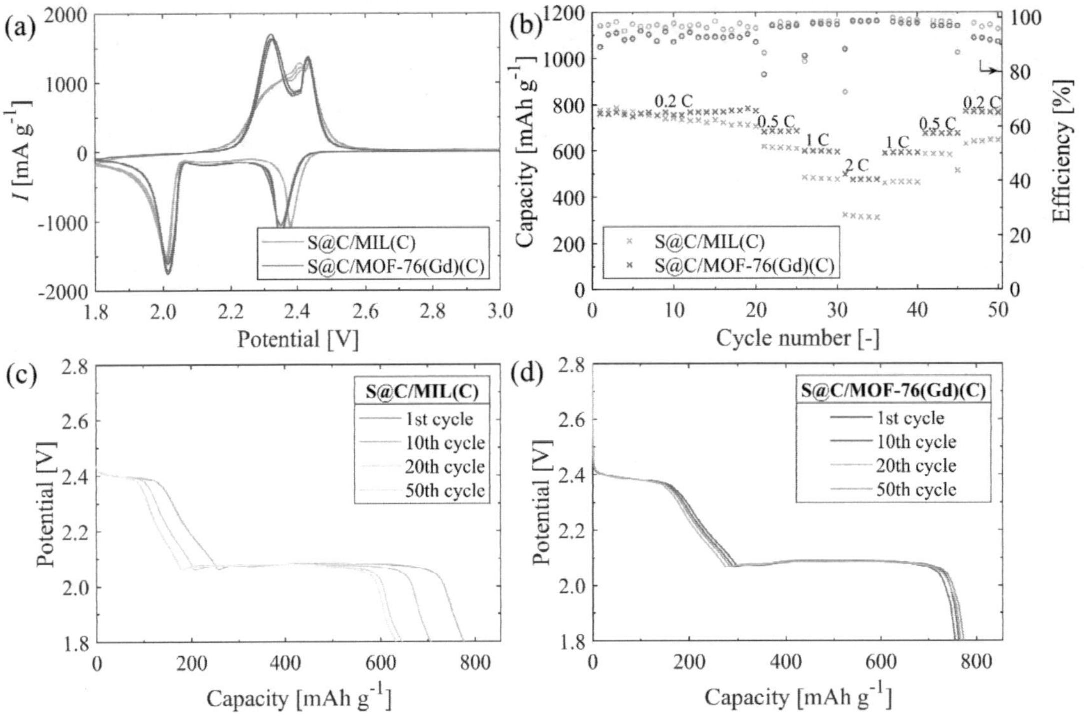

Figure 5. (a) CV curves of the Li-S cell with S@C/MIL(C) and S@C/MOF-76(Gd)(C) cathodes at a scan rate of 0.1 mV s^{-1}. (b) Cycling performance of the S@C/MIL(C) and S@C/MOF-76(Gd)(C) electrodes at various C-rates. The discharge curves despite cycling at 0.2 C of the (c) S@C/MIL(C) and (d) S@C/MOF-76(Gd)(C) electrodes.

An overview of the electrochemical parameters of all investigated electrodes with activatedand carbonized MOF materials is summarized in table 1.

Table 1. The electrochemical performance of the S@C/GaTCPP(AC), S@C/MOF-76(Gd)(AC), S@C/MOF-76(Gd)(C), and S@C/MIL(C) despite multi-current and long-term cycling.

	Multi-current cycling				
	Capacity [mAh g^{-1}]				Coulombic efficiency [%]
	0.2 C 1st cycle	0.2 C 20th cycle	2 C 35th cycle	0.2 C 50th cycle	
S@C/GaTCPP(AC)	667	677	378	699	93
S@C/MOF-76(Gd)(AC)	684	774	325	767	91
S@C/MOF-76(Gd)(C)	763	789	476	767	94
S@C/MIL(C)	775	706	312	646	97

	Long-term cycling				
	Capacity [mAh g^{-1}]		Coulombic efficiency [%]	Capacity retention [%]	Fading rate per cycle [%]
	0.5 C 1st cycle	0.5 C 200th cycle			
S@C/GaTCPP(AC)	601	517	95	86	0.07
S@C/MOF-76(Gd)(AC)	521	502	93	81	0.10
S@C/MOF-76(Gd)(C)	658	610	94	93	0.04
S@C/MIL(C)	705	476	95	68	0.16

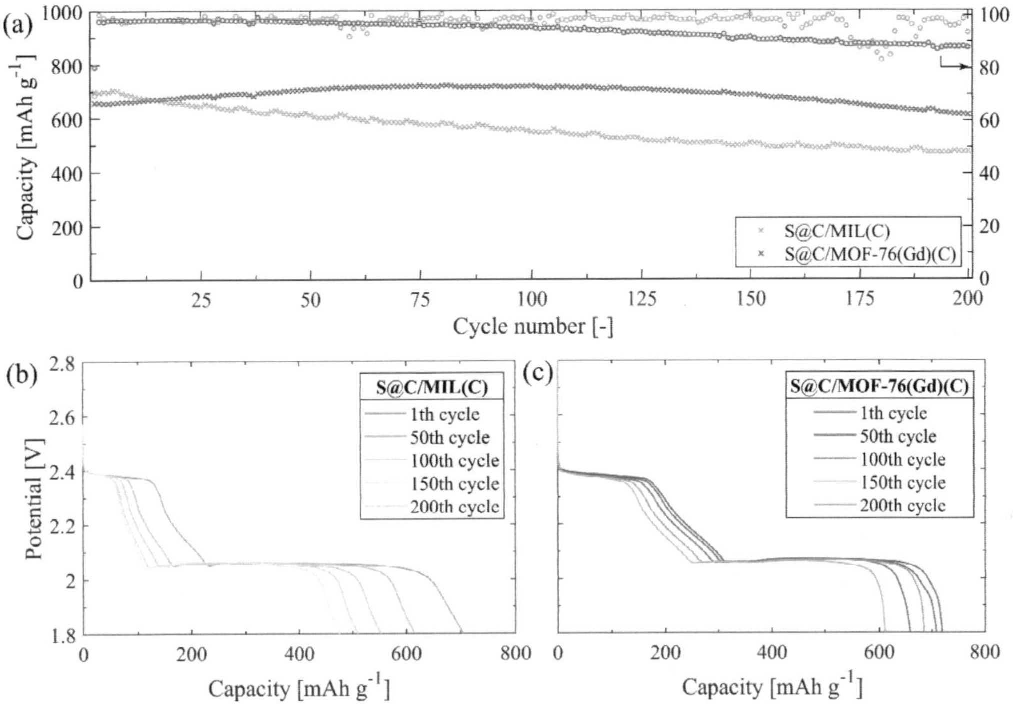

Figure 6. (a) Long-cycle stability at 0.5 C for 200 cycles of the S@C/MIL(C) and S@C/MOF-76(Gd)(C) electrodes. The discharge curves during cycling at 0.5 C for the (b) S@C/MIL(C) and (c) S@C/MOF-76(Gd)(C) electrodes.

4. Conclusion

High-efficiency MOF materials were synthesized and applied as a host for sulphur in the cathode material for Li-S batteries. MOF materials were prepared in an activated form (GaTCPP(AC) and MOF-76(Gd(AC)) or in a carbonized form (MIL-101(Fe)-NH_2 and MOF-76(Gd)(C)).

In order to explore the influence of the MOF structure on the electrochemical properties of the sulphur-based cathode material, two different activated MOFs were investigated. Comparatively, high specific capacities were observed for the MOF-76 and GaTCPP materials (684 and 667 mAh g^{-1} at 0.2 C, respectively). GaTCPP MOF showed a better ability to confine polysulfides and suppress the polysulfide shuttle resulting in high capacity retention (86 % for GaTCPP and 81 % MOF-76(Gd)) and low capacity fading rate per cycle (0.07 % for GaTCPP and 0.10 % for MOF-76(Gd)) during long-cycling stability at 0.5 C. It can be concluded that porphyrin-based material (GaTCPP) is more stable and beneficial as a host for sulphur than MOF-76(Gd).

The enhancement of the properties was demonstrated by the carbonization of MOF-76(Gd), wherein in an activated form the initial specific capacity at 0.2 C was 684 mAh g^{-1} and in a carbonized form the capacity increased up to 763 mAh g^{-1}. The capacity retention despite long-cycling stability at 0.5 C was improved for the electrode containing MOF-76(Gd)(C), where it reached 93 %, and the capacity fading rate per cycle was only around 0.04 % compared to 81 % and 0.10 % for S@C/MOF-76(Gd)(AC). Moreover, carbonized MIL-101(Fe)-NH_2 was also applied in cathode material for Li-S batteries. The S@C/MIL electrode achieved a specific capacity of 775 mAh g^{-1}, similar to S@C/MOF-76(Gd)(C). However, the capacity retention and fading rate per cycle after 200 cycles at 0.5 C were less satisfactory, 68 % and 0.16 %, for the S@C/MIL cathode. It can be concluded that the structure of MOF-76(Gd) is more stable and beneficial for the Li-S battery than the structure of MIL-101(Fe)-NH_2. The carbonization process may improve the electrochemical properties of the material, but it requires an additional step in the cathode material preparation and increases the cost.

Acknowledgments
This research was financially supported by the following projects: iCoTS No. 313011V334, Innovative Solutions for Propulsion, Power and Safety Components of Transport Vehicles; specific graduate research of the Brno University of Technology No. FEKT-S-20-6206; No. LTT19001, Development of New Electrochemical Energy Storage Devices – Batteries supported by the Ministry of Education, Youth and Sports of the Czech Republic under project; APVV-20-0138, Development of Novel 3D Materials for Post Lithium Ion Batteries with High Energy Density; APVV-20-0111, Towards Lithium Based Batteries with Improved Lifetime; and VVGS-2022-2123, Porous coordination polymers for environmental applications.

References

[1] Zheng Y, Zheng S, Xue H and Pang H 2019 Metal-organic frameworks for lithium-sulfur batteries *J. Mater. Chem. A* **7** 3469–3491
[2] Knap V and Stroe D-I 2021 Effects of open-circuit voltage test and models on state-of-charge estimation for batteries in highly variable temperature environments: Study case nano-satellites J. *Power Sources* **498** 229913
[3] Yu Z, Wang H, Kong X et al. 2020 Molecular design for electrolyte solvents enabling energy-dense and long-cycling lithium metal batteries *Nature Energy* **5** 526 – 533
[4] Capkova D, Kazda T, Cech O et al. 2022 Influence of metal-organic framework MOF-76(Gd) activation/carbonization on the cycle performance stability in Li-S battery *J. Energy Storage* **51** 104419
[5] Kazda T, Capkova D, Jasso K et al. 2021 Carrageenan as an ecological alternative of polyvinylidene difluoride binder for Li-S batteries *Materials* **14** 5578
[6] Capkova D, Kazda T, Strakova Fedorokova A 2019 Carbon materials as the matrices for sulfur in Li-S batteries *ECS Trans.* **95** 19-26
[7] Capkova D, Kazda T, Cudek P and Strakova Fedorokova A 2020 Binder influence on electrochemical properties of Li-S batteries *ECS Trans.* **99** 161-167
[8] Capkova D, Kazda T, Almasi M et al. 2021 Metal-organic framework for stable cyclability of Li-S batteries for space missions *Proceedings of the International Astronautical Congress* **C3** 177587
[9] Capkova D, Almasi M, Kazda T et al. 2020 Metal-organic framework MIL-101(Fe)-NH_2 as an efficient host for sulphur storage in long-cycle Li-S batteries *Electrochim. Acta* **354** 136640
[10] Ma G, Wen Z, Wang Q et al. 2015 Enhanced performance of lithium sulfur battery with self-assembly polypyrrole nanotube film as the functional interlayer *J. Power Sources* **273** 511
[11] Zhao M Q, Zhang Q, Huang J Q et al. 2014 Unstacked double-layer templated graphene for high-rate lithium-sulphur batteries *Nature Communications* **5** 3410
[12] Yao W, Zheng W, Xu J et al. 2021 ZnS-SnS@NC heterostructure as robust lithiophilicity and sulfiphilicity mediator toward high-rate and long-life lithium-sulfur batteries *ACS Nano* **15** 7114-7130
[13] Garg A, Almasi M, Bednarcik et al 2022 Gd(III) metal-organic framework as an effective humidity sensor and its hydrogen adsorption properties *Chemosphere* **305** 104419
[14] Almasi M 2021 A review on state of art and perspectives of Metal-Organic frameworks (MOFs) in the fight against coronavirus SARS-CoV-2 *J. Coord. Chem.* **74** 2111-2127
[15] Kim D, Liu X and Lah M S 2015 Topology analysis of metal-organic frameworks based on metal-organic polyhedral as secondary of tertiary building units *Inorg. Chem. Front.* **2** 336-360
[16] Zheng S, Li X, Yan B et al 2017 Transition-metal (Fe, Co, Ni) based metal-organic frameworks for electrochemical energy storage *Adv. Energy Mater.* **7** 1602733
[17] Demir-Cekan R, Morcette M, Nouar F et al 2011 Composite cathodes for Li-S batteries via the use of oxygenated porous architectures *J. Am. Chem. Soc.* **133** 16154-16160
[18] Kiraly N, Capkova D, Almasi M et al 2022 Post-synthetically modified metal-porphyrin framework GaTCPP for carbon dioxide adsorption and energy storage in Li-S batteries *RSC Adv.* **12** 23989-24002

IOP Publishing
2382 (2022) 012011 doi:10.1088/1742-6596/2382/1/012011

Reversible photoelectrochemical cell produced by using of 3D print for the accumulation "solar" hydrogen

S V Chivikov[1*], I A Rusetskyi[1], S S Fomanyuk[1], M O Danilov[1], V O Smilyk[1] and G Ya Kolbasov[1]

[1] V.I.Vernadskii Institute of General and Inorganic Chemistry of the NAS of Ukraine, 32/34 Academic Palladin Avenue, Kyiv, 03142, Ukraine.

[*]Corresponding Author E-mail Address: s.v.chivikov@gmail.com

Abstract. A reversible photoelectrochemical system with using of accumulated hydrogen has been developed. Reversibility was ensured by introducing a third electrode based on Cu_2S, onto which a metal hydride electrode based on MH alloys was discharged. The photoaccumulator made it possible to obtain a maximum current of 97 mA and Vo.c. ≈ 0.4 V, during the discharge by the MH electrode without illumination. A new feature of the cell design was construction in 3D. Almost all the details of the cell were printed on a 3D printer. It was found that the peak of current during cell illumination was several times higher than the maximum current obtained by the 2-electrode scheme. For consumers with uneven power consumption, such a system is equivalent to a solid state system with more power.

1. Introduction

Recently, the problem of finding new alternative energy sources is of particular relevance. This is due to the fact that the situation has become absolutely obvious when the further active development of modern energy and transport can lead civilization to a global environmental crisis. Along with this, fossil fuel reserves are declining every year, which further contributes to the development of alternative energy sources. One of the rapidly developing directions in this area is hydrogen energy. It is a renewable and environmentally friendly source of energy. The use of hydrogen as an energy carrier makes it possible to consider and resolve energy problems in close connection with environmental ones. Favorable opportunities are being created to reduce the formation of solid waste, harmful gas emissions and eliminate the greenhouse effect.

The development of efficient systems for the evolution, accumulation and use of hydrogen as the most environmentally friendly fuel is one of the urgent tasks of modern energy. For the evolution and accumulation of hydrogen, photoelectrochemical systems with photosensitive semiconductor electrodes can be used.

Currently, research continues in two main directions. The first is further research and modification of already known photoelectrodes in order to optimize their electrophysical and photoelectrochemical characteristics and the development of new schemes of photoelectrochemical cells and the improvement of old ones. The second direction of research is related to the search, synthesis and development of technologies for obtaining new photosensitive semiconductor materials and systems. Such works are carried out in several scientific centers in the USA, Europe, Japan, China, etc.

Electrochemical reactions occur on the surface of the electrodes and are associated with the process of charge transfer through the interface. For their study and improvement, electrochemical cells are used - specially designed devices with the help of which conditions are created for carrying out one or another electrochemical process.

At present time, additive technologies are increasingly used in industry and technology. The ability to print complex-shaped parts by using of the 3D printer can be used to manufacture electrochemical cells.

We have developed a layout of a fully reversible photoelectrochemical cell using stored hydrogen (Figure 1). Reversibility was provided by the introduction of the third electrode to the anode part of the cell, on which the metal hydride electrode based on the alloys AB_5 was discharged. The necessity of introducing the third electrode is due to the fact that on photoelectrodes based on cadmium chalcogenides, the polysulfide ion reduction reaction practically does not proceed at the potential at which MH electrodes operated in the absence of illumination.

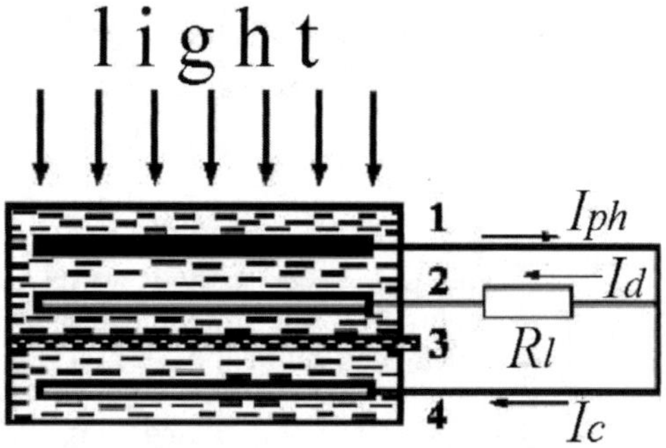

Figure 1. Scheme of a reversible photoelectrochemical system: 1-photoelectrode in a polysulfide solution, 2-counter electrode, 3- cation-exchange membrane, 4- the storage electrode in an alkaline solution. R_l - load resistance. Iph - photocurrent, Ic - charge current, Id - discharge current.

Copper, nickel, cobalt, lead, ruthenium, and molybdenum sulfides can be used as counter electrodes. These electrodes have a low overvoltage for the $2S^{2-} + 2p^{+} \rightarrow S_2^{2-}$ reaction.

The results of some prior investigation [1] let us to develop three electrodes photo-electrochemical cell. New feature of the design of the cell was construction in 3D. Almost all details of the cell were printed on 3D printer. New additive technology make possible to change design of details of the cell according to the results of experiments. It is create fundamentals for quick improvement of electrochemical characteristic of the cell.

2. Experimental

For the printing anode and cathode cell chambers of photoelectrochemical cell (Figure 2) we used polypropylene, material for other parts was ABS polymer.

The TiO_2 ($NT\text{-}TiO_2$) nanotubes on the Ti substrate were obtained by the method of potentiostatic anodic polarization of titanium foil. The electrolyte contained 70 % glycerol, 30 % water and 1.5 wt % NH_4F. Titanium foil was previously degreased in the acetone. The process was carried out at 30 V for 1 hour, followed by annealing at 500 °C for 3 hours [2, 3]. Under laboratory conditions, it was found that the formation of a layer of TiO_2 nanotubes leads to an increase in the efficiency of charge carrier accumulation on $NT\text{-}TiO_2/CdSe$ electrodes. This is due to the increase in the flat band potential, which determines the magnitude of the electric field at the surface of the semiconductor that affects the separation rate of photogenerated carriers, thereby reducing recombination in the space charge region (SCR) [4].

Figure 2. Reversible photoelectrochemical cell.

CdSe films were obtained by the painting method on previously prepared Ti substrate with surface area 15.8 cm^2. For the deposition following suspension was prepared: to the 1 g of CdSe and 0.25 g of $CdCl_2$ was added 0.5 g of an aqueous solution of polyethylene glycol (1 % by weight). The starting CdSe powder was milled in a ball mill with agate balls for 32 hours and then treated for 2 hours with an ultrasonic disperser UD-11 (Techpan, Poland): the oscillator frequency was 22 kHz, and the power was 100 watts. The films of CdSe were applied in 2 layers. Each layer was annealed in an air atmosphere at 535 °C for 30 minutes. After that, the surface was activated in an aqueous solution of 0.5 M CrO_3 for 45 seconds at room temperature and washed in distilled water. The surface of the CdSe semiconductor film was examined by using of an SEM JSM 6700F electron microscope. X-ray diffraction (XRD) was performed on a DRON-4-07 X-ray diffractometer.

As a hydrogen absorbing electrode (cathode), a commercially produced active mass MH from industrial Ni-MH batteries were used. The capacity of the MH cathode was determined by cycling in the galvanostatic mode at a charge/discharge current of 3.2 mA and was up to 25 mA h. The counter electrodes for the photoaccumulator were prepared by several methods with using of nickel foam (d= 0.6 cm) as a base.

The Cu_2S electrode was prepared by dipping copper-coated nickel foam into a polysulfide solution. Copper was deposited by the electrochemical method from a solution containing $CuSO_4$ 200 g/l H_2SO_4 50 g/l by the current density 3 A/dm^2. Copper in a polysulfide solution gradually corrodes, forming a Cu_2S layer. CoS counter electrodes were prepared by pressing CoS powder and CoS powder with conductive acetylene black with PTFE emulsion onto nickel foam under a pressure of ~150 kg/cm^2. For comparison, the volt-ampere characteristics of electrodes in a polysulfide electrolyte based on carbon graphite and titanium foil of the VT1-0 brand were taken. On the titanium foil, the currents in the selected potential range did not exceed 50 ÷ 100 μA, and the carbon-graphite electrode showed characteristics similar to CoS in terms of electrocatalytic properties.

The photoelectrode and counter electrode were placed in the anode space filled with polysulfide electrolyte 1M KOH + 1M Na_2S + 1M S, and the hydrogen sorbing electrode was placed in the cathode chamber filled with an alkaline solution (30% KOH).

The anode and cathode chamber are separated by an ion exchange membrane MF-4SK. The MF-4SK membrane had the following characteristics: thickness 0.034 cm; exchange capacity, referred to 1 g of swollen polymer (Q) - 0.78 mol / g swollen polymer; capacity moisture (W swollen.) - 15.6%; equivalent polymer weight (EW) - 1100 dry g of polymer./mole; specific moisture (n) - 11.1 mol H_2O / mol SO_3^-; frequency of weaving threads (number of threads on 10 cm) are 100.

3. Results and Discussion

In microscopic examinations with using of SEM, it was found that the average particle size of CdSe deposited by the painting method was about 5 μm (Figure 3).

Figure 3. SEM micrograph of the CdSe electrode surface.

An X-ray diffraction analysis of the surface of a CdSe electrode deposited on a titanium substrate showed the formation of a hexagonal CdSe phase (Figure 4).

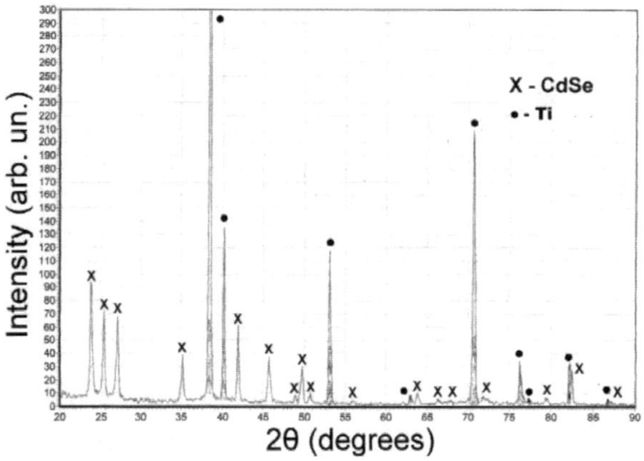

Figure 4. X-ray diffraction pattern of the CdSe electrode surface.

The electrocatalytic activity of the studied electrodes in a polysulfide electrolyte is shown on the graph in the current-potential coordinates (Figure 5).

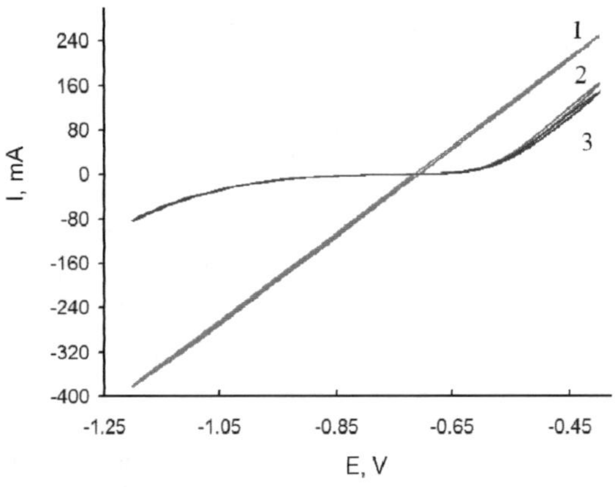

Figure 5. Cyclic polarization curves of various electrode materials in 1M KOH + 1M Na_2S + 1M S electrolyte at 20 °C (5 mV/s, ref. Ag/AgCl): Cu_2S (1); CoS (2); CoS+C (3).

As can be seen from the given current-voltage characteristics, the material based on Cu_2S has the highest electrocatalytic activity. Similar results were obtained in [5].

When studying the parameters of a 3-electrode photoelectrochemical cell, it was found that the open-circuit voltage of such a photoaccumulator in the absence of illumination is about ~0.4 V, and the peak of current during the discharge of such a battery under illumination was at least three times greater (depending on illumination) than the maximum current generated by the photoanode and was ~ 150 mA (Figure 6).

In this case, the recorded current is the sum of the MH discharge current and the photocurrent. The current recorded when the photoanode is illuminated in a pair with the counter electrode (Cu_2S) is greater than the current when the MH is charged so there is no voltage drop across the membrane. The voltage drop across the membrane was 66 mV at a current of 10 mA.

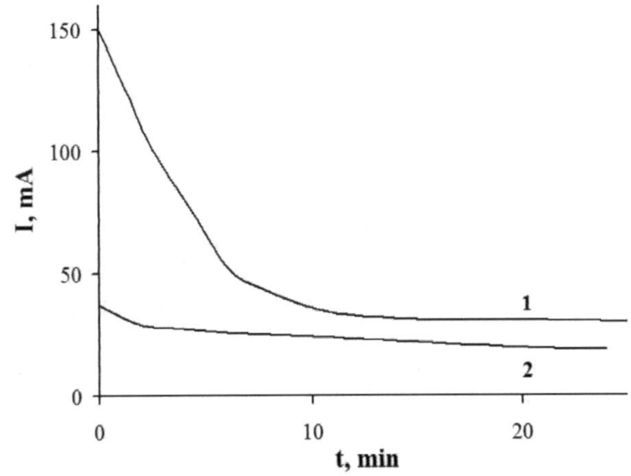

Figure 6. Dependence of current - time under an illumination for a 3-electrode photoelectrochemical cell (1) and for a 2-electrode cell (2).

At the same time, the photoaccumulator made it possible to obtain a maximum current of 97 mA during the discharge of the MH electrode without illumination (Figure 7).

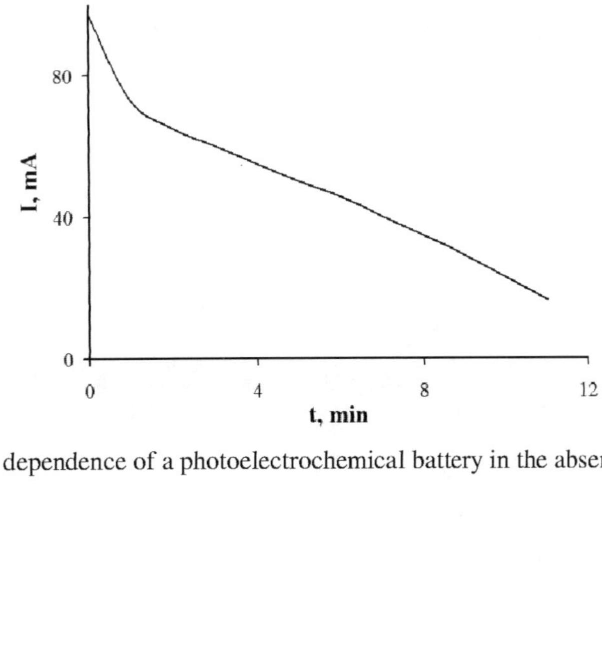

Figure 7. Discharge dependence of a photoelectrochemical battery in the absence of illumination.

4. Conclusions

We have developed a reversible photoelectrochemical system by using of the accumulated hydrogen. Reversibility was ensured by introducing a third electrode based on Cu_2S, onto which a metal hydride electrode based on MH alloys is discharged. The need to introduce a third electrode is due to the fact that on photoelectrodes based on cadmium chalcogenides, the reduction reaction of the polysulfide ion practically does not proceed at the potential that is established on MH electrodes in the absence of illumination.

The photoaccumulator made it possible to obtain without illumination a maximum current of 97 mA and Vo.c. ≈ 0.4 V during the discharge of the MH electrode.

This creates the basis for rapidly improving the electrochemical performance of the cell. It was found that the maximum current when the cell was illuminated were at least three times greater than the maximum current generated by the photoanode and amounted to ~150 mA. For the consumers with uneven consumption of electricity, such system is equivalent to a solid state system with the more power.

A new feature of the cell design is construction in 3D. Almost all the details of the cell were printed on a 3D printer. New additive technologies make it possible to change the design of cell parts based on the results of experiments.

References

[1] Rusetskii I A, Shcherbakova L G, Danilov M O, Slobodyanyuk I A, Kolbasov G Ya, Fomanyuk S S and Solonin Yu M 2018 *ECS Trans.* **87** 335
[2] Rusetskyi I A, Danilov M O, Fomanyuk S S, Slobodyanyuk I A, Vorobets V S and Kolbasov G Ya 2019 *Res. Chem. Intermed.* **45** 4121
[3] Macak J M, Hildebrand H, Marten-Jahns U and Schmuki P 2008 *J. Electroanal. Chem.* **621** 254
[4] Pleskov Yu V and Gurevich Yu Ya 1986 *Semiconductor Photoelectrochemistry* (New York: Consultants Bureau)
[5] Hodes G, Manassen J and Cahen D 1980 *J. Electrochem. Soc.* **127** 544

Design and optimization of solar-powered irrigation system

K Jandova[1] **and R Stranak**[1]

[1] Department of Electrotechnology, Brno University of Technology, Faculty of Electrical Engineering and Communication Technologies, Technicka 10, 616 00 Brno, Czech Republic

jandovak@vut.cz, 220856@vut.cz

Abstract. Currently, there are several commercially produced solar-powered irrigation systems that allow irrigation to be used even in the areas and applications where piped water supply is not available. However, these are mostly systems that are only basic and do not allow for further improvements. This contribution describes the design and manufacture of a custom solar-powered irrigation system that includes, for example, a liquid fertilizer reservoir for better plant growth or a moisture meter that prevents waterlogging of individual crops as well as protection against self-destruction. This system reflects not only the needs of the operator, but also the needs of the plants and, thanks to the use of a solar module; it is one of the most environmentally friendly sources of energy.

1. Introduction

With the advent of more efficient and affordable photovoltaic panels, a new sub-category of their utilization called solar irrigation is emerging in agriculture. This sub-category offers an energy-independent solution with minimal future costs. An electric pump controlled by a timer switch offers the advantage of eliminating human factor and replacing it with a time-controlled water-saving system. Precise timing is achieved by the dependence of the solar cells power generation on the intensity of the sun radiation. At high intensities, the batteries take less time to charge. The plants receive irrigation at the exact moment when the soil beneath them is dry and needs watering the most. Water savings compared to general irrigation are achieved by using drip irrigation, where water is not sprayed fan-shaped onto the soil but drips from central holes in the hose directly to the root of the plant, thus avoiding watering unused areas and thus weed growth [1, 2].

2. Design and optimization of solar-powered irrigation system

The actual design of the solar-powered irrigation system was based on the primary components of the system. The first component of the system analyzed was the solar panel. A 12 V/10 W polycrystalline solar panel was selected. Lithium-ion batteries type 18650 were selected for the energy storage function with a nominal voltage of 3.7 V and a capacity of 2600 mAh. The next step involved a diaphragm water pump. The FL-3308 type was chosen, with an input of 70 W at 12 V and a flow rate of 4-6 l/min. Lastly, a microcontroller was selected as the control unit, specifically a Raspberry Pi Pico [2].

The next step in the development of the whole device were calculations that reflected the volume of water that needs to be pumped to irrigate an average garden when watering repeatedly several times a day. The design continued with the creation of a block diagram, which at the top level consists of the power management, controller and output peripherals.

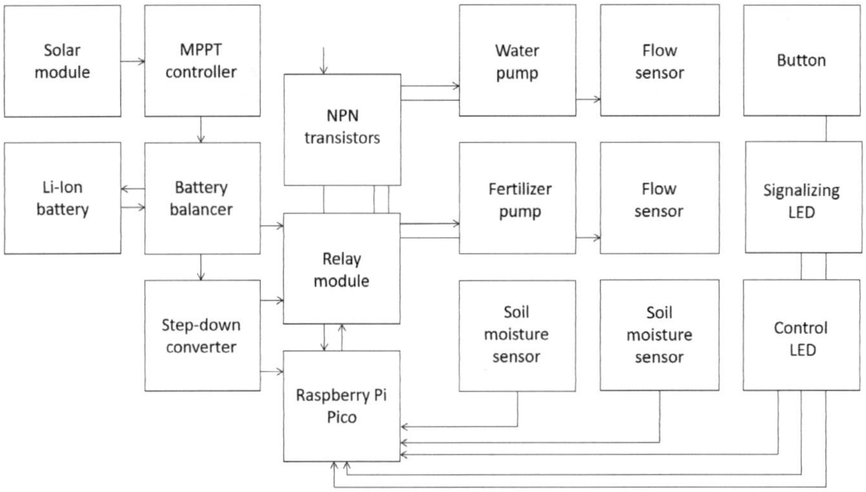

Figure 1. Solar-powered irrigation system layout block diagram [2]

The electric design is based on a block diagram, which is divided into three parts. The left part of figure 1 can be called the power management. This part is responsible for processing electric energy from solar panel, equal charging of three serial connected Li-ion batteries and making separated voltage circuits for the rest of the device. The device consists of three main voltage circuits, first, 12 V circuit is used for battery charging and powering a main water pump, second, 5 V circuit powers almost all the electrical components and the last 3.3 V powers the rest of the electrical components and indicator LED diodes.

A control unit of the device is placed in the middle part of Figure 1. The heart of it is a microcontroller from the company Raspberry, Raspberry Pi Pico, which is programmed in language C. Input/output pins of the microcontroller control and manages transistors and relay modules, which allows the microcontroller to switch on and off high-power demanding components, as such the main water pump. Rest of the pins are used as data readers from output peripheries, the monitored variables include, for example, actual battery voltage.

The last part consists of output peripheries. Output peripheries use the energy contained in batteries either for doing physical work (water pump) or for obtaining important data from the device (water flow sensor). In addition to the already mentioned components, such as water pump or water flow sensor, output peripheries contain soil moisture sensors and two types of LED diodes. Thanks to signaling LED diodes, the device is capable of communication with the user, while control LED diodes are hidden inside the box and serve as a control of each electrical circuit in the event of a malfunction.

Subsequently, after completing the electrical design, which also includes production of a few PCBs, a 3D model of the device was created in SolidWorks and consists of an external and an internal part. In the process of 3D modeling, the name of the device, IRIS, was created. Simplicity and compatibility were emphasized in the design. The inner part consists of the base panels, which contain all the electrical components, and a container for fertilizer, which is located at the rear of the unit. The protection function is primarily on the outer part that protects IRIS from natural phenomena such as heavy rain, hail or even insects.

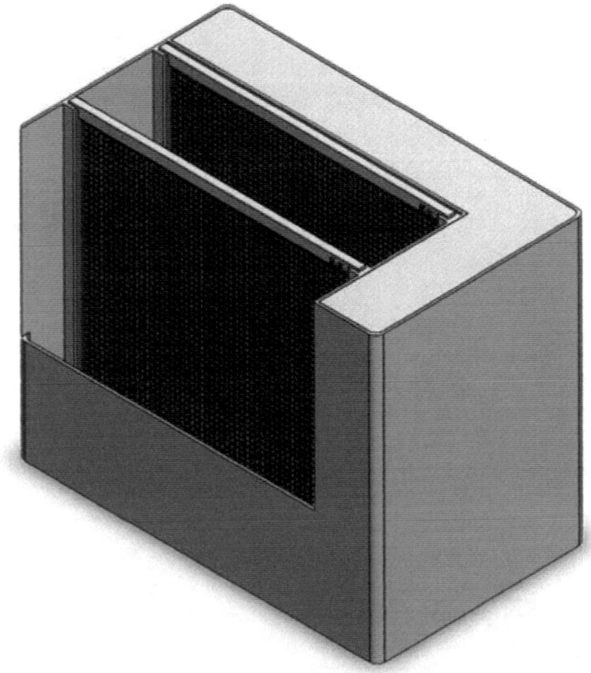

Figure 2. The inner part of IRIS [2]

The inner part is a solid entity mainly because of the plug-in connections between base panels and inner part structure, which made it mechanical and vibration durable. A cut-out is created from the front of the device, main reason is easier placement of components, as well as possible repairs in the future.

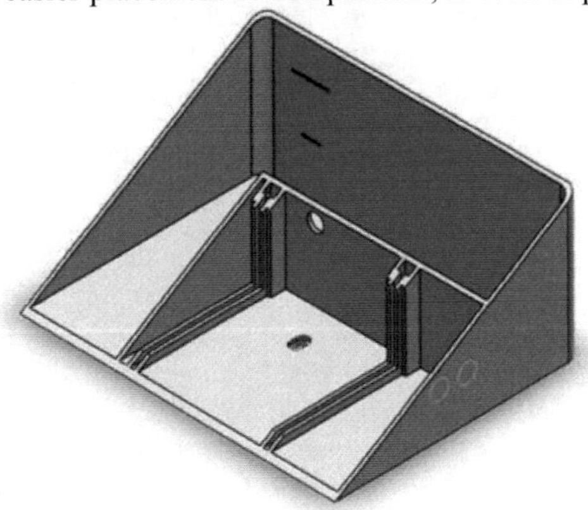

Figure 3. Cutaway view of the inner part [2]

Cutaway view in Figure 3 shows the fertilizer tank with indication lines, the narrower line represents every 0.5 l and wider line every 1 l of liquid. The total volume of the tank is approximately 4 liters and thanks to these indication lines it will be easier for the user to estimate the remaining volume of fertilizer. Figure 3 also shows the location of signaling LEDs on the front of the unit.

Individual parts were gradually printed on a 3D printer Creality Ender 5 plus, which main advantage is relatively big printing space of 350 x 350 x 400 mm.

Figure 4. Shot from 3D printing of the outer part [2]

PET-G material was used for printing, while PET is commonly used in the production of bottles, cans, and caps. Modification G consists in adding a modified glycol to the material composition, which makes the resulting product more resistant to impact, together with a reduction in brittleness and an increase in gloss. Another reason for choosing this material is its ability to be recycled, which increases the environmental friendliness of the entire product.

Figure 5. The front of the assembled device [2]

For compatibility with end environment, two colors were chosen for 3D printing. The outer cover makes up the majority part of the device, so grass green color was chosen. The inner part is mint green, making the rest of the device stand out from greenery on the ground.

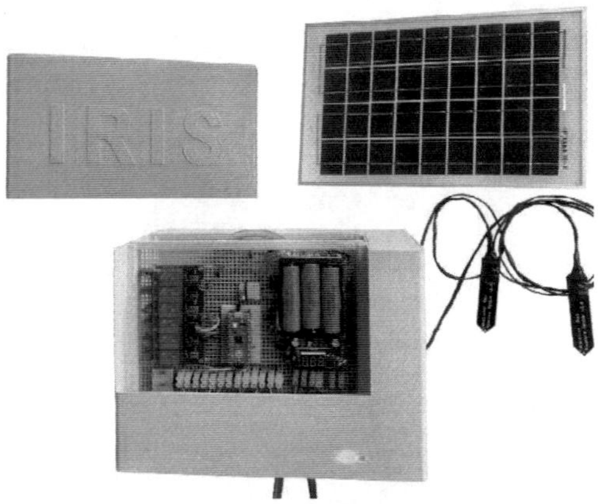

Figure 6. Designed solar-powered irrigation system [2]

IRIS is controlled through a single button, which is located on the underside of the inner part. One button control was chosen mainly to maintain simplicity for the end user. The algorithm is set so that after pressing the button for 6 seconds, the device turns on, which will be signaled by the triple flashing of both signaling diodes. After pressing the button again for less than 6 seconds, the fertilization process will be initialized for the next watering cycle, which will be signaled by one flash of the green LED for 2 seconds. If the fertilizer ran out in the previous fertilization cycle, the red LED lights up instead of the green one.

After turning on the device, the battery voltage is cyclically measured and possibly recharged, and after reaching 12.3 V the soil moisture measurement will begin. If the soil moisture is low, the irrigation process starts, which last approximately 12 minutes, during which sixty liters of water are pumped. Thanks to the flow sensor, the water flow is controlled throughout the irrigation period. A decrease in the flow from the original four lit/min may indicate a blockage in the inlet or outlet hose or too high pressure in the outlet hose caused by an insufficient amount of droppers or other water dispensers. When a low flow rate is detected, the device goes into "fault" mode, in which the red LED blinks constantly and the device is shut down until restarted, which will initialize a device check by the user. When the watering completes without error, the green LED flashes six times and the device is switched to a “charging“mode, where it remains for at least one hour. If the battery does not recharge within one hour, for example due to a cloudy sky, then irrigation will wait until fully recharged state, after which the soil moisture will be measured again.

The functionality of this system was verified by several tests. The first test was a classic device functionality test, during which one irrigation period took place. After the successful completion of the first test, the fertilization test followed. The fertilization test consists primarily of one irrigation period, but the fertilization function was also initialized, during which the device successfully pumped such a volume of fertilizer into the outlet hose that a satisfactory ratio of fertilizer volume to water was maintained after the end of irrigation. The reason for these tests was to verify the functionality of the device without exposing it to crisis "fault" situations. This was followed by an important test of the functionality of the main pump protection, during which the equipment was exposed to a critical situation.

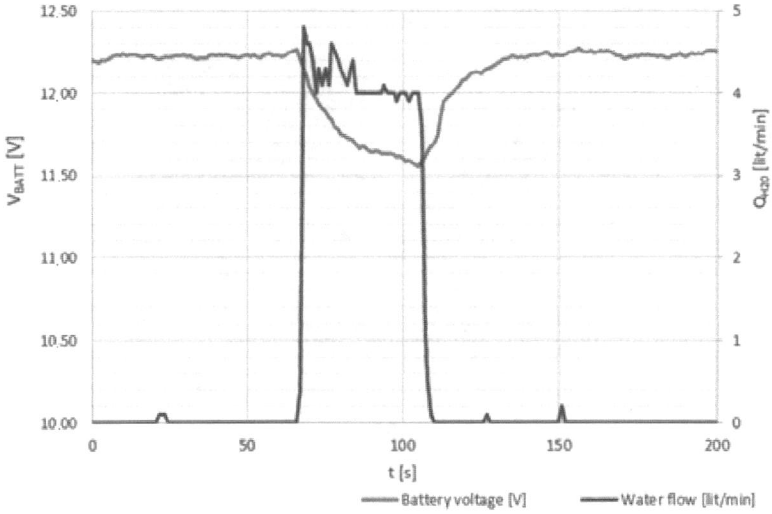

Figure 7. Water pump protection test [2]

In Figure 7, it can be observed that the test started with a nearly charged battery, while the desired value was reached in 62 seconds from the start of the measurement. After a few seconds, the flow meter recorded the first values of the water flow. The inlet hose was submerged until the 105th second from the start of the test when it was pulled out of the water, causing the flow rate to drop fairly quickly to zero liters per minute. As the red curve shows, the device correctly recognized the situation and turned off the main pump, which resulted in a decrease in current flow and a gradual increase in battery voltage. This test simulated the clogging of the inlet/outlet hose with unwanted material, when the pump would not be able to pump water from the inlet to the outlet and the flow meter will show very low flow values.

The last test was the test of protection against overwatering, which is critical from the point of view of plant protection, mainly in cases of heavy rain or other similar situations, when the soil is sufficiently moist and does not need to be additionally watered. The test started with charged batteries, while the humidity sensors were gradually immersed in water, which resulted in the measurement of 100% humidity on both green curves in the graphic dependence in Figure 8. Irrigation started when both sensors were taken out of the water, i.e. by measuring humidity lower than 50 %.

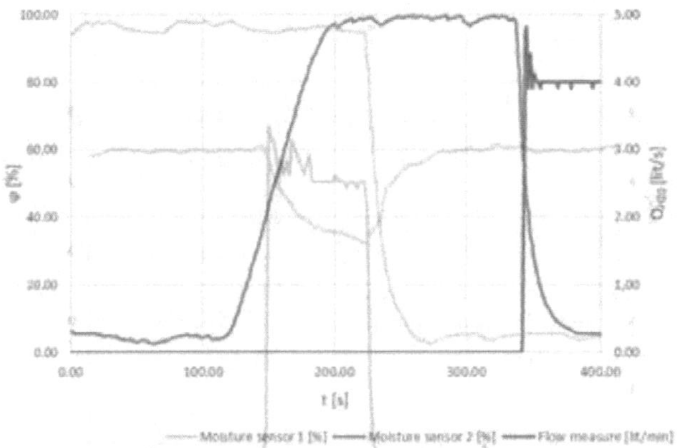

Figure 8. Soil moisture sensor tests [2]

3. Conclusion

The paper focuses on the description of an automatic solar-powered irrigation system that was developed following the identified shortcomings of commercially produced irrigation systems. Powered by a photovoltaic panel, the system is independent of grid connection, and, thanks to the built-in pumps, no source of pressurized water is required. The major benefits of this system are its environmental friendliness, water conservation and automated operation. The addition of a fertilizer tank ensures even better plant care. Similarly, the addition of protective elements ensures a longer lifetime of the entire system.

4. Acknowledgments

This work was supported by the specific graduate research of the Brno University of Technology No. FEKT-S-20-6206.

References

[1] Solar irrigation [online]. [cit. 2021-5-9]. Available from: https://ecoprodukt.cz/c/vsechno-solarni-zavlazovani-1880?gclid=CjwKCAjwkN6EBhBNEiwADVfyawnaR3AwnD8bx3I_I7AZlhqytbmz-NzKudAAGT2SO68ZSXYBvLO-DhoCqA0QAvD_BwE.

[2] R. Straňák, K. Jandová: Solar irrigation, Bachelor Thesis, Brno University of Technology, 2022.

Overview of tribological properties of UHMW polyethylene under rotation

Z. Studeny[1], D. Dobrocky[1], J. Viliš[1], J. Adam[2]

[1] Department of Mechanical Engineering , University of Defence, Brno, Czech Republic
[2] Department of Weapons and Ammunition, University of Defence, Brno, Czech Republic

corresponding author: zbynek.studeny@unob.cz

Abstract. The paper works with ultra-high molecular weight polyethylene (UHMW). This polyethylene is widely used in technology, in engineering or its applications are also in the field of medicine in prosthetics in the lining of joint replacements. This wide use is mainly due to its properties. It is mainly its abrasion resistance and very low coefficient of friction (COF), so it has excellent sliding properties. It is these properties that are the subject of research on UHMW polyethylene. The investigation of these properties is always linked in tribological tests to the material in the friction pair, which consists of the material under test (UHMW) and the indenter. For the purpose of tribological testing, 6.35 mm diameter balls of ZrO2 ceramic material and a cobalt-chromium alloy known as Stellit 6 were chosen as indentors. Practical applications are the above mentioned fields of technology, engineering, etc.

1. Introduction

The material dealt with in the article belongs to the polyethylene group. It is Ultrahigh Molecular Weight (UHMW) polyethylene (PE). This material was created in a structure of extremely long and highly intertwined molecular chains in the early 1960s [1]. This structure of UHMWPE gives it a high wear resistance. This polyethylene is also very tough, has a low coefficient of friction and is abrasion resistant. It is these characteristic properties of UHMW polyethylene, such as low coefficient of friction and wear resistance, that are the focus of this work. These properties are used in the application of UHMWPE in engineering, machine components or structures. Examples include the use of UHMWPE for the manufacture of ropes in maritime transport or in lifting equipment. Particularly suitable is the use of this material in wet environments where there is a significant risk of corrosion, such as bearings. For bearings, this material is also advantageous in terms of vibration damping capability.

The above mechanical properties combined with low weight are also useful in the field of ballistic protection, both for technology and for people, as described by Kartikeya [2] or Vilis [3]. An equally important application of UHMWPE is in the field of implantology. Due to the combination of good mechanical properties, in particular the already mentioned low coefficient of friction and wear, and biocompatibility in the human body, this material is also widely used for joint linings as reported by Kurtz [4].

With regard to the above mentioned properties, it is essential to know the tribological properties of the material under study. However, the tribological properties are evaluated for decades, as reported by Klapperich [5]. However, in the field of tribological evaluation, the bonding of the pair of the material under study and the indenter material is important. For the purpose of the tribological properties analysis, the material ZrO_2 was chosen to the pair of the studied material UHMWPE and the indenter material or ball as investigated by Ge [6]. The work is subsequently extended to the material cobalt-chromium alloy known as Stellit 6.

The experiments were carried out using the Pin-on-Disk method on a Bruker UMT-3 Tribolab. Subsequent evaluation of the tribological test marks was carried out on a TalySurf CLI1000 profilometer.

2. Material and Methods

2.1. Materials Under Investigation

Ultrahigh Molecular Weight Polyethylene (UHMWPE) is a subset of thermoplastic polyethylene, with the chemical designation $H(CH_2CH_2)_nH$. This material used for the experimental work is odorless, tasteless and non-toxic. As already mentioned, it has high abrasion resistance. Compared to structural steel, it is approximately 15 times more resistant. Its coefficient of friction is significantly lower than that of nylon and PLA and is comparable to polytetrafluoroethylene (PTFE). However, the UHMWPE material studied has better abrasion resistance than PTFE [7, 8].

This material consists of polyethylene chains. Compared to the common chains of molecules in other materials, UHMWPE has these chains of molecules very long and oriented in the same direction. This orientation gives the material as a whole the mechanical properties mentioned above. The van der Waals bonds between the individual molecules are relatively weak. Since the overlap of the individual molecules is long, they have the ability to transfer greater shear forces from one molecule to another. This increases the total amount of bonding between molecules and thus the magnitude of the forces transferred.

The melting point of UHMWPE is between (130 - 136)°C, due to the deteriorated mechanical properties it is not advisable to exceed (80 - 100)°C in use, and it will also become brittle at temperatures below -150°C [9]. Due to the chain structure of the molecules, UHMWPE does not absorb water on the surface of the fibres. It does not interact with the skin. As UHMWPE does not contain chemical groups (e.g. esters, amides or hydroxyl groups) which are susceptible to attack by aggressive substances, it is very resistant to water, moisture, most chemicals, UV radiation and micro-organisms.

In addition to its wide application in medicine, especially for lining knee or hip implants, this material has applications in ballistic protection. Companies such as Dyneema or Spectra produce this material under different brand names, e.g. Ticona, Braskem, DSM, Teijin (Endumax) [3], Celanese or Mitsui. The UHMWPE material under discussion is commercially available in the form of fibres or in the form of sheets or rods.

The indenter materials chosen were ZrO_2 ceramic and then a cobalt-chromium alloy known as Stellit 6. These materials are again closely related to the use of the UHMWPE material tested, both in machinery applications and in contact in biological environments. The balls made from both materials had a diameter of 6.35 mm for testing purposes. Their basic properties are given in Table 1.

The ZrO_2 ball material is a highly durable material capable of operating at high temperatures and in aggressive environments of organic solvents or acids. It has high resistance to abrasion and corrosion. The second ball material used for the experiment was Stellite 6.

Table 1. Properties of the balls used for the Pin on Disk test

Ball material	Diameter (mm)	Hardness (HV10)	Temperature resistance (°C)	Standard
ZrO_2	6.35	1240	to 2400	ISO 3290 G5-G40
Stellit 6	6.35	527 - 612	to 600	ISO 3290 G10-G40

It is a chromium-cobalt alloy composed of complex carbides that is resistant to wear, mechanical damage and corrosion even at very high temperatures. It was invented by Elwood Haynes in 1907. It also resists impact and erosion cavitation. These versatile properties are due to the CoCr matrix. The name "Stellite" is derived from the Latin word "stella" (star) because of its high lustre. Nowadays, the name Stellite is a trademark of Deloro Stellite, part of the Kennametal group [10]. High-cobalt alloys do not have any exact chemical formula because the different types of alloys are made up of a combination of several elements in different proportions. Therefore, the alloys are designated by a number or a letter, such as Stellite 6, which is the most widely used [11]. The chemical composition of Stellite 6 is given in Table 2.

Table 2. Chemical composition of chromium-cobalt alloy Stellite 6 (10)

Element (wt%)	Co	Cr	W	C	Si	Fe	Ni	Mo
Ranges	54-55	30	4-5	1,2	< 2	< 3	< 3	< 1

2.2. Experiment Characterization

2.2.1. Tribological Tests. Tribological tests were performed using the Pin on Disk method [12]. The test results not only in the measurement of the coefficient of friction (COF) but also in the acquisition of a trace that can be further evaluated in terms of wear. The Pin on Disk method was performed on experimental specimens made of UHMWPE materials.

The specimens were disk shaped with a diameter of 70 mm and a thickness of 7 mm. The requirement for flatness of the samples is 0.01 mm to avoid measurement error due to harmonic pulsation of the sample during rotational motion. The specimens were completed with a central hole for mounting the specimen on the stage. A 3 mm diameter hole is required at the 12 mm radius to accommodate the pin of the drift.

This test was performed on a Bruker UMT-3 Tribolab. The Pin on Disk test is one of the basic methods for observing material surface abrasion [13, 14]. The principle of the method is to load a ball-shaped indenter that moves vertically and exerts a load force F_N. The indenter was loaded with a constant force. The sample stage under the indenter rotates at a defined angular velocity. This of course depends on the radius of the track and the chosen path to be traversed by the indenter on the sample. The penetration of the indenter into the sample creates a trace which is further evaluated.

The reciprocal motion was carried out on the selected path of 100 m at constant speed. The load during the test on one track was also always constant. However, different load levels were investigated. Each track was created at a different radius on the sample. The radiuses at which the measurements were made were chosen to be (20 and 25) mm, taking into account the design capabilities of the machine and especially the rotary table. As the radius of the track increased, the distance per revolution was also increased, thus adequately reducing the total test time while maintaining constant speed. The loading force was set sequentially at two levels (20 and 50) N for the ZrO_2 ball (tracks 2 and 3) and the Stellite 6 material (tracks 4 and 5), see Figure 1 or Figure 3. Track 1, which was used for calibration, was not evaluated further.

Friction is created between the surfaces being evaluated, in this case between the sample and the bead. The value of friction is expressed by the coefficient of friction μ (-). This coefficient is the ratio of the friction force to the normal force. The normal force F_N (N) corresponds to the load applied perpendicularly across the ball to the sample. The frictional force F_t (N) is measured during motion and expresses the force required to overcome the resistance to motion. These forces are expressed in relation (1). The coefficient of friction can then be expressed [12]:

$$\mu = \frac{F_t}{F_N} \tag{1}$$

The aim of the measurement is then to determine the actual values of the coefficient of friction between the two materials. Depending on the relative movement of the functional surfaces, the friction coefficient can be divided into a rolling friction coefficient and a sliding or shear friction coefficient. Important note: the friction force is negatively oriented with respect to the motion and its absolute value is less than the force required to set the body in motion.

2.2.2. Measuring traces of wear. The evaluation of wear marks can be approached in several ways, one can use the measurement of the 3D texture of the surface with the calculation of the volume of the wear marks, as stated, for example, in [15, 16]. Another possibility is the measurement of 2D profiles, from which the content of the surface of the wear mark in a cross section, together with the width and depth of the mark is determined [17, 18].

The measurement of traces of wear was performed on a Talysurf CLI 1000 device, using an inductive touch sensor with a measuring tip radius of 2 μm. The method of measuring 2D wear trace profiles, or their cross-sections. The measurement methodology was as follows:

- measurement of the 2D profile of the wear trace,
- levelling of profile,
- form removal,
- zoom of the wear trace,
- analysis of the maximum depth and the area of wear trace,
- analysis of the width of wear trace.

The measurement of the 2D profiles of wear traces was always carried out at five locations for a given indenter load, the individual measured locations were rotated by 60° relative to each other. Fig. 1 shows the evaluated experimental sample, on the first side, testing was performed at three loads, on the second side of the sample, two loads were selected.

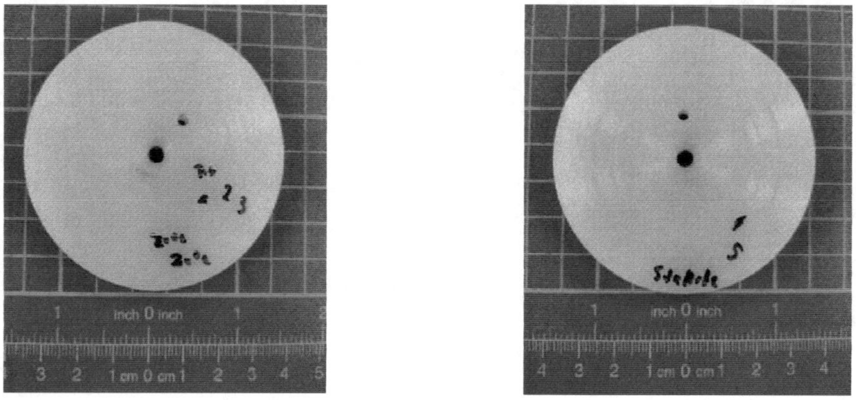

Figure 1. Experimental sample

3. Results and Discussion

3.1. Evaluation of Tribological Testing

The result of tribological tests are records of the friction coefficient over time. The comparative record is documented in Figure 2. For each waveform, the indenter material and the loading normal force F_N are listed for clarity. The specific friction coefficient values for each friction pair and load are given in Table 3. Standard deviations are given for the friction coefficient values in this table. From both the COF value plots in Figure 2 and the numerical values in Table 3, it is clear that the experimentally tested friction pairs UHMWPE vs. Stellite balls or ZrO_2 balls generally have very low COF values of less than 0.1. In tests, it was shown that the ZrO_2 ceramic ball achieves demonstrably lower COF values than the Stellite 6 chromium-cobalt alloy ball at both normal load levels. Of course, it depends on the specific application whether there will be a requirement for hardness or ductility of the material in addition to the requirement for low COF. This property is then better guaranteed by the chromium-cobalt alloy material Stellite 6 than by the ceramic material ZrO_2.

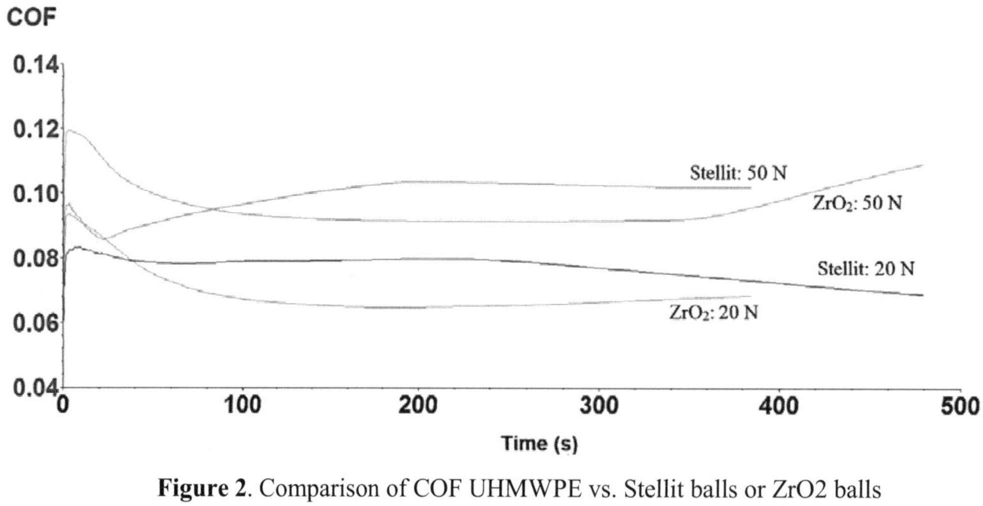

Figure 2. Comparison of COF UHMWPE vs. Stellit balls or ZrO2 balls

Table 3. Resulting values of COF (coefficient of friction) for UHMWPE vs. Stellit balls or ZrO2 balls

Normal Force (N)	20		50	
	COF	Std. Dev.	COF	Std. Dev.
UHMWPE vs. Stellit balls	0.0768	0.00357	0.0992	0.00525
UHMWPE vs. ZrO_2 balls	0.0690	0.0066	0.0099	0.00572

3.2. Evaluation of traces of wear

A more detailed overview of wear traces, for individual measurements, is documented in Fig. 3. Images of wear traces on the surface of the experimental sample were taken with an Olympus MVX10 optical measuring microscope. Overview images were created at 6.3x magnification, images of wear traces are at 25x magnification. Wear trace 1 corresponds to loading with normal force $F_N = 20$ N, trace 2 $F_N = 20$ N, trace 3 $F_N = 50$ N, trace 4 $F_N = 20$ N a trace 5 $F_N = 50$ N.

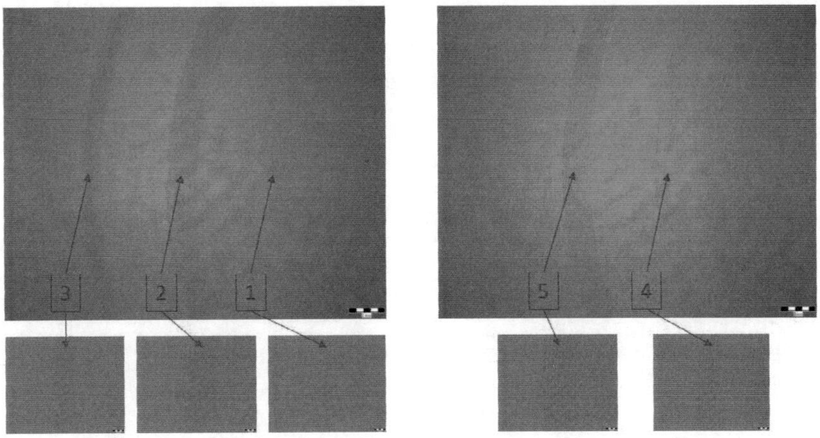

Figure 3. Traces of wear on the surface of the experimental sample

The profile of the wear track, together with the display of the analysis of its maximum depth, area and width, is shown in Fig. 4.

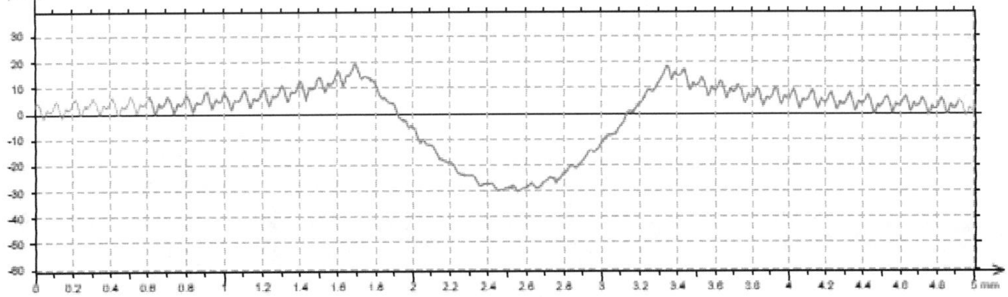

Wear trace profile

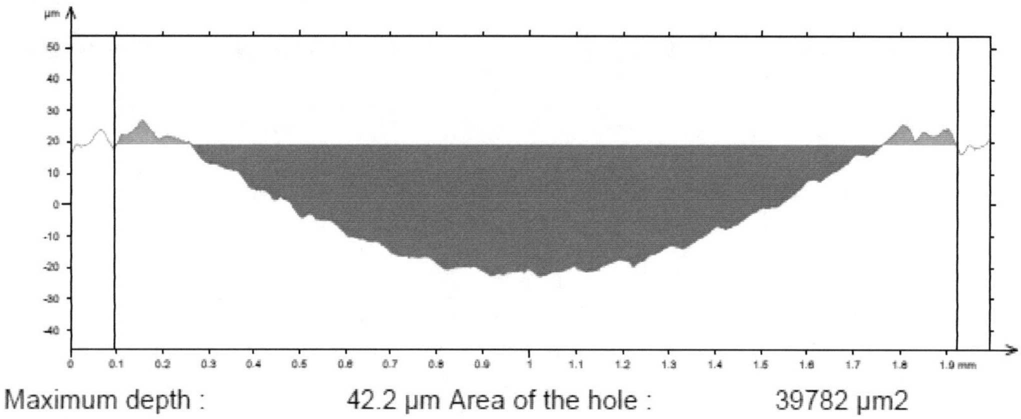

Determining the maximum depth and area of wear trace

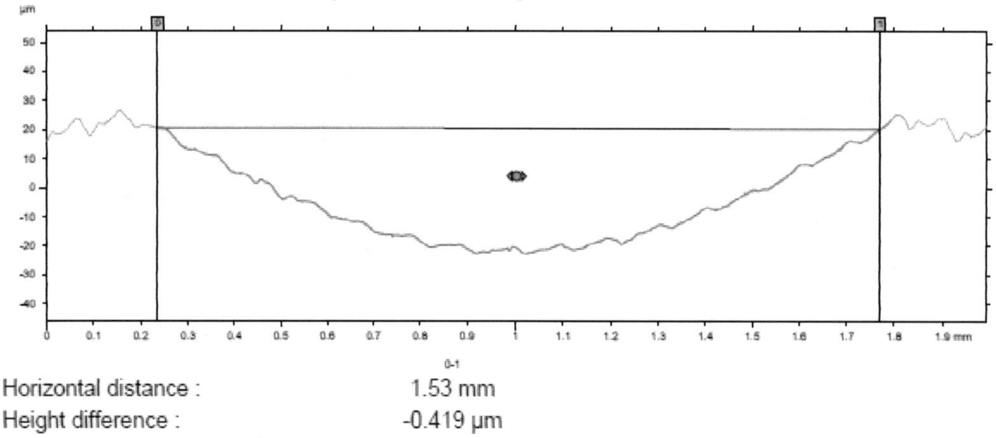

Determining the width of the wear trace

Figure 4. Wear trace analysis

The measurement results of selected geometric parameters of wear traces No. 1 to No. 5 are summarized in Table 4.

Table 4. Wear trace measurement results

Trace number	Normal force F_N (N)	Max. depth [µm]	Area of the hole [µm^2]	Width [µm]
1	20	9.83 ± 0.22	317.28 ± 37.87	67.12 ± 11.65
2	20	43.22 ± 0.75	41173.80 ± 1051.40	1550.00 ± 23.45
3	50	13.24 ± 0.62	6832.40 ± 586.33	936.40 ± 49.14
4	20	13.96 ± 1.06	6172.20 ± 352.07	961.40 ± 48.04
5	50	39.30 ± 1.82	35466.40 ± 2563.97	1522.00 ± 55.41

From the results, it can be seen that the smallest surface wear of the experimental sample was achieved in the case of wear trace No. 1 (load 20 N, ball material PTFE - not further evaluated). The highest wear values were found for wear track No. 2 (20 N load, ZrO2 bead material). High values were also achieved for wear track No. 5 (50 N load, Stellite 6 bead material). Wear tracks No. 3 (50 N load, ZrO2 bead material) and No. 4 (20 N load, Stellite 6 bead material) achieved almost identical results for wear trace geometric parameters.

4. Conclusions

The subject of the experimental work was to determine the tribological parameters of Ultrahigh Molecular Weight Polyethylene (UHMWPE), which is a subset of thermoplastic polyethylene, with the chemical designation $H(CH_2CH_2)_nH$. It is a material with a wide range of applications. The use of this material is in machinery components, bearings, gears, etc., as well as in the medical field. Due to its biocompatibility with the human body and good sliding properties, it is used in artificial joint parts. Another large area is, for example, ballistic protection of persons or technology. Due to its properties, it is used in combination with hard and durable materials.

As part of the experimental work, the tribological properties were measured on a Bruker UMT-3. Balls made of the ceramic material ZrO2 and the chromium-cobalt alloy Stellite 6 were used in a friction pair with the above material.

Based on the Pin on Disk test, the better sliding properties of the ZrO2 ceramic material were clearly demonstrated. This hypothesis is also consistent with the results from the wear marks measured after the tribological tests. The same wear values were achieved for the ZrO2 ceramic material at higher loads than at lower loads in the case of the Stellite 6 ball. Their practical use, beyond the above, will therefore depend on other strength, biocompatibility, economic and other aspects in specific applications.

Acknowledgments

The work presented in this paper was supported by the specific research project 2020 “SV20-216” at the Department of Mechanical Engineering, University of Defence in Brno and was supported by the Project for the Development of the Organization “VAROPS (DZRO VAROPS) Military autonomous and robotic assets” by the Ministry of Defence of Czech Republic.

References

[1] Slouf M. et al. Comparison of macro-, micro- and nanomechanical properties of clinically-relevant UHMWPE formulations. *Journal of the Mechanical Behavior of Biomedical Materials* **120**, p. 104205, 2021, doi.: 10.1016/j.jmbbm.2020.104205)

[2] Kartikeya K.et al. Ballistic evaluation of steel/UHMWPE composite armor system against hardened steel core projectiles. *International Journal of Impact Engineering* **164**, p. 104211, 2022, ISSN 0734-743X, doi.: 10.1016/j.ijimpeng.2022.104211

[3] Vilis J Pokorný Z Zouhar J. and Jopek M. Ballistic Resistance of Composite Materials Tested by Taylor Anvil Test. *Manufacturing Technology*, 2022.

[4] Kurtz S. M Rimnac C MPruitt L Jewett C W Goldberg V and Edidin A A. The relationship between the clinical performance and large deformation mechanical behavior of retrieved UHMWPE tibial inserts. *Biomaterials* **21**, Issue 3, p. 283-291, 2000. doi.: 10.1016/S0142-9612(99)00178-7

[5] Klapperich C M Komvopoulus K and Pruitt L A. Tribological properties and microstructure evolution of ultra-high molecular weight polyethylene. *J Tribology* **121** (1999), pp. 394-402

[6] Ge S et al. Friction and wear behavior of nitrogen ion implanted UHMWPE against ZrO2 ceramic. *Wear* **255**, Issues 7–12, 2003, P. 1069-1075, doi.: 10.1016/S0043-1648(03)00269-2

[7] Tong J Ma Y Arnell R D and Ren L. Free abrasive wear behavior of UHMWPE composites filled with wollastonite fibers. *Composites Part A: Applied Science and Manufacturing* **37**, Issue 1, P. 38-45 2006, doi: 10.1016 / j.compositesa.2005.05.023
[8] Brožek.M. Selected Plastics Wear Resistance to Bonded Abrasive Particles Compared to Some Ferrous Materials. *Acta Univ. Agric. Silvic. Mendel. Brun.*, **63**(2), 387-393. 2015. doi: 10.11118/actaun201563020387
[9] Lewin. *Handbook of Fiber Science and Technology Volume 3*: High Technology Fibers. CRC Press. 1996. ISBN 9780824794705.
[10] Gilbert N. Stellite Alloys - Chemical Composition, Mechanical Properties and Common Applications. AZO Materials [online]. b.r. [cit. 2018-12-03]
[11] Stellite 6 Alloy: Technical data [online]. In: . Koblenz, Germany: Deloro, wearsolutions, b.r., s. 2
[12] ASTM G 99 95a Standard Test Method for Wear Testing with a Pin-on-Disk Apparatus
[13] Studeny Z Pokorny Z Dobrocky D Joska Z Prochazka J. Tribological properties of DLC coating for parts of weapons. In: Kazda T., Klein F., Sedlarikova M. *ECS Transactions* ***99***. Brno; Czech Republic: IOP Publishing Ltd, No. 1, p. 297-307. ISSN 1938-5862. ISBN 978-160768539-5. (2020).
[14] Barényi I Majerík J Pokorný Z et al. Material and technological investigation of machined surfaces of the OCHN3MFA steel. *Metallic Materials* **57**(2), 2019, 131-142. ISSN 0023-432X.
[15] Liang Ch et al. *Journal of Alloys and Compounds* **911**, (2022).
[16] Haus L Wildfeuer M et al. *Wear*, 488-489, (2022).
[17] Nguyen N Ch Svoboda E Procházka J et al. Effects of Hybrid Surface Treatment Composed of Plasma Nitriding and CrN Coating on Friction-Wear Properties of Stainless Steel. *ECS Transactions*, 2021, **105**(1), 309-318. ISSN 1938-5862. doi:10.1149/10501.0309ecst
[18] Doan T V et al. Friction and Wear Behaviour of 42CrMo4 Steel Treated by Tenifer, Hard Chrome and Plasma Nitriding Technologies. *Manufacturing Technology,* 2017, **17**(2), 168-174. ISSN 1213-2489.

Investigation of material properties using acoustic emission

T Binar[1], J Zimáková[1], P Šafl[1]

[1]Brno University of Technology, Technická 10, 616 00 Brno, Czech Republic

zimakova@vut.cz

Abstract. This paper discusses the possibility of using acoustic emission to evaluate the effect of corrosion in grey cast iron. Experimental samples were fabricated and subjected to corrosion attack in condensation chamber for 7 days. Acoustic emission recordings were taken throughout the corrosion exposure period. The results are complemented by metallographic analysis.

Introduction

The acoustic emission (AE) method is nowadays a commonly used non-destructive test because it can be used in many practical areas. The acoustic emission of corrosion processes reliably shows the instantaneous rate of progress of corrosion and associated processes. In particular, the use of multiple acoustic emission sensor is advantageous, which can very accurately locate corroding parts of a structure, especially in inaccessible areas for internal corrosion modes.

The automotive and aerospace industries are some of the most widespread areas where the acoustic emission method can be used to investigate materials. Aerospace aluminium and magnesium alloys have been selected as examples of materials of interest. Another material where the acoustic emission method could be applied is cast iron with flake graphite. Due to their biocompatibility there, these materials can also be used in medicine. However, pitting corrosion appears on this alloy when in contact with an electrolyte containing chloride anions [1]. Investigations of magnesium alloys by means of electrochemical noise have been carried out, for example, by Z. Zhang et al. AZ91D alloy was immersed in an alkaline chloride solution and investigated using the electrochemical noise method. By analyzing the time records, three corrosion stages, including anodic dissolution processes associated with hydrogen bubble growth, absorption and desorption, pitting corrosion and inhibition process could be clearly distinguished [1].

Another one was by Wang et. al. who monitored pitting in AZ31 alloy in 4 simulated solutions using wavelet analysis of electrochemical noise signals. They found that the local corrosion development process in all four environments can be divided into two phases, and the electrolyte environment can influence the dominant corrosion phenomena in each phase [1]. Magnesium alloys have also been studied by Kim and his team on AZ31 Mg under potentio-dynamic polarization. They found that isolated hydrogen bubbles produced by cathodic reaction and bubbles produced by hydrolysis in pitting corrosion formed sources of AE. For this reason, the investigation also involved the AE method. The monitored electrochemical noise and acoustic emission parameters were reported and related to different corrosion stages. [1] The sample was 15 x 15 20 mm in size and its functional surface was fixed in epoxy resin sanded with 2 000 grid silicon carbide paper, than cleaned by ultrasonication in ethanol and inspected by scanning electron microscope. The electrolyte was prepared from sodium chloride solution in deionized water at room temperature.

AE in corrosion of aluminium alloy – here the corrosion process on aluminium alloy aircraft sheet metal was investigated. Accelerated corrosion testing of this material was investigated, for example, by J. Kober and Z. Převorovský form CAS. They had 2 L-shaped samples which were first exposed for 140 hours in weaker solutions of sea salt with hydrochloric and nitric acid, and then in stronger solution for about another 500 hours. The samples were 100 mm long and 15 mm wide at the sides. Their surface was completely unmodified except for minor oxidation products typical of this material. The Dakel XEDO analyser recorded the counts of emission over 2 threshold levels (count 1 near the noise floor and count 2 for about half the range) and the effective AE signal at the corrosion test site. First peak lasted about 2 hours and it estimated to correspond to the rapid formation of a passivation layer on the surface of the sample after the test was interrupted and of the corrosion layer cleaned. The second peak had a duration about 5 hours and then always decreased in time, although its height varied during the test. This may be the progression of the corrosion products and the possible depletion of the aggressiveness of the solution. The AE here thus revealed the mechanism of corrosion damage, but to determine it accurately, it was still necessary to construct a corrosion model using additional physico-chemical analyses. For the purpose, in this case, the NWMS nonlinear ultrasonic method measurements were used several times during the corrosion test. [2, 3]

1. Experiment

1.1. Test samples

LLG/GJL or grey cast iron with the designation GJL-300 was selected for corrosion testing using AE. Test specimens with a circular cross-section of 3 cm and plate width of 1 cm were prepared form base material (see figure 1). The chemical composition of the constituent elements of the GJL-300 cast iron, as a percentage of the volume present in the material, are shown in Table 1.

Table 1. Chemical composition of the constituent elements of the GJL-300 cast iron

Cast iron	Basic elements of cast iron [%]				
	C	Si	Mn	P	S
GJL - 300	3.061	1.833	0.971	0.073	0.079

Six sets of 3 specimens and one comparison set of specimens that were subjected to the corrosion test were created. This set served as a source of the original properties of material. To place the material in the condensation chamber, a special fixture was designed and created using 3D printer to hold the samples, which is shown in Figure 2. It was necessary to attach AE sensors to the samples.

Figure 1. Prepared samples for corrosion testing

Figure 2. Sample placement jiq for condensation chamber (left – unfolded, right – with sample clamped)

1.2. Condensation chamber

Testing was carried out in a Liebisch Constanco condensation chamber. This is a 300 litre tabletop chamber that can be operated manually or programmed for automatic operation. It is suitable for testing either in a pure condensation atmosphere of for test using suplhur dioxide (so-called Kesternich tests), in a temperature range from -50°C to +70°C. For tests with suplhur dioxide, it can optionally be equipped with a manual or automatic external SO_2 gas dosing device with o dosing capacity of 200 – 2000 ml of gas per chamber volume. It is also suitable for salt test at temperatures up to 50 °C. [4]

The corrosive action on the samples was continuous in this chambre for 168 hours (7 days) for both uprotected and coated samples. A constant relative humidity of 100 % created by distilled water, and a constant temperature of 50 °C were maintanined in the chambre. Throughout this test, AE data were recorded by sensors attached to the samples. A view of the samples places in the condensation chambre is shown in Figure 3.

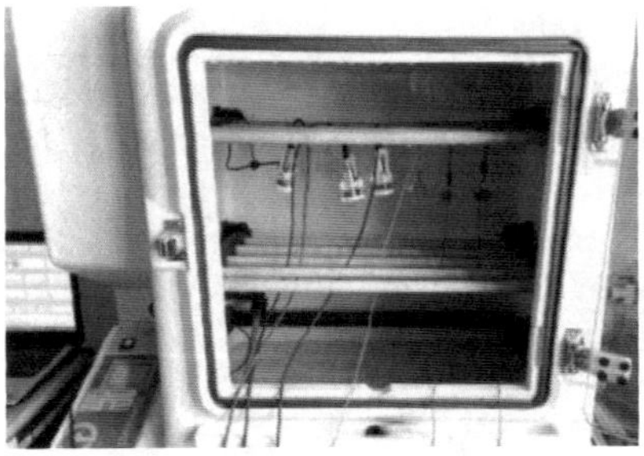

Figure 3. Samples placed in the condensation chamber

1.3. Coating for corrosion test

The specimen were at first exposed to corrosion without corrosion protection and then with corrosion protection in the form of a selected coating with corrosion inhibitors under the designation VpCI-368D from Cortec Corporation. VpCI-368 is a temporary preservative coating that provides corrosion protection to various types of metals (carbon and stainless steel, copper, aluminium and cast iron) in outdoor climatic conditions. The VpCI-368D version is a factory diluted version of the previous version of the above mentioned coating suitable especially for spry application.

2. Results of the experiment

2.1. Metallography

2.1.1. Sample preparation
Samples for metallographic analysis were cut approximately halfway around the circular area, perpendicular to the rough cutting grooves, and one half of the cut was always used for evaluation. In the case of samples not protected by coating, an oxide layer formed on their surface, which made it impossible to press the samples well. The oxide layer rolled out and caused the polishing wetting agents to come out of the gaps after rolling, so it had to be dewetted first. To do this, a solution of 3.5 g of hexamethylene tetramine powder (also urotropine of hexamine), 500 ml of hydrochloric acid and 500 ml of water was mixed, with a pickling time of about 2 minutes. The samples could then be pressed into plastic and standardly sanded and polished.

2.1.2. Metallographic analysis
During metallographic analysis, the depth of corrosion of the samples was also measured with a counting module at 200x magnification. The results of the measurements were statistically processed using STADAT software, where the parameters of the samples were compared with each other using a pairwise test (F-test for variances and T-test form means). A Neophot 32 metallographic microscope and a Canon EOS 4000D digital camera were used for observation and photodocumentation. Surface evaluation is performed for both reference and corrosion-exposed samples, both completely unprotected and with a protective coating.

2.1.3. Metallography results after 7 days
After 7 days, there was a slight increase in the local appearance of surface irregularities on the perimetr (flank) of the specimens not protected by the corrosion coating, which can be compared to pitting corrosion, e.g. in stainless steels (Figure 4). The specimens protected by the coating were also exposed to corrosion for 7 days. It can be assumed that samples coated with corrosion inhibitors will be less affected by corrosion than unprotected samples. The analysis showed that after 7 days of corrosion exposure, the samples coated with the protective coating have a similar surface condition as the initial samples. This is illustrated in Figure 5.

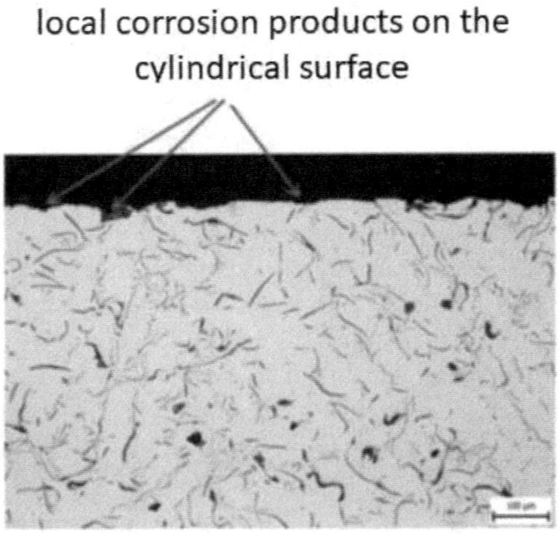

Figure 4. Surface of the sample after 7 days exposure – sample without protective coating

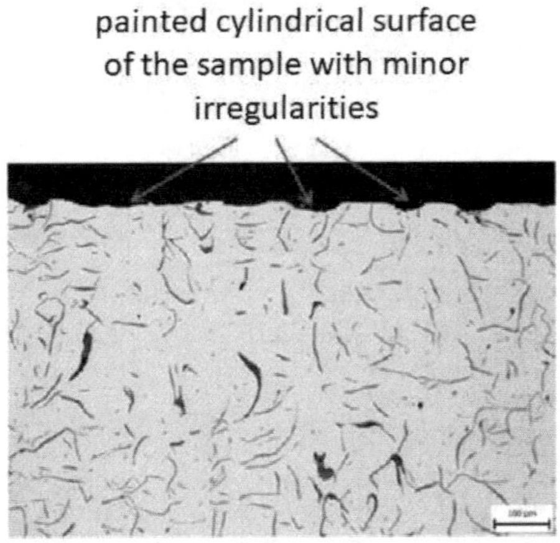

Figure 5. Surface of the sample after 7 days exposure – sample with protective coating

2.2. Acoustic emission results

Accelerated corrosion tests were monitored with an AE diagnostic tool. The measurements were performed with a DAKEL type XEDO AE diagnostic tool. The measuring system is equipped with a total of 6 channels. The formation of a corrosion-fatigue crack in cast iron is accompanied by a slightly increased AE activity. As the number of oscillation increases, i.e. as the crack grows, the frequency of AE events and their amplitude increases.

2.2.1. Monitored parameters

Measured data from the AE diagnostic system were evaluated form a total of 3 sensors. The following parameters were monitored to evaluate the AE signals:

- Count 1 (c1) – the number of overshoots over the set threshold level expressed on a logarithmic scale (left y-axis) for a specified time interval
- Count 2 (c2) - the number of overshoots over the set threshold level expressed on a logarithmic scale (left y-axis) for a specified time interval
- RMS (Root Maen Square) – root mean square level of the detected signal in mV (right y-axis)

2.2.2. AE of samples without corrosion protection.

Figure 6 shows the evaluated graphical record of the AE of test specimen 2 without corrosion protection after 7 days of measurement. In the first phase (indicated by number 1 in the graph), an immediate increase in AE activity, expressed by RMS level, can be observed during first 12 hours of testing. The second phase (marked as 2 in the graph) expressed the acoustic activity record typical of graphite deposition and the formation of local graphitic corrosion in the form of flakes and pitting. The charring is accompanied by the reaction of iron with water to form iron oxide in the form of uniform corrosion. Thus it can be assumed that the AE noise signal is caused by the reaction of Fe with H_2O to form FeOH2·2H_2O and discontinuous AE signal is caused by local graphitic corrosion, flaking and pitting. In this respect, the signal in cast iron is different form that in steel corrosion, for example, since in steel we would only obtain a signal of continuous nature. These corrosion mechanisms are also demonstrated by the results of metallographic analyses.

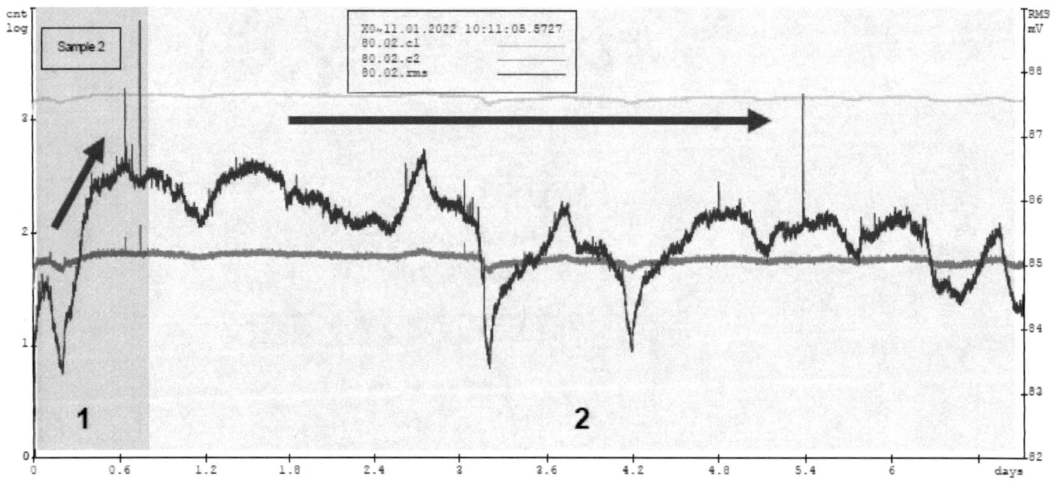

Figure 6. AE of specimen without corrosion protection after the 7 days

2.2.3. AE of samples with corrosion protection

Figure 7 shows a graphical record of the AE test specimen with corrosion protection in the form of a coating after 7 days of corrosion exposure. In the first phase, indicated by number 1 in the graph, a gradual decrease of the AE activity can be observed for the RMS values and the overshoots over the set threshold, indicated as count 1. Compared to the specimen without corrosion protection, the course of the AE activity in this observed region is reversed. This phenomenon can be attributed to the fact that the initial major corrosion events occur after the corrosion protection layer has been breached. Since "each corrosion protection layer is only as effective as its weakest part", it is always necessary to verify under realistic conditions where the critical points with minimum layer thickness are located. These can be edges and folds in particular, as confirmed by metallographic photographs. The so-called undercorrosion occurs, which has its own specifics – from the point of view of the cast iron, graphitic selective spongiosis occurs under the corrosion layer, which is enhanced by the presence of trace amounts of sulphur in cast iron. A porous structure of graphite is formed and selective dissolution of iron occurs.

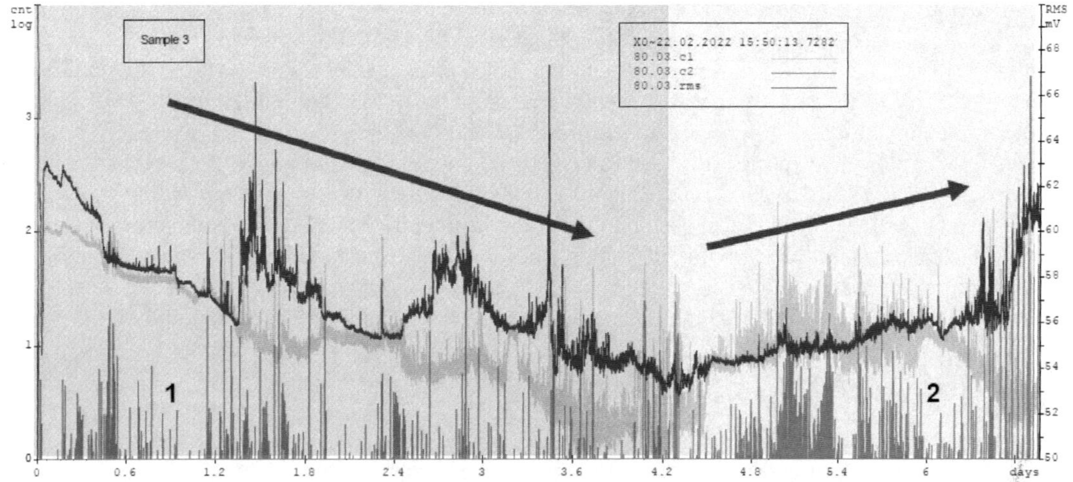

Figure 7. AE of specimen with corrosion protection after 7 days

2.2.4. Number of acoustic emission events

The graphs below show the number of AE events, again comparing samples without and with corrosion protection. From the graph for the number of events for the samples without corrosion protection in figure 8, it can be seen that the number of AE events for these samples is in the order of tens of thousands and the phases of the corrosion process are quite sharply delineated with a linear curve. The graph for number of events for the samples with corrosion protection in figure 9 shows that the number of AE events for these samples is only in the order of thousands compared to previous case, and the original relatively sharply defined individual phases of the corrosion process show a somewhat different shape. The curve is linear with slight variations.

Figure 8. Number of AE events – without protection

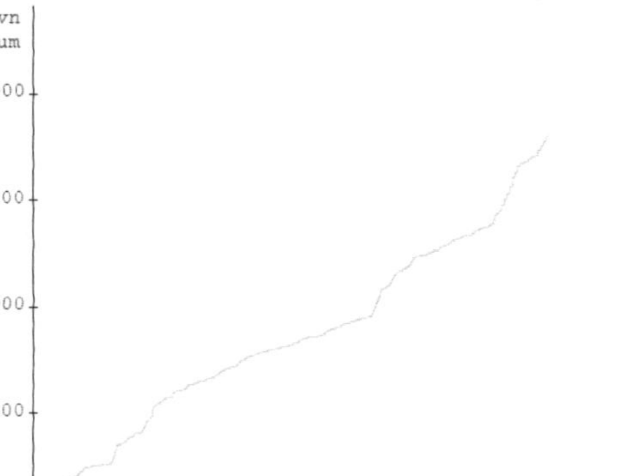

Figure 9. Number of AE events – with corrosion protection

3. Conclusion

It is quite obvious that for unprotected samples subjected to accelerated corrosion test in a condensation chamber, corrosion will gradually increase from the initial phase of the corrosion test. The number of events in the order of tens of thousands shows a sharp linear increase in process stage 1. The metallographic images also show corrosion from day 1 and document the gradual increase. Thus, for unprotected samples, the two methods used to investigate corrosion already differ somewhat. As far as the number of AE events is concerned, for protected samples the values are only in the thousands. Moreover, a fairly linear increase can also be found here in stage 1. While AE reveals corrosion activity in the protected samples already in the first stage, where corrosion can reach below protected surface at the point of weakening (i.e. edges or crevices) can subsequently propagate under the protective surface, no corrosion products are visible on the surface in the protected samples for metallography.

References

[1] Zhen Z, Zhanyonng Z, Peikang B, Xiaofeng L, Bin L, Xinqiang W 2021 In-situ monitoring of pitting corrosion of AZ31 magnesium alloy by combining electrochemical noise and acoustic emission techniques. *Journal of Alloys and Compounds*, ISSN 0925-8388. Available at: doi:160334

[2] Kober J, Převorovský Z 2011 Evaluation of the course of corrosion damage by methods of acoustic emission and nonlinear ultrasound spectroscopy. Faculty of Nuclear and Physical Engineering CTU: Institute of Thermomechanics AS CR NDE for Safety / DEFEKTOSKOPIE 2011, Harmony Club Hotel, Ostrava Czech Republic: Czech Society for Nondestructive Testing, 2011 [cit. 2021-4-28].

[3] Binar T, Švarc J, Dostál P, Šustr M, Tippner J 2017 The Spread of Corrosion in Cast Iron and its Effect on the Life Cycle of Transportation Vehicle. *ACTA UNIVERSITATIS AGRICULTURAE ET SILVICULTURAE MENDELIANAE BRUNENSIS* Brno 65(2), 383-389

[4] Liebisch *LABIMEX CZ* [online]. Praha [cit. 2022-05-14]. Available at: https://labimexcz.cz/laboratore/produkty/korozni-komory-liebisch/zakladni-rozdeleni-komor-liebisch/

Journal of Physics: Conference Series **2382** (2022) 012015 doi:10.1088/1742-6596/2382/1/012015

Failure mechanism in the pearlite structure with lamellar graphite due to temperature

J Zimáková[1], T Binar[1], P Šafl[1], P Čudek[1]

[1]Brno University of Technology, Technická 10, 616 00 Brno, Czech Republic

zimakova@vut.cz

Abstract. Critical operating temperature can be determined for the machine components that form the key functional part of the device. This article describes a change of the failure mechanism in the material of an engine, where the material structure consists of fine pearlite layers with lamellar graphite equivalent to grey cast iron. The mechanical properties of this material influence not only the layers formed by the perlite, but also the lamellar graphite acting as a stress concentrator. To assess the failure mechanism in the material, a fractographic analysis of the fracture areas was performed after the impact test. The range of test temperatures at which the impact test was performed on the test specimens with notch ranged from -80°C to +180°C. The occurrence of brittle fracture was detected at the low temperatures by cleavage along and over the planar layers of the lamellar graphite. The brittle fracture initiation at the higher temperatures was observed by cleavage in the basic layer of the perlite. Depending on the failure mechanism of the test specimens fracture area, critical temperatures were determined, in which the measured value of the notch toughness decreased. Regression of temperature dependence of notch toughness allows very precise quantification of this decrease.

1. Introduction

There are a number of factors which influence the failure mechanism of materials from which the combustion engine cylinder insert, in which the piston moves, is made. One of the significant factors that can cause degradation processes in the material due to external stress is the ambient temperature or temperature generated during the engine operation [1, 2]. Determining the critical temperatures that initiate a failure mechanism in the material allows us to determine, for instance, the optimal operating temperature of the engine. Due to the material composition of the engine cylinder insert, which is the basic phase of a fine pearlite with lamellar graphite, a brittle fracture appears to be the most dangerous. Because of its nature and consequences, the brittle fracture represents the most serious and the most dangerous material failure which is the cause of failures of machine parts and structures. Based on this fact, extensive research has been carried out to define the underlying causes and process of brittle fracture formation and to assess the impact of external and internal factors involved in this degradation process [3, 4]. The early results of these studies, continuing to date, have enabled us to understand the causes and mechanism of the brittle fracture. Knowledge has been gained to identify the different types of sources from which fragile fractures occurs. From the results was also evaluated the role of internal stresses and the influence of thermal effects. From a macroscopic point of view, the brittle fracture can be characterized as a failure of cohesion during which did not occur macroscopic deformation in the fracture area.

Fracture occurs at a lower nominal stress than is the macroscopic slip strength. Depending on the mechanism of violation, the brittle fracture area is usually formed by cleavage, i.e. by breaking the interatomic bonds in the cleavage planes of the individual grains (transcrystalline fracture) or separating the individual grains by breaking the interatomic bonds in the grain boundary (intercrystalline fracture). In the base phase of fine pearlite there is lamellar graphite, on which it can be assumed that the field of elastic deformations with a high density of deformation energy is concentrated and thus the conditions for brittle fracture are created. Grey cast iron is a very brittle material and may contain many cast defects [5]. Both facts lead to a wide distribution of the mechanical test results that require statistical processing. Their dependence on external parameters usually is hard to interpret without regression using a suitable regression function. For this purpose, it is possible to use the MS Excel Spreadsheet add-in as it will be shown in the case of the regression of the impact toughness and temperature

2. Experimental material, measuring methods and laboratory instruments

2.1. Specimen preparation

To prepare the test specimens for the impact bending test, an electro-erosion wire cutting method was used, which produces only a slight heat-affected area and does not require any further treatment of the samples. The test specimens were sized to 7.0 x 7.5 x 55 mm, with a V-notch of 2 mm depth of 45 ° angle, and a notch base radius R = 0.25 mm, surface roughness 3.2. Using optical emission spectrometry, a chemical analysis of the engine cylinder material (Table 1) was carried out by bulk analysis - "BULK" on the LECO SA-2000 spectrometer at a laboratory temperature of 25 °C. A total of 300 samples were produced and tested. In this work, illustrative examples of representative samples are given. The numbers for each sample correspond to the numbering in the test set.

Table 1. Chemical composition of experimental specimen

Chemical composition	C	Mn	Si	P	S	Cr	Ni	Cu
Amount (wt. %)	2.880	0.794	1.512	0.296	0.056	0.084	0.050	0.148

2.2. Analysis of the microstructure of the material under investigation

The Neophot 32 microscope and the Olympus C-5050 digital camera were used for metallographic analysis of the base material. From the notch toughness specimens, cross-sections were removed from both halves of the test pieces at distance about 7 mm from the fracture and, after being pressed into a plastic, the samples were polished in a standard way. The microstructure was developed with 2 % nitaletch. The microstructure consists of a mixture of evenly distributed lamellar graphite [6] (ČSN 42 0464) and fine perlite with numerous clusters with a phosphide eutectic (steatite) (Figure 1, Figure 2, Figure 3) which is formed at a phosphorus content of about 0.3 %. In Figure 1, Figure 2 and Figure 3 small areas of the excluded ferrite in the area of the lamellar graphite are visible

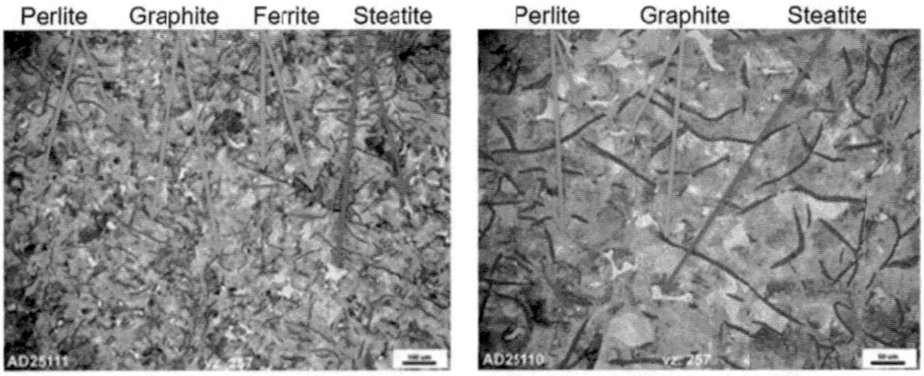

Figure 1. Microstructure of cast iron sample 257 after impact test at 22 °C

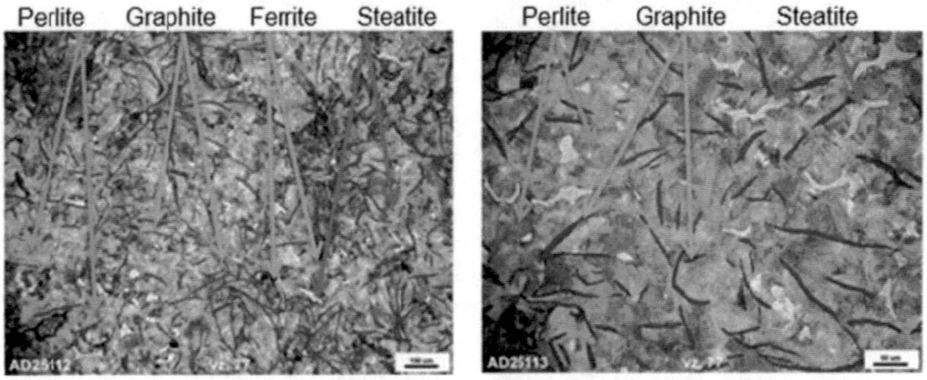

Figure 2. Microstructure of cast iron sample 77 after impact test at -80 ° C

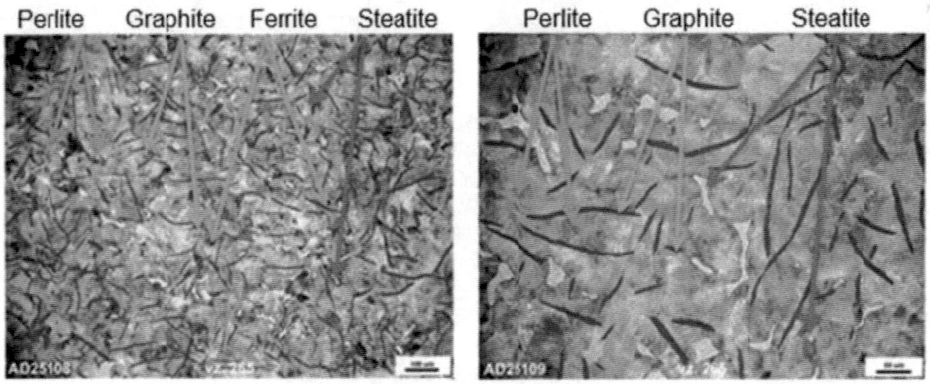

Figure 3. Microstructure of cast iron sample 257 after impact test at 180 °C

Microstructure assessment of selected specimens was performed for the upper and lower test temperatures, where different values of the measured notch toughness quantity were measured, and a different occurrence of the brittle fracture failure mechanism was observed in the examined material.

The microstructures of all the analysed specimens are identical and consist of a metal matrix of fine perlite and a mixture of lamellar graphite, see Figure 1, Figure 2, Figure 3.

The influence of carbon, silicon, and phosphorus on the cast iron structure can be determined by the degree of eutecticity Se according to:

$$S_e = \frac{\%\, C}{4.3 - 0.3(\%\, Si + \%\, P)} \tag{1}$$

Where at $S_e = 1$, the cast iron is eutectic, at $S_e < 1$ the cast iron is sub-eutectic and at $S_e > 1$ the cast iron is super-eutectic.

Using the values of the chemical composition of the experimental material from Table 1 in the expression (1), it was determined that the material was a sub-eutectic cast iron, meeting the condition $S_e < 1$.

$$S_e = \frac{2.880}{4.3 - 0.3(1.512 + 0.296)} \tag{2}$$

$$S_e = 0.7665 \tag{3}$$

3. Results and discussions

3.1. Evaluation of the test material resistance against brittle fracture

The evaluation of the specimens in terms of susceptibility to brittle fracture was performed by an impact bending test at a temperature range of -80 °C to +180 °C on the Pendulum Impact Tester RKP 450 Zwick with a nominal energy of 300 J [7 - 10]. The impact speed of the hammer was 5.52 m/s. A Lauda Proline RP 1290 refrigerated circulating water bath was used to cool and heat samples for the low and elevated temperature experiments.

For the regression analysis of the temperature dependence and the notch toughness, a hyperbolic tangent regression function was used [11 - 14].

$$KCV = \frac{L+H}{2} + \frac{H-L}{2} tanh \frac{2(t - t_{tr})}{\Delta t} \tag{4}$$

Having parameters whose significance is closely related to the characteristics of the transit curve: L is the lower and the H is the peak level of notch toughness, t_{tr} is the transit temperature, and Δt is the width of the transit area. The decrease in KCV for two experimental temperatures is expressed by a decrease in ΔKCV. The resulting regression is shown in Figure 4 and Table 2.

The lower toughness level is 37.2 J/cm^2, the top level is 64.6 J/cm^2, the transit temperature is 80 °C and the transit area width is 157 °C. The impact toughness decrease is at temperatures of 120 and 140 °C equal to 12.7 J/cm^2. The regression analysis of the experimental dependence is relatively successful (especially with regard to the large variance of the measurement results), only at the lowest temperatures there appears to be a further decrease in KCV, which, however, the abovementioned simple regression function is not able to take into account appropriately.

Table 2. numerical expression of average values of KVC and standard deviations for individual temperatures

t [°C]	-80	-70	-60	-50	-40	-30	-20	-10	0	25	40
Ø KCV [J/cm2]	35.45	37.07	37.77	37.91	38.06	40.11	40.47	41.16	41.46	42.51	43.94
standard deviation	0.00	0.58	0.40	0.21	0.24	0.60	0.33	0.34	0.38	1.08	1.03
t [°C]	50	60	70	80	90	100	120	140	160	180	
Ø KCV [J/cm2]	44.35	45.77	47.85	50.80	54.23	55.02	44.57	46.89	61.68	61.90	
standard deviation	0.97	1.33	1.71	1.15	0.91	1.16	0.55	1.80	1.35	0.61	

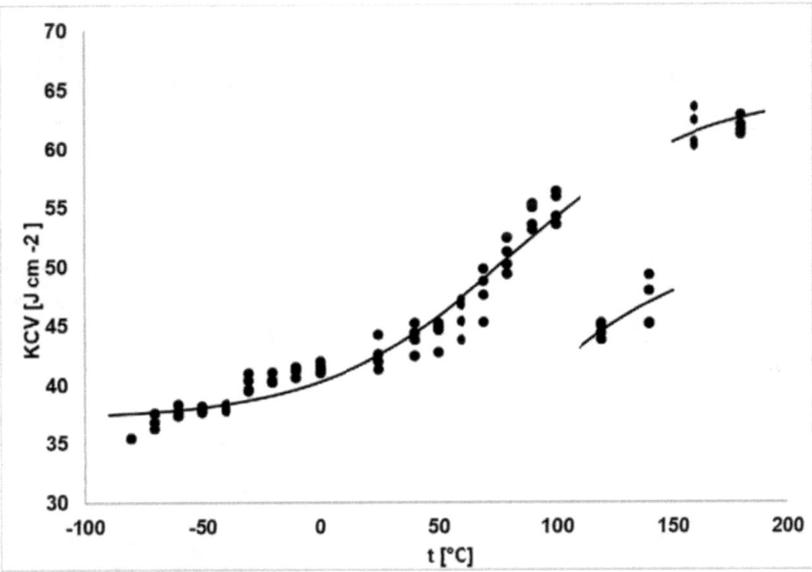

Figure 4. Temperature dependence of notch toughness

3.2. Evaluation of the fractographic analysis of fracture areas after impact bending test

The Tescan Vega 3 XMU Electron Scanning Microscope equipped with the LaB6 Cathode was used to evaluate the fracture areas following the impact bending test. To obtaining a finer surface structure, all images were taken in vacuum at a 5 kV accelerating voltage using the Everhart-Thornley scintillation detector. A fracture area with a magnification of 1000x was documented.

3.2.1. Temperature range from -80 °C to 0 °C

At the test temperature of -80 °C (Figure 5), comparing with the test temperature +160 °C (Figure 9), the fracture morphology after impact bending test is in the form of straight triangular shaped facets and at the boundary of the grain are visible opened facets. The failure mechanism has transcrystalline form with the transcrystalline cleavage regions shown in Figure 5, Figure 6, Figure 7 and Figure 8.
From a test temperature of -30 °C and lower, the fracture morphology (Figure 7) can be seen as a failure of the cohesiveness at both the grain boundary and between the lamellar graphite and the basic

matrix. The measured notch toughness values (Figure 4) in the temperature range of -40 °C to -80 °C are lower than at more positive temperatures. The most critical test temperature for assessing the effect of temperature on the notch toughness values is -80 °C. From -40 °C to -80 °C, a fine cracking of the surface layer of the facets on the fracture surface due to a the low test temperature can be observed. Lamellar graphite, which forms a sharp notch and is evenly distributed in the metal matrix, represents a stress focus that initiates a brittle fracture. In the low temperature range, the interlayer cleavage affects the lamellar graphite and nucleation of the brittle fracture failure mechanism results from its cleavage and then from the matrix of material.

In the pictures from low temperature studies, the larger openings of the lamellar graphite faces are seen comparing with the fracture surface at the above-zero temperatures. Also, the presence of steatite in the matrix, which is relatively tough and very brittle material, affects the measured notch toughness properties in the lower temperature range, and can be assumed an influence in the initiation of a brittle fracture in the material. At minus zero temperatures, larger openings may be observed than at the above-zero temperatures. The failure mechanism is intercrystalline with transcrystalline cleavage regions.

Figure 5. Fractionation of specimen 77 after a shock test at -80 °C

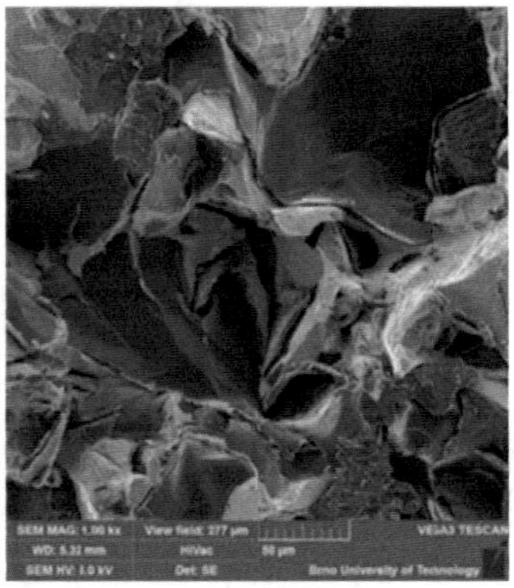

Figure 6. Fractionation of specimen 51 after a shock test at -60 °C

Figure 7. Fractography sample 37 after a shock test at -30 °C

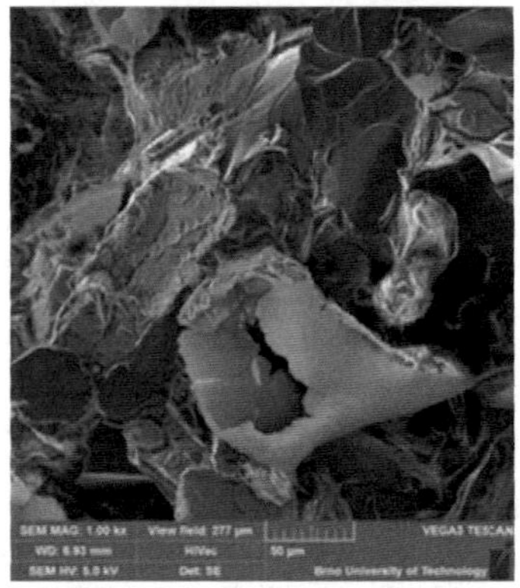

Figure 8. Fractionation of sample 32 after a shock test at -20 °C

3.2.2. Temperature range from +25 °C to +180 °C

From +25 °C to +100 °C (Figure 12, Figure 13, Figure 14) a linear increase in the measured notch toughness characteristic (Figure 4) is evident. Due to the increased test temperature, although the fracture morphology is noticeable, the opening of the facets and the lamellar graphite are separated from the core matrix, but the KCV values are higher compared to those at the lower test temperatures. The failure mechanism is transcrystalline in the temperature range under investigation and the fracture spreads longitudinally. This fracture failure mechanism is evident even at the test temperature of +100 °C and (Figure 12) this temperature can be considered a boundary both due to the failure mechanism (transcrystalline) and to the KCV notched toughness value. At the test temperatures of +120 °C and +140 °C, the morphology of the fracture is a mixed fracture with a majority of the transcrystalline failure mechanism with a sign of plasticity (Figure 10, Figure 11). The fracture morphology (Figure 9, Figure 12) shows a wider boundary between the borders forming the fracture surface, which caused a decrease in the notch toughness values between the test temperatures of +100 °C and +160 °C. The failure mechanism at +160 °C is a mixed transcrystalline, but the thermal energy was sufficient (adhesion of the lamellar graphite at break compared with +120 and 140 °C) and there was no significant reduction in the measured notch toughness value. Figure 9 shows a better adhesion to grain than at +100 °C (Figure 12) resulting in a higher KCV value. From the point of view of the operating temperatures at which the engine cylinder is operating, the test temperatures of +120 °C and +140 °C can be set as critical for assessing the influence on the toughness values, not only that there is a partial decrease in the KCV quantity, but also a change in the failure mechanism of fracture areas. In the field of positive test temperatures, the initiation of a brittle fracture occurs in the matrix.

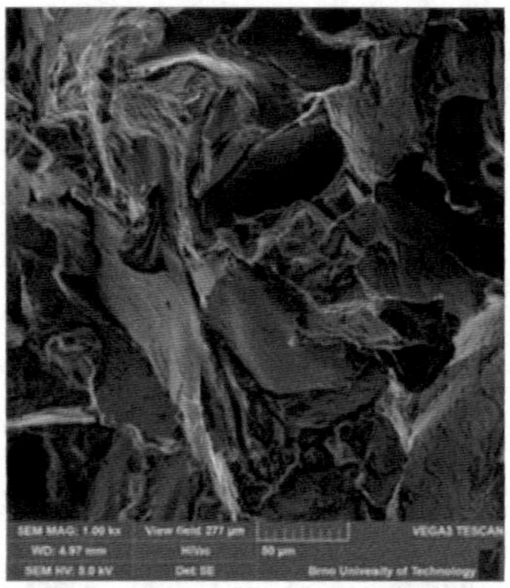

Figure 9. Fractography of sample 261 after a shock test at +160 °C

Figure 10. Fractography of sample 257 after a shock test at +140 °C

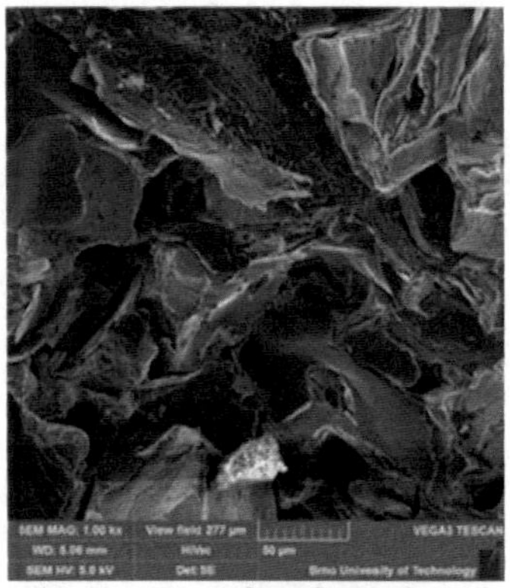

Figure 11. Fractography of sample 255 after a shock test at + 120 ° C

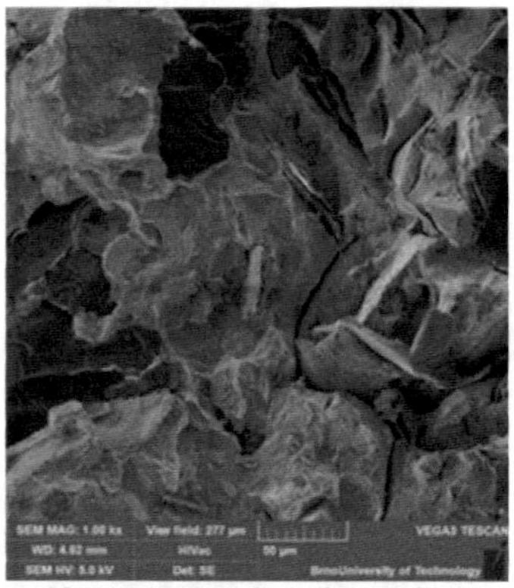

Figure 12. Fractography of sample 88 after a shock test at +100 °C

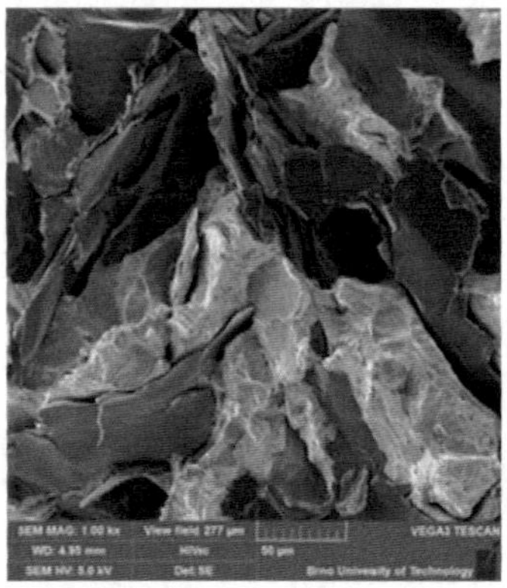

Figure 13. Fractography of sample 71 after a shock test at + 80 °C

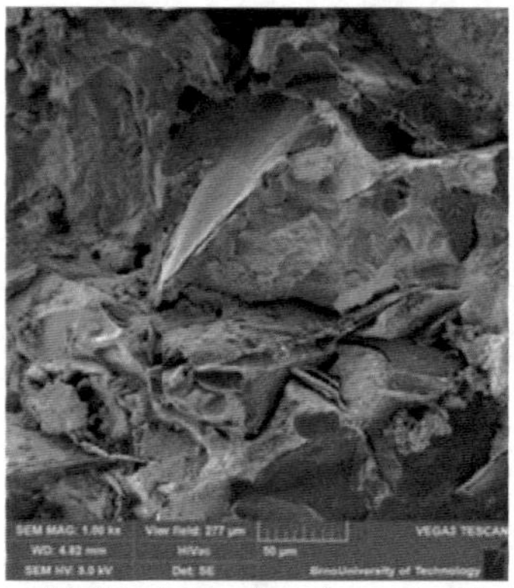

Figure 14. Fractography of sample 57 after a shock test at + 60 °C

3.2.3. Elemental analysis by EDS

For EDS elemental analysis, a Bruker Xflash 6|10 spectroscope was used. The aim of the applied method was to explore the chemical composition of a particular fracture area (Figure 15) after the impact bending test and to determine the amount and distribution of the elements carbon, manganese, iron and silicon (Figure 16). Because the specimens were stable in the electron micrograph vacuum and not vulnerable to electron beam damage, acceleration voltage of 30 kV was used. From the elemental distribution of the studied area shown in Figure 15, it is evident that ferrite sites appear in the lamellar graphite. An increased occurrence of graphite together with ferrite worsens the mechanical properties, which in the area of reduced test temperatures has been implicated, and a decrease in the measured notch toughness characteristic and the nucleation of the failure mechanism by cleavage of laminated lamellar graphite.

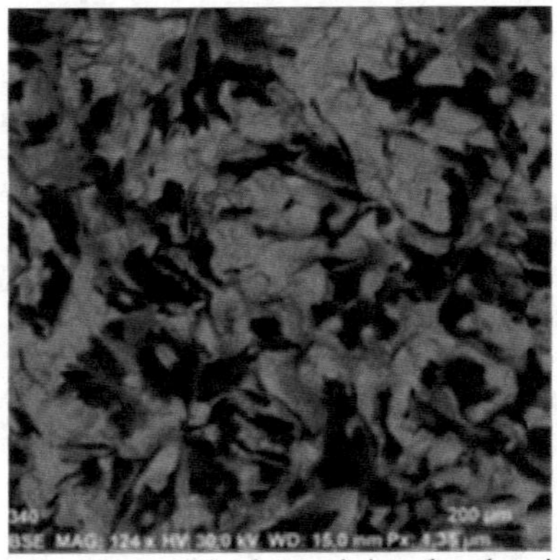

Figure 15. Fracture area used in determining the chemical composition

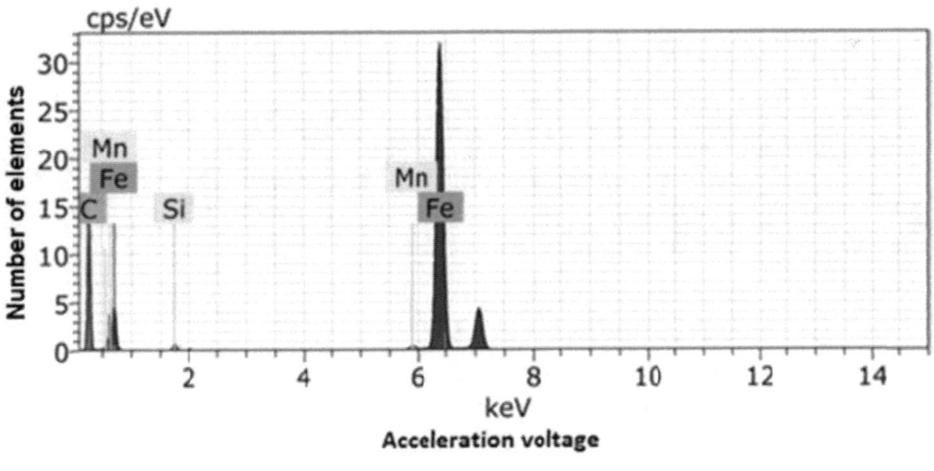

Figure 16. Elemental analysis by EDS

4. Conclusion

The effect of the test temperature on the mechanism of failure sustained during an impact bending test in the material with a fine pearlite structure, where ferrite clusters were formed in the area of the lamellar graphite, was investigated. In the investigated microstructure, structures of steatite have been identified, which are extremely fragile and hard. In the region of low test temperatures of -80 °C to 0 °C, a failure mechanism was observed on the fracture surfaces with transcrystalline cleavage regions (zones). The nucleation of the failure mechanism occurs by splitting the laminated graphite and then passes through the grains of the basic metal matrix. The most critical test temperature for assessing the effect of temperature on the rating of the resistance properties of the material against brittle fracture was determined to be -80 °C.

In the temperature range of above-zero test temperatures from +25 °C to +180 °C, a linear increase in the notched toughness characteristic was observed, and the failure mechanism was transcrystalline and longitudinally spreading at the facets of graphite. In the temperature range of +120 °C and + 140 °C, the morphology of the fracture is composed of a mixed fracture, a proportion of the transcrystalline disruption and a sign of plasticity, and in that temperature interval a reduced measured notch toughness was observed compared to +100 °C and +160 °C. In the above-zero temperature range, a brittle fracture failure mechanism occurred in a matrix consisting of fine perlite and graphite facets cleavage.

Various bent planes of lamellar graphite observed in the fractographic analysis of the fracture were probably caused by greater overheating of the melt and its rapid cooling.

The transient temperature of the studied material determined on the basis of the temperature-dependence of the notch toughness regression is 80 °C. This is relatively high value, however, this comes into play only during engine start-up and the time when the engine is warmed up, which is fairly short time. The notch toughness decreases at temperatures of 120 and 140 °C of nearly 13 J cm-2 is explained by a change in the failure mechanism.

Acknowledgements

The publication was created as part of a specific research project at BUT (No. FEKT-S-20-6206) and project number FW01010609.

References

[1] Sukáč J, Binar T, Kazda T and Šilinger K 2015 *ECS Transactions*, 70(1) 167-175. doi:10.1149/07001.0167ecst, Retrieved from www.scopus.com
[2] Binar T, Kadlec J, Rejzek M, Vlkovsky M and Hruby V 2011 *Strength of Materials*, 43(5), 537–542
[3] Sukáč J, Binar T, Sedlaříková M and Šilinger K 2015 *ECS Transactions*, 70(1), 177–185
[4] Binar T, Dvořák I, Kadlec J, Rolc S and Křest'an J 2014 *Advances in Military Technology*, 9(2), 33–39
[5] Švarc J, Binar T, Dostál P and Tippner J 2017 *Acta Universitatis Agriculturae et Silviculturae Mendelianae Brunensis*, DOI: 10.11118/ACTAUN201765041295
[6] ČSN EN ISO 643 Steels - Micrographic determination of the apparent grain size
[7] ISO 14556:2015 Metallic materials – Charpy V-notch pendulum impact test – Instrumented test method
[8] ISO 148-1:2016 Metallic materials – Charpy pendulum impact test – Part 1: Test method
[9] ČSN 42 0382 (420382) Metal testing. Charpy impact test at low temperature. (U-notch and V-notch)
[10] ČSN 42 0383 (420383) Metal testing. Charpy impact test at elevated temperatures
[11] Shin H S, Park J S and Lee H M 2008 *International Journal of Modern Physics B*, vol. 22, no. 9/11, p. 1496-1503.

[12] Rahmanifard R, Farhangi H and Novinrooz A J 2016 *Journal of Alloys and Compounds*, vol. 657, p. 646-654.
[13] Kohout J and Věchet S 2001 *International Journal of Fatigue*, 23(2), pp. 175–183
[14] Kohout J 1999 *International Journal of Fatigue*, 21(8), pp. 813–821

New possibilities of using cold kinetic deposition

P Šafl[1], J Zimáková[1], T Binar[1], P Čudek[1]

[1]Brno University of Technology, Technická 10, 616 00 Brno, Czech Republic

zimakova@vut.cz

Abstract. This work deals with the properties of the protective layer applied to the base material using the cold spray method. Aluminum plate, on which a thin layer of copper was applied, served as the basic material. The samples prepared in this way were subjected to accelerated corrosion tests. Sensors for measuring the acoustic emission signal were attached to the samples. The recording was an acute emission and was taken during the entire period of corrosion. The properties of the deposited layer were also verified using metallography.

1. Introduction

Cold spray (CS) belongs to the spray family. It exhibits properties that have made hot spray inappropriate in some applications and hence a great deal if research and testing is currently being devoted to it. An important observation is that the coating itself is carried out at very low temperatures compared to incandescent methods and at relatively high (supersonic) propelant gas velocities. The actual bonding of the substrate to the material to be deposited is by a mechanism different from the glow methods.

Cold spray, thanks to its unique properties, enables the spraying of various types of materials such as metal, ceramics, composite materials or cements or polymeric substances. The CS process can be used to create anti-corrosive layers, layers with increased wear resistance or higher mechanical resistance. It can also be used for more specific applications such as creating components with different thermal and electrical properties than the substrate. This can create different insulating layers, conductive pathways in polymer structures or ultrathick protective coatings. Very interesting applications are the refurbishment or repair of special application is in biomedicine, where, due to the high desired surface porosity, special coatings can be applied to prostheses or implants.

The first experiments and applications of Cold spray technology were carried out at the turn of the twentieth and twenty-first century. Since then, there has been a development and a huge expansion of this technology for various applications.

Applications in the energy sector – cold spray coating and surface formation is most often used to create a high temperature resistant coating in combination with abrasion resistance or chemical resistance to halide elements, generally against high temperature corrosion. According to [1], the coating created by Cold spray method is used on the heat exchanger tubes inside the fireplace for protection of the superheater against high temperature corrosion, when the substrate is a classic boiler steel (SA516) covered with a layer with a spray concentration of 77 – 88 mg/cm^2. The coating was created by a mixture of chromium, nickel and spinel.

Applications in biomedicine – Cold spray method is widely used in biomedical practice due to properties. It is used for spraying biocompatible materials on implant and prostheses. Thermal spray processes such as HVOF or VPS are often used for the production of bioactive materials with a titanium surface. Titanium or its alloys are among the materials that are suitable for these applications, mainly due to the promotion of cellular activity, protein adsorption or reactions with tissue. Some other thermal spray processes are used to deposit layers of tantalum (Ta) or silver (Ag). Tantalum is seen a very promising material for the future and is the subject of various studies. Biocompatible materials were the first to be used for Cold spray deposition in combination with high grade stainless steel. The whole process was adapted to achieve a high surface porosity for good fusion of the osteoimplant with the bone. [2]

Good results are achieved using ceramic materials such a TiO_2. Coatings produced in this way have excellent mechanical properties while maintaining a high level of biocompatibility. [2]

Applications in additive manufacturing – one of the interesting applications of Cold spray in additive manufacturing is the creation of components from IVAR 36 alloy. This nickel-iron based alloy is characterized by its thermal stability or very low coefficient of super-precision tools or in optics. Although this material has excellent thermal properties, it is characterized by poor mechanical resistance. For example, high sensitivity to plastic deformation. This is a very limiting factor for furthers processing and production. [3]

These findings lead to the fact that it is suitable to produce components form IVAR alloy by additive manufacturing. Conventional additive manufacturing uses a laser and achieves similar results to casting. However, additive manufacturing of the component using the CS method is beginning to play an important role. From the results obtained by [3], the theoretical assumption of using the CS method for this application if confirmed. Low surface porosity and high density of the formed layer are achieved. A similar conclusion was reached in the case of the evaluation of the microstructure of the specimen, where the desired bonding of the layers occurred under elastic deformations. Due to the high impact velocities and elastic deformations, there is always a higher internal stress in the substrate material. This can be eliminated by heat treatment of the sample. [3]

The cold spray method is more than suitable for additive manufacturing as low porosity achieved. The porosity is influenced by the carrier gas pressure and must therefore be adapted to different production needs. It is fully confirmed that phase changes of the material or oxidation occur sporadically within the layer.

Applications in repair – from a production sustainability perspective, the question is when to repair a part and when to replace it. This is largely influenced by cost of repair and the frequency of repair. It is the Cold spray process that can help optimize both of the above. [4]

CS is most often used in dimensional repair where the same material is applied as the substrate. It is most often used for repair in the aerospace industry, and is also indispensable in the military. Magnesium-based alloys are often used in these applications. In Aerospace, additive repair is used, for example, to repair parts of aluminium alloy panels that are worn out by constant clamping or by airflow in flight. Magnesium alloys used in roto gearboxes can also be repaired in this way. These parts are, of course, very expensive and therefore their eventual renewal by CS is desirable. Another suitable application is for the repair of bearing and camshaft washers in large individually cast engine blocks. If a mistake is made in the manufacture of such an atypical block, it is not necessary to throw away the original block and make a new one, but only to repair it with CS. [5]

Corrosion test for verification to examine coating formed by spray methods are most often performed in an accelerated mode. Their aim is to expose the test article to an environment that the test article will withstand in a real process and to simulate its behaviour after the article has been exposed to external influences for a longer period of time. In real practice, this can hardly be verified. This is due, for example, to the speed of product launches, where it is economically impossible to carry out test in situ and in a real time. In the course of work with various sources, the observation according to [6] regarding the suitability of testing deposited materials in salt chambers has been established. Many methods can be used for the evaluation and examination of materials, ranging from those are common and time-tested in practice to new and modern methods thar are beginning to be widely applied in particular fields, such as acoustic emission (AE).

2. Materials and methods of sample preparation

In this work, we focus on the creation of a copper coated aluminium substrate sample using the Cold spray method. The copper layer is intended to serve as a corrosion protection. In general, corrosion can be prevented in various ways. Recently, manufactures are focusing on the precaution of products with anticorrosion layer made by spraying for its simplicity and effectiveness. In the case of copper, the most common method is electroplating. Here, however, a limitation is encountered, where either the entire surface must be plated or, in the case of a requirement to plating, the need mask is eliminated, saving time and material. This result is considerable cost savings, especially in series production. Thermal spray methods are limited in particular by their high deposition temperature and the associated oxidation of the deposited material and mechanical stress on the substrate during deposition. Therefore, CS, which does not suffer for these disadvantages, appears to be a suitable process.

Samples were created using CS and prepared by water jet. Subsequently, the prepared samples were fitted with an acoustic emission sensor and placed in a condensation chamber where the corrosion formation was evaluated. Corrosion testing was carried out in the condensing chamber under steady state conditions at 50 °C and 100 % humidity. The specimens were exposed to the internal environment of the chamber for 168 hours, with continuous AE measurements. A metallographic test was performed on each sample after the end of the test time cycle.

2.1. Substrate and application material

An aluminium plate with dimensions 120 x 100 mm and a thickness of 3 mm was used as a substrate for copper deposition. The copper powder chosen for CS was form Safina a.s.. The purity of the copper is 99,9 % according to the specification. The composition of the copper powder is shown in Table 1. The PSD parameter of the powder used in the range of 15 - 33 mm.

Table 1. Parameters Cu

Cu composition – supplier Safina a.s.
Cu
CuCr1Zr
CuA110
Ag

2.2. Cold spray process

The equipment for the CS process is from Impact Innivations GmbG, type Impact Cold Spray Systems 5/11 (see figure 1). This equipment operates at temperatures up to 1 100°C and pressures up to 50 bar. The heating power is 40 kW.

Nitrogen was chosen as the working gas. It is a commonly available gas that is internal, non-toxic, readily available on the market and less expensive than other types of working gas such as helium etc. With regard to the selected deposited material, i.e. copper and nitrogen carrier gas, the parameters of the CS process were set. The required temperature is 450 °C, working pressure 25 bar. The distance of the Laval nozzle from the substrate was 30 mm. At the same time, the parameters were set to achieve the required deposition rate for copper to aluminium.

Figure 1. Impact cold spray system 5/11 [7]

2.3. Resulting layer and further sample processing

Figure 2 shows the resulting sample prepared using CS. This was subsequently modified by waterjet cutting into additional, smaller samples which were subjected to condensation chamber testing.

Figure 2. Cu layer on Al substrate formed by CS

Since the sample will be tested in a condensation chamber, the cutting into smaller pieces was done by waterjet cutting. This procedure was chosen in order to avoid mechanical damage to the copper-aluminium interface or thermal stresses around the cutting area, as occurs when cutting with a diamond wheel, etc. A thermally degraded or damaged copper-aluminium interface would lead to distortion of the results of the measurements made in the condensation chamber.

2.4. Anti-corrosion coating

Since the corrosive environment inside the chamber affects the entire sample, not just the copper-sprayed surface, it was expected that corrosion would progress significantly faster at the copper-aluminium interface. Therefore, three more samples were prepared. Two of the edges were painted with a protective varnish to prevent corrosion.
Figure 3 shows a sample with protective varnish from Cortec corp. It is temporary water-based, non-stick coating of type VpCl-391. It is designed for medium to long term protection in both indoor and outdoor conditions. The varnish characterized by the manufacturer as highly resistant to salt spray, moisture and UV radiation. The chosen coating is recommended for protective coatings of metal

surfaces where a non-sticky surface and easy removal of the dried coating is required. The manufacturer states that the paint is particularly suitable for carbon, stainless and galvanized steel, as well as Al and cast iron.

Figure 3. Detail of prepared sample with protective coating

3. Measurements and tests performed in the condensing chamber

The prepared plates were fitted with an AE sensor and inserted into the condensing chamber in a vertical position. For the test in the condensation chamber, a time cycle of 168 h was selected. Using the AE, the extent and rate of corrosion was determined during the experiment. For each sample, a metallographic test was performed after removal from the condensation chamber to evaluate the corrosion cracks formed towards the copper surface towards the Al substrate at the edges.

3.1. Measurements and test performed in the condensing chamber

The test environment in the condensing chamber was at a steady temperature of 50 °C with a ± 5°C allowable variation at 100 % relative humidity. The water bath inside the condensation chamber consisted of distilled water with a conductivity less than or equal to 20 °C ± 2 °C and relative humidity of 65 °C ± 5 % after settling for at least one hour in accordance with DIN 038131 – Corrosion Test Condensing Chamber.

Figure 4 shows a sample marked E168h which was subsequently exposed to a corrosive environment for 168 h.

Figure 4. sample before placing in the condensing chamber

3.2. Measurement results in the condensing chamber

Figure 5 documents the status of each sample after removal form the chamber. Further test were performed on there samples. Continuous AE sensing was performed while the samples were placed in the condensation chamber. A similar procedure was followed for the samples that were coated with an anti-corrosion coating.

Figure 5. Sample E168 hours in the condensing chamber

After removal of sample E168h, which was exposed to the corrosive environment for 168 hours, corrosion attack is not visually apparent. In addition, a metallographic test was performed on sample E168h.

3.3. Measurement by AE

The aim of the AE measurements of the individual samples was to verify how the protective copper layer created by CS method affects the viability of the sample in the condensing chamber environment. The measuring system DAKEL with AE diagnostic tool type XEDO was equipped with a total of 6 channels. The measurements were performed on the following set-up:

- 3x piezoelectric sensor type MTR-15
- 3x 35 dB preamplifiers
- 3x sensor extension cables
- DAKEL Xedo diagnostic system
- Laptop (connection to Xedo)

The sensors operating on the principle of piezoelectric effect were attached to the samples by means of a "cage". Silica gel paste was used between the test sample and the contact of the sensor. The paste is used to transfer signals form the sample to the sensor.

The measured data from the AE diagnostic system was evaluated from three sensors. The following parameters were monitored:

- Count 1 – the number of overshoots over the set threshold level
- Count 2 - the number of overshoots over the set threshold level
- RMS – root mean square level of the detected signal

4. Measurements results

4.1. Acoustic emission results

Based on metallographic measurements, it was confirmed that no significant corrosion progression was observed on the samples. According to figure 6, an increased acoustic variability – RMS signal (blue waveform) is measured in section 1 (pink). This is a due to the direct reaction of water with the substrate material (Al). The difference in the AE signal of the coated and uncoated sample is determined by the fact that the interaction to the corrosive environment with the Al produces Al oxide Al2O3 (corrund), i.e. a white oxide layer on the surface of the material. The layer thus formed protects the base material to a certain extent. However, pores are formed in this layer, though which the layer is eroded and corrosion reactions penetrate deeper into the surface of the base material. Due to the non-uniformity of the pore formation in the oxide layer, a variability of the AE signal arises, the activity of which is higher in the fi phase of corrosion. In this phase the oxide layer is formed. Further non-uniform acoustic signals results from local disruption and the formation of porosity in this layer. This can be seen in the highlighted part 2 (green shading) in figure 6.

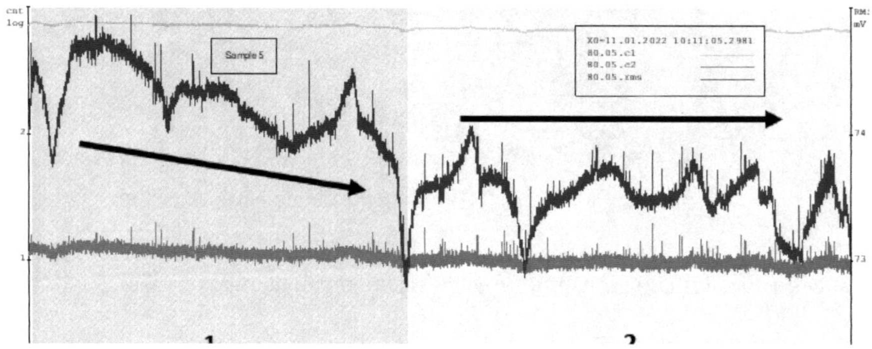

Figure 6. Record for a sample without protective coating

Figure 7 shows the acoustic activity of a sample coated with a protective coating. This sample had a relatively high signal activity. This part 1 (pink underpainting) and the period of approximately the first 48 hours. Part 2 form 48 hours onwards showed an uniform signal.

On the edges, where the protective anti-corrosion coating does not adhere sufficiently to the material, corrosion penetrates under the copper layer, which was confirmed by metallographic analysis. At these points, oxides begin to form on the base material. The oxide formation is what measures the increased the AE activity. However, from the point of view of the corrosion protection of the whole sample, this has a negligible effect. After evaluation of the AE, it can be said that the corrosion protection coating is very effective. This is further confirmer by metallographic analysis.

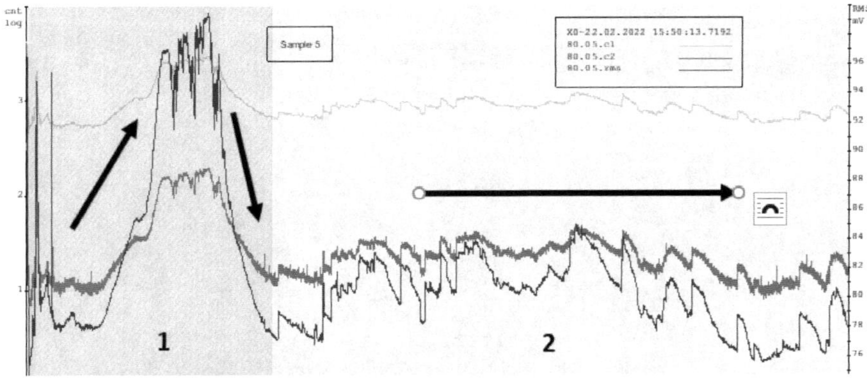

Figure 7. AE record for the sample with protective coating

4.2. Metallographic measurements

In order to perform the metallographic analysis, the samples had to be prepared as follows. The original sample was cut approximately in half. This produced two cuts of 25 mm in length. For the sample with the protective coating, the surface of the Cu coating was covered and two edges out of four were covered, i.e. on two edges there was an aluminium-copper-anti-corrosion coating composition and on the remaining edges there was a "clean" sample. The purpose was that for these samples, the halves with and without coating at the edge were evaluated separately. The samples for the metallographic test were carefully prepared to avoid delamination during cutting, which is essential for the evaluation. The samples were then pressed into plastic, ground and polished. All observations were carried out on a ZEISS Neophot 32 metallographic microscope. All images are labelled as MG for better clarity.

4.3. Metallographic analysis of a "clean sample"

Figure 8 shows section on an Al-Cu specimen without corrosion coating and without exposure to the corrosive environment in the condensing chamber at a lower magnification. From this figure, a good idea of the surface texture of the copper coating can be made.

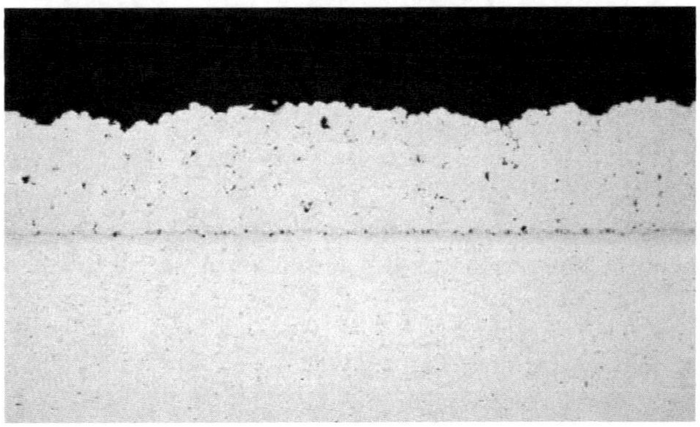

Figure 8. MG "clean" specimen without corrosion attack

Figure 9 shows a sample in detail of the Al-Cu interface in a higher magnification. Slight voids are observed, and oxide elements are also present but are negligible. Small pores are observed at the interface between the Al substrate and the deposited Cu.

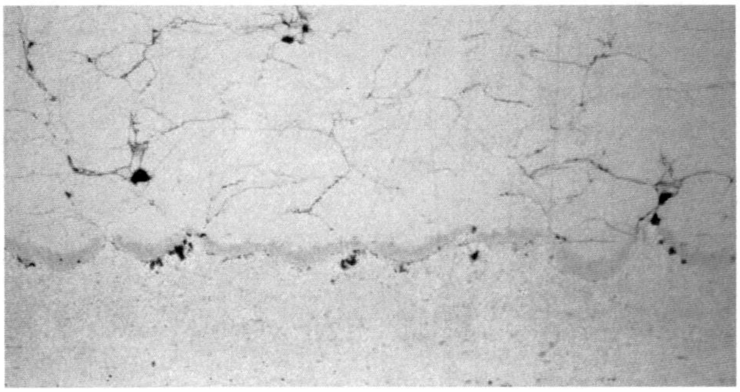

Figure 9. MG „clean" sample – detail of substrate-copper interface

4.4. Metallographic analysis after exposure in the condensing chamber

Further, the analysis was performed on a sample exposed to the condensing chamber environment for 168 hours. It can be concluded that the state sample E168h (figure 10) corresponds to the initial state.

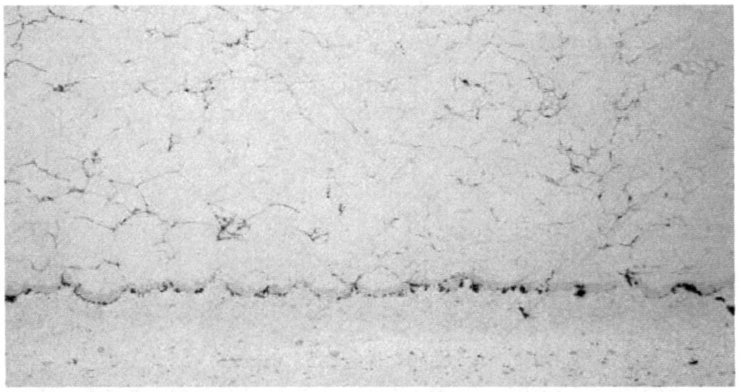

Figure 10. MG Sample E168h without protective coating

Figures 11 and 12 shows that locally thin continuous oxide enveloped begins to appear at the Al-Cu interface. However, after 168 hours in the condensing chamber, the sample is not yet stratified.

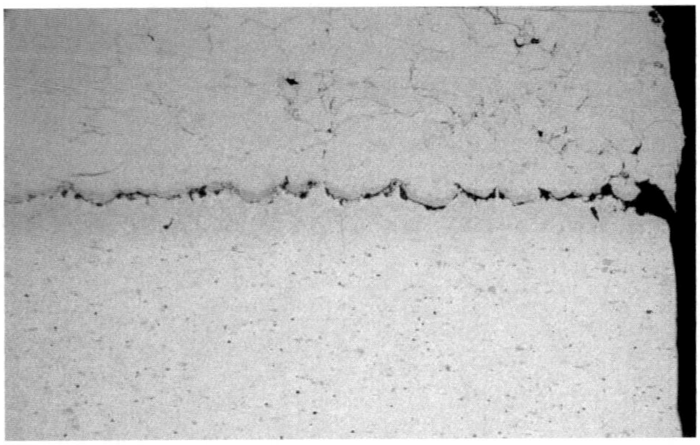

Figure 11. MG Sample E168h without protective coating

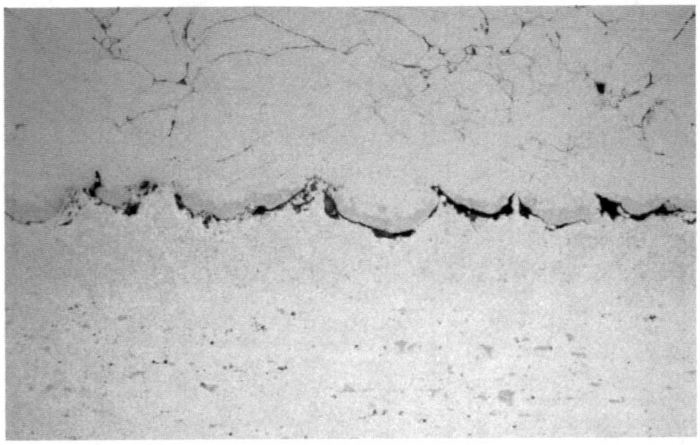

Figure 12. MG Sample E168h without protective coating

Figure 13 shows the occurrence of oxide envelopes at the boundaries of Cu blocks. This condition was also observed on the pure sample.

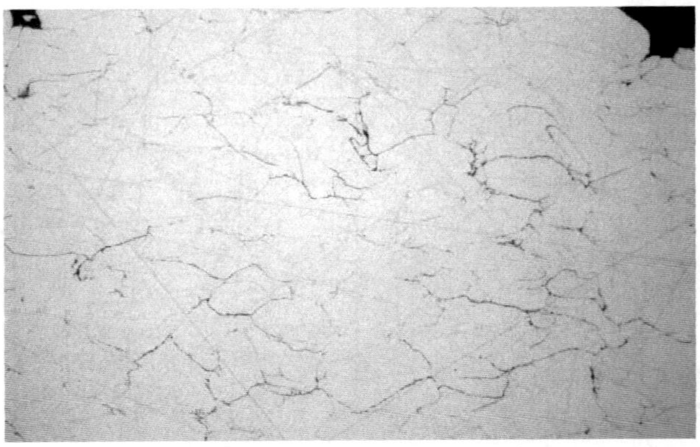

Figure 13. MG Sample E168h without protective coating

The following figures 14 and 15 shows the results after exposing the sample to a corrosive environment, one part of which has been coated with a protective coating. This sample, like all the others, was exposed to the condensing chamber environment for 168h.

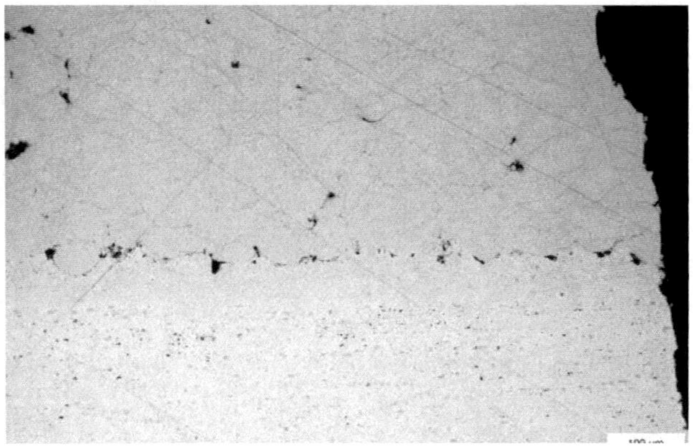

Figure 14. MG specimen E168h – coating, edge painted with anti-corrosion coating

Specimen with anti-corrosion coating was painted on only two edges out of four. Figure 14 shows an edge with this coating. Figure 15 shows an edge of the same specimen that was not coated.

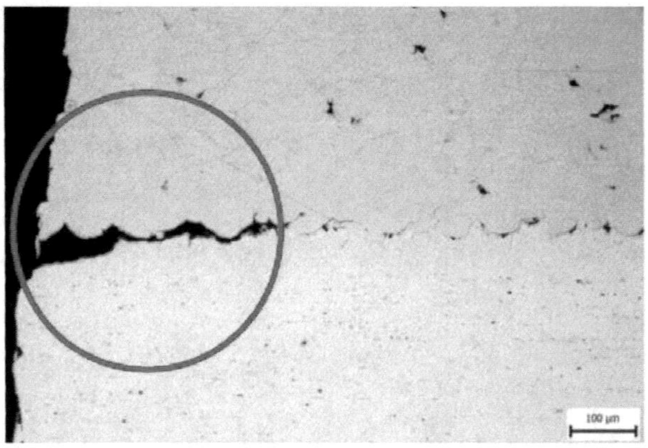

Figure 15. MG specimen E168h – coating, edge without anti-corrosion coating – detail of lay-up

In this case, delamination is observed on the specimen (figure 15) that has not been coated with a protective coating on the edge. On this specimen, an increased occurrence of oxide envelope at the Al-Cu interface is also observed. Metallographic analysis shows that the sample with protective coating does not show corrosion after 168h. This is a relatively short time point compared to the time stated by the manufacturer. However, it can be concluded that after such a short period of time the protective coating performs its function.
Both short and long oxides regions are presented in the sample. These oxides are mainly formed by corrosive effects inside the condensing chamber. Contrary to the assumption, the highest occurrence of oxides was measured in sample E168h – the coating on the comparison edge, which does not contain coating. This peculiarity can be attributed to the breakage of the individual layers during the waterjet cutting of the sample.

5. Conclusion

This work describes the production of a sample where a layer of Cu is deposited on an Al substrate using equipment form Impact Innovations GmbG as corrosion protection. The sample thus produced is ready for corrosion testing in a condensing chamber. The test in the condensing chamber took 168h. Six smaller samples were prepared from the initial sample using a waterjet. For three of them, a layer of anti-corrosion coating was applied to the edge. An AE sensor was attached to all the samples, which sensed the corrosion processes in the samples that were placed in the condensing chamber during the whole time. At the end of a given time cycle, the individual samples were removed and subjected to metallographic analysis. The individual results are discussed in the partial conclusions for each type of test.

The aim of the work was to assess the corrosion attack of the sample and how the protective layer of Cu affects the formation and development of corrosion. A significant finding was that Cu slows down the progress of corrosion. In addition, the samples that were protected with an anti-corrosion coating showed minimal corrosion development. Furthermore, the suitability of CS application as corrosion protection was confirmed due to its properties. The resulting layer has a high density of deposited material and, due to the relatively low deposition temperatures compared to Thermal spray processes, prevents the formation of corrosion within a layer and no further corrosion occurs when exposed to the corrosive environment inside. The most noticeable corrosion progression was always at the edge of the specimen, where the cutting of the original specimen broke the structure, even though the specimen were prepared using a waterjet. On the edges, the corrosion progression was already evident after 168h. However, it can be noted that the oxide corrosion layers always formed at the edge of the sample and did not progress further into the sample at the interface between the substrate and the Cu layer. Delamination of the protective layer from the specimen was always observed except for the specimen that had edges treated within an anti-corrosion coating.

Acknowledgments

The publication was created as part of a specific research project at BUT (No. FEKT-S-20-6206) and project number FW01010609.

References

[1] Kumar S, Kumar M, Neeru J 2020 Overview of cold spray coatings applications and comparisons: a critical review *World Journal of Engineering* 10.1108/WJE-01-2019-0021.

[2] Vilardell A M, Cinca N, Concustell A, Dosta S, Garcia C I, Guilemany J M 2015 Cold spray as an emerging technology for biocompatible and antibacterial coatings: state of art *Journal of Materials Science* 50. 10.1007/s10853-015-9013-1.

[3] Chaoyue Ch, Yingchun X, Longtao L, Ruixin Z, Xiaoli J, Shanqing L, Renzhong H, Jiang W, Hanlin L, Zhongming R 2021 *J. of Mat. Sci. & Tech.* 72, P 39-51, ISSN 1005-0302, https://doi.org/10.1016/j.jmst.2020.07.038. (https://www.sciencedirect.com/science/article/pii/S1005030220308148)

[4] Widener Ch A., Ozdemir O C, Carter M 2018 Structural repair using cold spray technology for enhanced sustainability of high value assets. Procedia Manufacturing. 21, P 361–368 ISSN 2351-9789. Available at: doi:10.1016/j.promfg.2018.02.132

[5] Champagne V and Helfritch D 2015 Critical Assessment 11: Structural repairs by cold spray. Materials Science and Technology. 2015, 31, P 627–634. ISSN 0267-0836. Available from: doi:10.1179/1743284714y.0000000723

[6] PROŠEK, T 2016 Accelerated cyclic corrosion tests: TECHNOLOGICAL INTERESTS AND ARTICLES FROM PRACTICE. Available from: doi:10.1515/kom-2016-0008

[7] Flame Spray Technologies: IMPACT COLD SPRAY SYSTEM 5/11. Flame Spray Technologies. Netherlands, 2018. Available from: fst.nl/thermal-spray-equipment/modular-thermal-spray-systems/cold-gas-spray-systems/impact-cold-spray-system-5-11.html

The mullite nanofibers for electrotechnical applications

J Kočí[1,2], M H Míka[1], N Klusoňová[1]

[1] Department of Glass and Ceramics, University of Chemistry and Technology Prague, Czechia
[2] Department of Electrotechnology, Faculty of Electrical Engineering, Czech Technical University in Prague, Prague, Czechia

JanKoci89@email.cz

Abstract. This paper is focused on the development and modification of inorganic refractory nanofibers for advanced applications, made from the most available ceramic raw materials (Al_2O_3 and SiO_2) in the right proportion for crystallization of mullite, commonly used high-temperature material. The intention is to create a modifiable refractory material for use in electrotechnical applications as electrodes, separators, or a solid electrolyte in a new type of safe solid-state batteries. For these reasons, the nanofibers are modified with different conductive materials, for example different types of spinels. Spinels with the formula AB_2O_4, specifically $CuCr_2O_4$, have been studied due to their significant electrical and magnetic properties, that are suitable for their intended purposes.

1. Introduction

The development and modification of inorganic refractory nanofibers for advanced applications is highly desirable at a time of growing economic and environmental demands. The intention is to create a modifiable material for use in electrotechnical applications as electrodes, separators, or solid electrolyte for the new types of safe solid-state batteries and many other applications. The most studied system are composite nanofibers made from the most available ceramic raw materials (Al_2O_3 and SiO_2) in the right proportion for crystallization of mullite, commonly used high-temperature material. [1]

Mullite has been used for high-temperature applications because of its outstanding thermomechanical properties at elevated temperatures e.g. creep resistance, thermal stability (i.e. thermal expansion and thermal shock resistance), low density, low thermal conductivity, high compressive strength and fracture toughness, or good corrosion resistance in severe atmospheres and others. [2, 3] The mullite fibers are attractive for their applications such as insulators, high-temperature filters, catalysts, etc. [4, 5]

For practical and economic reasons, the possibility of mixing solid particles (e.g. nanopowders) into the basic spinning solution was investigated. In this way, the chemical reaction leading to alumina can be omitted and instead the boehmite or pseudoboehmite (generated as a waste fraction in industrial production) can be used directly. [6] With this method, it will be possible to modify composite nanofibers with, for example, carbon nanopowder, and press electrodes for use in current Li-ion technology. This work focuses on the production of high-quality ultrafine and refractory mullite nanofibers from TEOS, industrially prepared boehmite nanoparticles with a diameter of about 5 nm and high molecular weight PVP, and modification of nanofibers with precipitated spinels from molten glass, for applications in safe solid-state batteries.

2. Composite modifiable nanofibers

Many publications focus on the production of pure alumina nanofibers due to their high temperature properties. However, most methods are based on chemical reactions of aluminum acetate, aluminum nitrate or aluminum chloride and others as an alumina precursor. [7-9] The principle of production consists in a chemical reaction, the formation of boehmite AlO(OH) from alumina precursors and the subsequent heat-treatment of the product, which leads to the synthesis of γ-Al_2O_3 and the gradual transition to high-temperature modification of α-Al_2O_3 with increasing temperature. [10] A significant part of the production also belongs to amorphous SiO_2 nanofibers produced from the TEOS precursor. They do not have as excellent thermal properties as alumina but are useful due to their optical and electrical properties. [11, 12]

2.1. Mullite nanofibers

The presented work is focused on the preparation of mullite nanofibers (Figure 1) from boehmite AlO(OH) nanoparticles, tetraethyl orthosilicate (TEOS) with a silica component and PVP polymer, which together with colloidal TEOS form fibrous templates for different types of nanoparticles. The high-voltage electrospinning method was successfully used to produce mullite nanofibers with a diameter in the range of hundreds of nanometers by the sol-gel technique. Mullite is one of the best and cheapest thermal insulation materials, e.g. in industrial, due to its refractory properties, high melting point (1840 °C) and economic availability. The nanofibers created from mullite have constant properties even above 1000 °C. [13]

2.2. Crystaline spinels

One of the main modifiers are different types of spinels due to their extraordinary properties. The spinels with the formula AB_2O_4 (Figure 2) and their various properties have taken significant roles in many applications such as electronics, lasers, and sensors. The structure of $CuCr_2O_4$ is normal spinel structure where Cu occupies the centers of tetrahedrally coordinated positions, Cr occupies the centers of octahedrally coordinated positions, and the anion sits at the polyhedral vertexes. Due to their magnetic and electrical properties, $CuCr_2O_4$ spinels show potential as suitable materials for the intended applications in batteries (conduction and storage of electrical energy). [14]

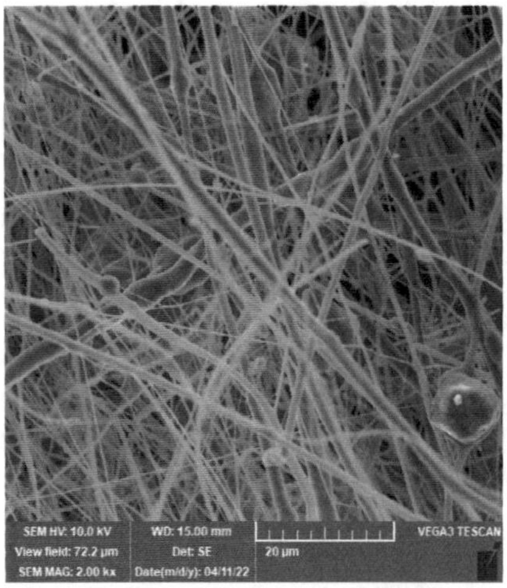

Figure 1. SEM image of composite ultra-thin nanofibers after heat-treatment

Figure 2. SEM image of spinels crystallized in molten glass after long heat-treatment.

3. Methods

3.1. Electrospinning

The process of spinning fibers using electrostatic forces known as electrospinning was developed in 1914 and today serves not only to spin various polymeric fibers, but also to produce fibers from many different materials in the submicron range. [7, 8] This method is very suitable for the formation of oxide and composite nanofibers, but there are many parameters that affect the size and morphology of fibers and thus the possibilities of their use. The influence of voltage, collector working distance, needle diameter, solution flow rate, viscosity and concentration of precursors was determined. [9, 10] For the production of nanofibers from various inorganic precursors, it is necessary to add polymers to the basic spinning solution. High molecular weight polymers such as PVB, PVP, PVA and PEO are most often used. [11, 15]

The basic electrospinning system consists of three parts: a high voltage source, spinning nozzle and collector plate. The high voltage is connected to the end of a capillary nozzle, usually a syringe needle containing the solution, and to a collector plate. The size, shape and properties of the resulting nanofibers are affected by many parameters: needle diameter, magnitude of electrical voltage, collector distance, flow rate and temperature of following calcination but the most important parameter that affects everything is viscosity of solution.

Table 1. Setting parameters of electrospinning

Parameter	Set up
Needle thickness	0.8 mm
Voltage	15 kV
Working distance	15 cm
Flow rate	1 ml/h
Viscosity	0.26 dPas

3.2. Precipitation of spinels from molten glass

The glass batch from pure oxides with composition of No. 1 (Table 2) was melted at 1400 °C for 2 hours and then by controlled heat-treatment process at 1100 °C and 800 °C each for 12 hours for 7 days in corundum crucible in superkanthal electrical furnace. The glass marked No. 2 was melted from the shards of glass No. 1 at 800 °C for 12 days in a corundum crucible in the same electrical furnace. The glass powder was milled to submicron particles in vibratory disc mill and mixed with ethanol and PVP for preparation of nanofibers by electrospinning.

Table 2. Glass composition after the first and second heat treatment.

	SiO_2	B_2O_3	Na_2O	MgO	Al_2O_3	Cr_2O_3	CuO	Fe_2O_3
No. 1	55.35		19.42	2.17	15.59	1.73	5.36	0.08
No. 2	52.06	6	17.48	1.78	15.59	1.60	0.10	0.28

3.3. Measuring devices

The resulting materials were characterized under the Confocal laser microscope LEXT OLS5000 SAF and scanning electron microscope TESCAN VEGA 3 LMU with energy dispersive spectrometer (EDX). Subsequently, they were analyzed on X-ray devices: XRF – spectrometer ARL 9400 XP Rh, and XRD – difractometer PANanalytical X´Pert PRO.

4. Experiments and results

4.1. Modifiable refractory nanofibers

The electrospun nanofibers were prepared from sol-gel solution consist of PVP (with average M_w 1 300 000) + TEOS solution (precursor of SiO_2) + boehmite AlO(OH) nanoparticles (precursor of Al_2O_3). The concentration of the polymer, and thus the viscosity of the solution, has the greatest effect on resulting fibers. The solution with the best properties for electrospinning had a viscosity of 0.26 dPas. It was found that the degree of polymerization of TEOS solution (TEOS + H_2O + HNO_3) increases over time. It is possible to use a solution up to one year old, but the older it is, the shorter is the workability or applying time. It is a fact, that fresh solution has 1 week workability and 1 year old only 1 day workability. The nanofibers created in this way have an average diameter slightly above 1 μm (Figures 3 and 5). Their morphology is not destroyed even by a very long heat treatment process (10 hours at 1200 °C) and after this process their diameter is reduced by up to half to about 500 nm (Figures 4 and 6).

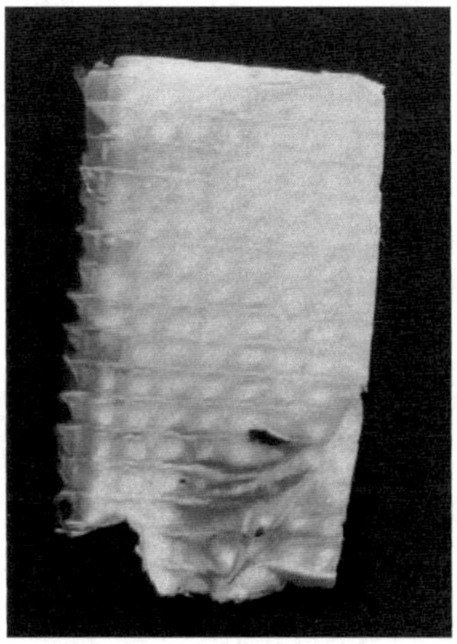

Figure 3. Image of electrospun nanofibers before heat-treatment.

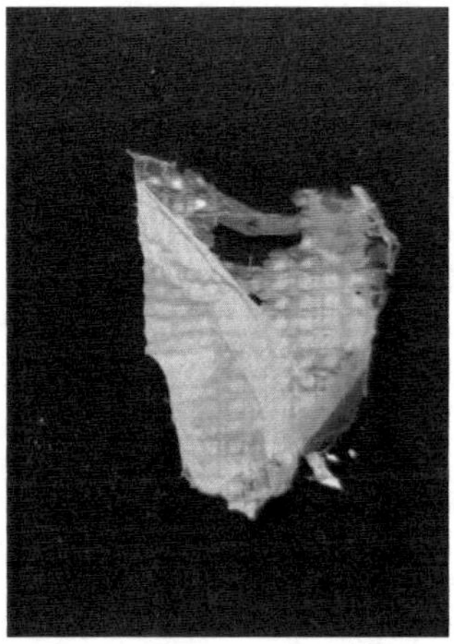

Figure 4. The electrospun nanofibers after 10 hours at 1200 °C.

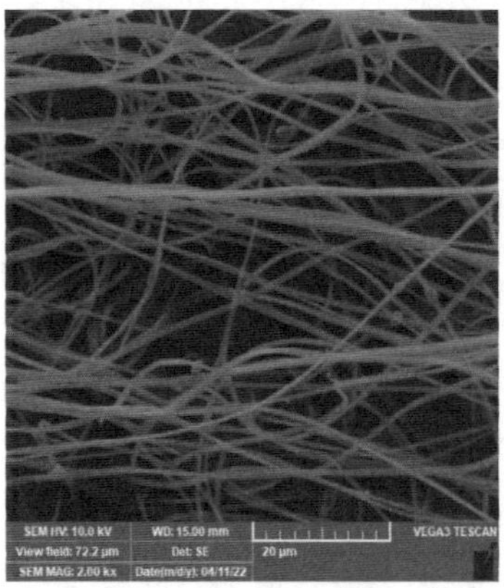

Figure 5. SEM image of as-spun nanofibers before heat-treatment.

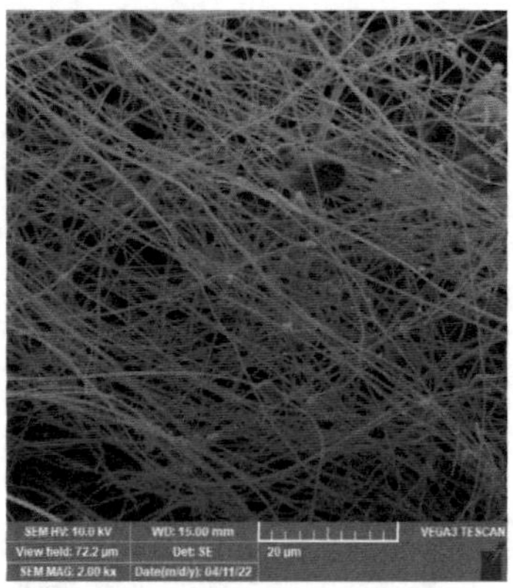

Figure 6. SEM image of nanofibers after 10 hours at 1200 °C.

4.2. Crystalization of mullite in nanofibers

The nanofibers were prepared by electrospinning from sol-gel solution with viscosity of 0.26 dPas, and subsequently were heat treated at 1000 °C and 1200 °C for 10 hours. When silicon oxygen tetrahedra and aluminum oxygen octahedra combine in the correct ratio at high temperatures, it is called mullitization. The mullitization into flake-like form occurs at a temperature of about 980 °C. Recrystallization into a needle-like shape appears around our tested temperature of 1200 °C. The melting point is incongruent between 1840 and 1890 °C.

It was found, that for the needs of efficient production of fibrous structures, mullitization depends more on temperature than time, and 10 hours at a temperature of 1000 °C was not enough for mullitization (Figure 7). After 10 hours at a temperature of 1200 °C, mullite recrystallization was already achieved (Figure 8).

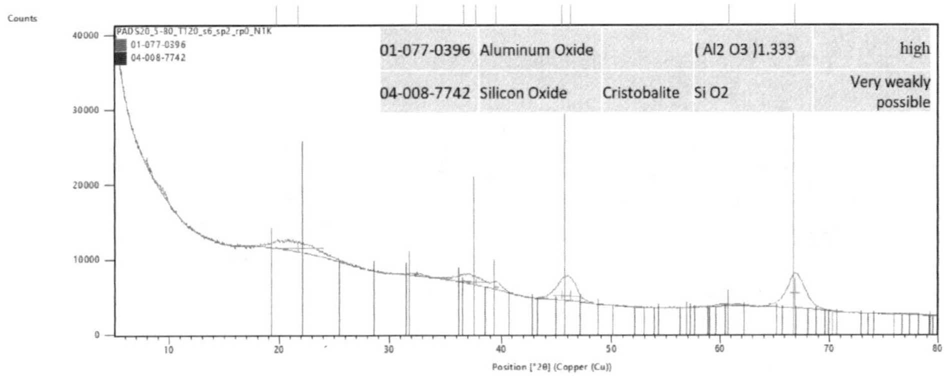

Figure 7. XRD of nanofibers after calcination at 1000°C for 10 hours

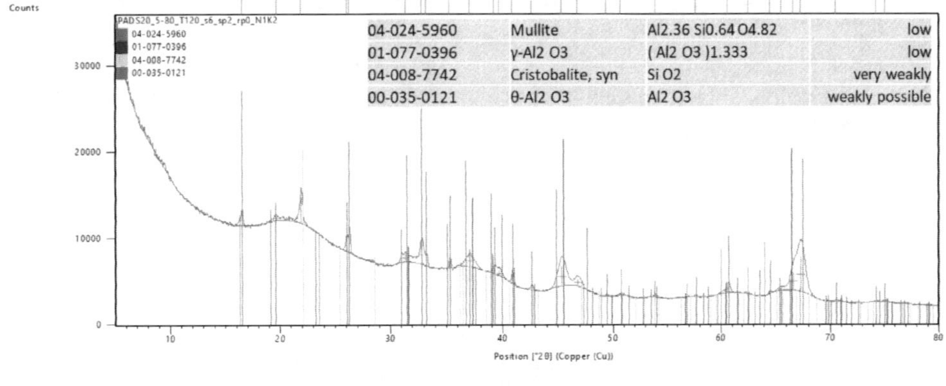

Figure 8. XRD of nanofibers after calcination at 1200°C for 10 hours

4.3. Preparation of crystalline spinel $CuCr_2O_4$

The crystalline spinels were precipitated from glass (Figure 9) prepared with the correct stoichiometric composition (No. 1) after prolonged thermal exposure (No. 2) at lower temperatures (800 °C, 12 days). The composition of the prepared spinel in glas No. 2 after heat-treatment (800 °C, 12 days) is Cu-Cr-Mg-O and was confirmed by XRD analysis (Table 3). Spinels of a different composition were also confirmed. Crystals of the correct composition (according to EDX) were also visually confirmed under a scanning electron microscope (Figure 10). The precipitation from glass has been found to be an effective method for preparing spinel crystals and may have other uses.

Table 3. XRD analyses of glass No. 2 after heat-treatment

Compound name	Formula	Wt %
Copper magnesium chromium oxide	$Mg_{0.46}Cu_{0.54}Cr_2O_4$	49
Silicon oxide - cristobalite	SiO_2	10
Magnesium copper chromium oxide	$Mg_{0.6}Cu_{0.4}Cr_2O_4$	22
Silicon oxide	SiO_2	8
Copper chromate	$CuCr_2O_4$	11

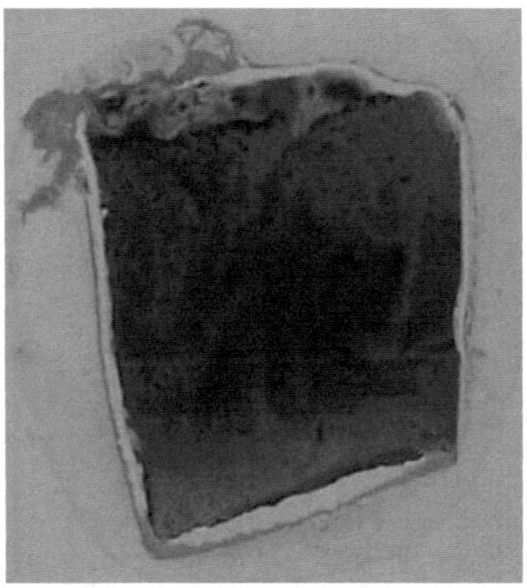

Figure 9. The glass of composition No. 2 after long heat-treatment (12 days at 800 °C)

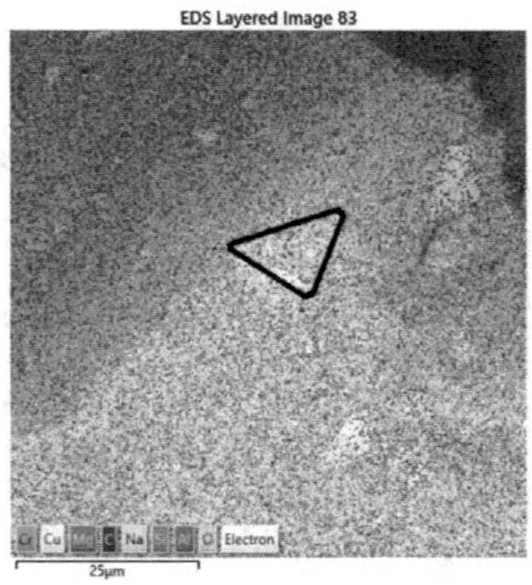

Figure 10. SEM image of crystalline spinel in glass shred after long heat-treatment

5. Conclusion

The production process of composite nanofibers was optimized for the needs of efficient production of highly resistant and refractory fibrous structures with the possibility of their modification with nanopowders of various origins. It was found that a temperature of at least 1200 °C is needed to initiate the recrystallization of mullite in the fibers. With a longer time of heat-treatment, the proportion of mullite in the structure increases. Crystallization of spinels in molten glass requires a lower temperature (only 800 °C) but a long time (at least 12 days) and the correct composition.

New possibilities of filling organic (PVP) templates with inorganic nanoparticles (boehmit, spinels, grinded glass nanopowders) and sintering into a fibrous structure are not yet as effective and fast as older and more expensive methods, but they give the possibility to produce ultra- fine nanofibers from any material that can be grind to the required small (submicron) fractions. One of the main modifiers are different types of spinels, with the formula AB_2O_4, due to their extraordinary properties. By combining these materials, it will be possible to quickly produce different electrically active parts of batteries, focusing mainly on a solid, inorganic, non-flammable electrolyte, separators, or electrodes.

References

[1] Lecomte-Nana G L and Hammas A 2021 Mullite: Structure and Properties *Encyclopedia of Materials: Technical Ceramics and Glasses* (Elsevier: Oxford) p 59-75
[2] Schneider H and Komarneni S 2005 *Mullite* (Wiley)
[3] Schneider H Okada K and Pask J A 1994 *Mullite and Mullite Ceramics* (Wiley)
[4] Hou Z 2013 Fabrication and properties of mullite fiber matrix porous ceramics by a TBA-based gel-casting process *Journal of the European Ceramic Society* **33**(4) p 717-725
[5] Zhu W 2019 Mechanical evaluations of mullite fibrous ceramics processed by filtration and in situ pyrolysis of organic precursor *Journal of the European Ceramic Society* **39**(4) p 1329-1335
[6] Wu J 2010 Synthesis and Characterization of Electrospun Mullite Nanofibers *Advanced Engineering Materials* **12**(1-2) p 71-74
[7] Liu Y 2008 Preparation and characterization of electrospun SiO_2 nanofibers *Journal of nanoscience and nanotechnology* **8**(3) p 1528-1536
[8] Mahapatra A Mishra B G and Hota G 2011 Synthesis of ultra-fine α-Al_2O_3 fibers via electrospinning method *Ceramics International* **37**(7) p 2329-2333
[9] Sharifi L 2017 Effect of Voltage and Distance on Synthesis of Boehmite Nanofibers with PVP by the Electrospinning Method International *Journal of Advanced Science and Technology* **98** p 63-74
[10] Dong X 2017 Electrospun mullite nanofibers derived from diphasic mullite sol *Journal of the American Ceramic Society* **100**(8) p 3425-3433
[11] Chen Z 2015 Electrospun mullite fibers from the sol–gel precursor *Journal of Sol-Gel Science and Technology* **74**(1) p 208-219
[12] Maneeratana V and Sigmund W M 2008 Continuous hollow alumina gel fibers by direct electrospinning of an alkoxide-based precursor *Chemical Engineering Journal* **137**(1) p 137-143
[13] Takei T 2004 Crystallization Kinetics of Mullite in Alumina–Silica Glass Fibers *Journal of the American Ceramic Society* **82** p 2876-2880
[14] Mika M Hrma P Schweiger M 1999 Rheology of spinel sludge in molten glass *Ceramics - Silikaty* **44**
[15] Zadeh M A Keyanpour M M and Ebadzadeh T 2013 Synthesis of mullite nanofibres by electrospinning of solutions containing different proportions of polyvinyl butyral *Ceramics International* **39**(8) p 9079-9084

Characterization of precursor sols for preparation of silica based nanofibers

V Procházka[1], M Havlík Míka[1], J Kočí[1]

[1]Department of Glass and Ceramics, Institute of Chemical Technology, Prague, Czech Republic

vaclav.prochazka@vscht.cz

Abstract. Objective of this study was to prepare and characterize electrospun silica-based nanofibers. The goal of this research was to create amorphous to semicrystalline nanofiber mats, which are easily modifiable and usable in electrochemical applications such as separators or electrodes in amorphous glass solid state batteries. We investigated stability and workability of precursor TEOS based sols, effect of sol ageing on fibre morphology and effect of modification on sol workability and spinnability. Sol-gel reaction mechanism and speed was investigated by liquid state ^{29}Si NMR spectroscopy.

1. Introduction

Nanofibrous structures shows great potential for applications in all industries such as tissue engineering, energy storage technologies, fuel cells composite materials, coating materials or filtration membranes. Electrospun nanofibers have diameters between 1 nm and 1 μm and their length ranges in meters. Due to this, nanofibers have extremely large length to diameter ratio resulting in higher mechanical strength, flexibility, surface area to volume ratio, porosity, gas permeability, and structural stability than microfiber or bulk counterparts [1, 2]. Their diameters generally range from 100 nm to 800 nm, however it is possible to mass produce fibres with diameter 36 nm and in small scale 2 nm [3, 4].

A large number of methods have been reported to prepare nanofibers, among these techniques most versatile is considered electrospinning. Electrospinning is a relatively simple and inexpensive method of preparing fibres with submicrometric diameters. During the spinning process, constant jet of spinning solution is ejected and elongated with strong electric field towards a collector. Electrostatic charging of the fluid at the tip of the nozzle results in formation of Taylor cone, from which a single fluid jet is drawn. After certain distance with stable jet, it is subjected to bending instability. The jet changes its path and is stretched and thinned in large ratios by electrical forces, solvent is evaporated, and fibres are deposited on collector. Typically used material in electrospinning are various polymeric solutions or polymer melts. It is possible to process large number of materials, easily control fibre diameter, fabricate multicomponent configurations an even nanotubes, nanorods or nanowires. There are many materials that are difficult or even impossible to process individually, however it is possible with right configuration i.e., multicore configuration or with use of carrier polymer [5]. Electrospun fibres are relatively easy to align, assemble and process into application with relatively low cost [6]. Electrospinning is one of few viable methods of silica-based amorphous nanofibers preparation. Other viable possibility is deemed melt-blowing technique which usually results in fibres s with diameters over 800 nm [4]. Besides the large fibre diameters special glass can easily crystalize if not cooled correctly after remelting.

 2382 (2022) 012018 doi:10.1088/1742-6596/2382/1/012018

Resulting product is influenced by large number of factors often divided into 3 categories: material properties (concentration, viscosity, conductivity, surface tension), processing parameters (voltage, flow rate, collector distance, spinner configuration) and ambient condition (mainly temperature and humidity) [4, 7].

The electrospinning process is affected by various forces such as surface tension, viscoelastic force, Coulombic force, electrostatic force of the external electric field or gravitational force. Stretching of solution jet and formation of nanofiber is mainly governed by Coulombic and electrical forces of the external field. Due to this fact, significant decrease in diameter can be observed after increase of conductivity. Therefor fibres with smallest diameters can be produced from solutions with highest conductivity. In literature, it has been reported that morphology of the fibres can be modified by addition of salts which reduces beads formation [7].

TEOS condensation reaction is illustrated in equations 1 and 2. Silanes condense into various linear, branched, or cyclic structures as shown in figure 1. Due to this fact several studies have been focused on characterizing and modifying reaction mechanisms to promote specific polymer structures.

$$Si(OH)_4 \xrightarrow{H_3O^+} (OH)_3Si\text{-}O\text{-}Si(OH)_3 + H_2O \tag{1}$$

$$(OH)_3Si\text{-}O\text{-}Si(OH)_3 \xrightarrow{H_3O^+} (OH)(Si(OR)_2O)_{n+1}Si(OH)_3 + H_2O \tag{2}$$

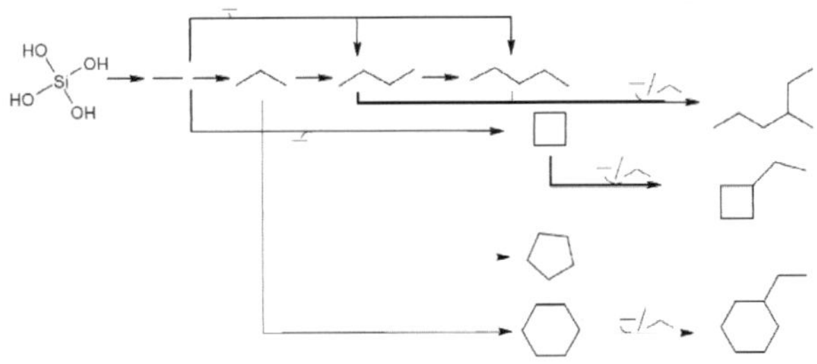

Figure 1. Teos condensation schematics

2. Experiment

Materials used in this work were Tetraethyl orthosilicate (TEOS)(>99,0%, Aldrich), Poly(vinyl pyrrolidone) (PVP) (Mw = 1300000, Aldrich), Ethanol, Hydrochloric acid (HCl), aluminium chloride (J.T.Baker Chemical), Boehmite. Deionized water was used throughout experiment.

The solutions for electrospinning were prepared by controlled condensation of TEOS precursor with or without PVP solution. For B and D sols, TEOS was dissolved in ethanol and then HCl, water and ethanol optionally with $AlCl_3$ were added dropwise while solution was continuously stirred at laboratory temperature. During mixing, solutions were water-cooled to stay within 20 °C and 30 °C. Then PVP-ethanol solution was added in chosen samples, and sols aged overnight and further on at room temperature. Remaining samples were modified with PVP-ethanol solution to increase viscosity prior spinning. The aged solutions were used for electrospinning immediately after ageing and further on as long as it was possible to process it.

Condensation reactions during aging within sols were analysed with ^{29}Si NMR spectroscopy. Spectra were recorded with Bruker Avance III (11,75 T). Resulting spectra were 1H decoupled (dept 45). Conductivity of sols was determined using two probe impedance method with stainless steel electrode 0.81 mm thick.

Morphology of resulting fibres were imagined using Confocal laser microscope LEXT OLS5000 SAF and SEM TESCAN VEGA 3 LMU.

End of spinnability and workability was determined by time when it was no longer possible to feed the sol through the nozzle or sediment formation.

Electrospinning setup was in horizontal configuration with horizontally placed rotating aluminium collector with diameter 7 cm covered with glass fibre mesh. Voltage vas set to 20 kV, nozzle diameter was 0.9 mm, spinner to collector distance was 12 cm and flow rate was set to 0.5 ml/h and 1 ml/h. Collected fibrous mats were calcinated in air at 600 °C and 650°C for 2 h. Figure 2 illustrates setup used for electrospinning experiments.

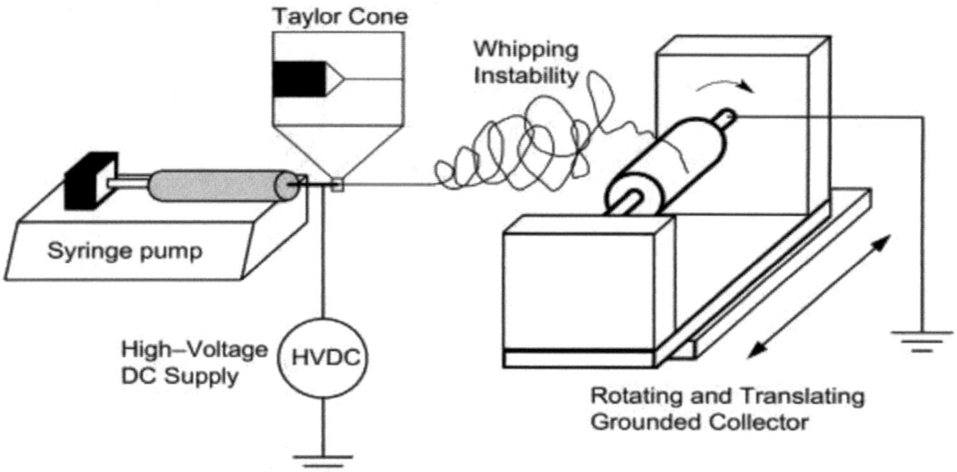

Figure 2. Schematics of apparatus with rotating cylinder [8].

3. Results and discussion

3.1. Conductivity

During the ageing during the short period after preparation sols keep their conductivity on steady levels as was reported, however in longer time intervals conductivity starts to decrease as shown in figure 3. The probable cause to this is decreasing ion mobility within solution due to growing siloxane clusters.

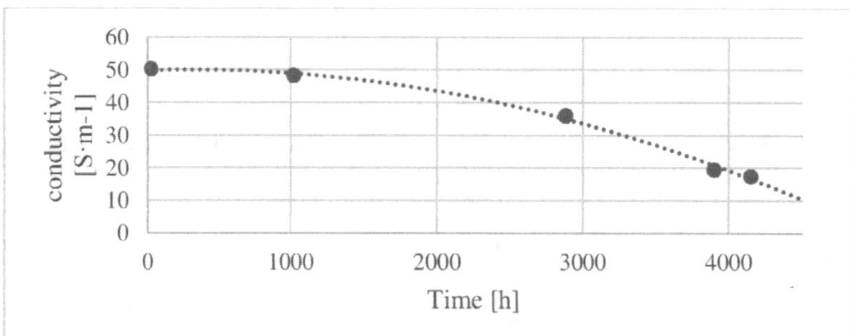

Figure 3. Conductivity of pure TEOS sol.

3.2. $_{29}$Si NMR

Acquired ^{29}Si NMR spectra recorded during the ageing of sols revealed presence condensed species Q_1 (dimers and end groups), Q^2 (linear chains and rings) and Q^3 (branched chains or rings). Q^0 structures were not detected due to the relatively high reactivity of monomer species. From spectra analysis is apparent that TEOS B series sols have highest relative content of Q^2 structures after initial ageing. After 2 days relative content is roughly the same as shown in Figure 4. Content of cyclic and linear oligomers was at 1:1 ratio for Q^2 motives and 1:0.9 cyclic to open branched structures for Q^3 motives. Modification of pure TEOS sol led to decrease of total conversion, however promoted formation of linear structures. Q^2 motives were more abundant than in pure B sol and cyclic to linear structures ratio increased to 1:1.1 and branched Q^2 motive structures had always higher relative concentration than Q^3 motive structures as shown in figure 5. This increase of linear structures relative concentration can be explained by interaction of siloxane species with PVP carbonyl groups as shown in figure 6 and figure 7.

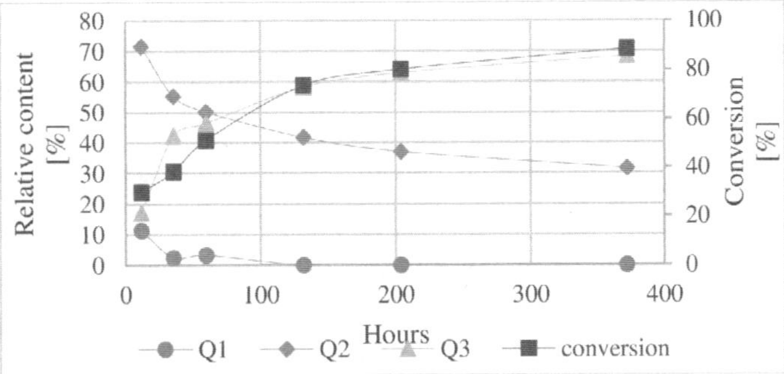

Figure 4. Relative content of Q motives in pure B sol and relative conversion of TEOS to condensed species.

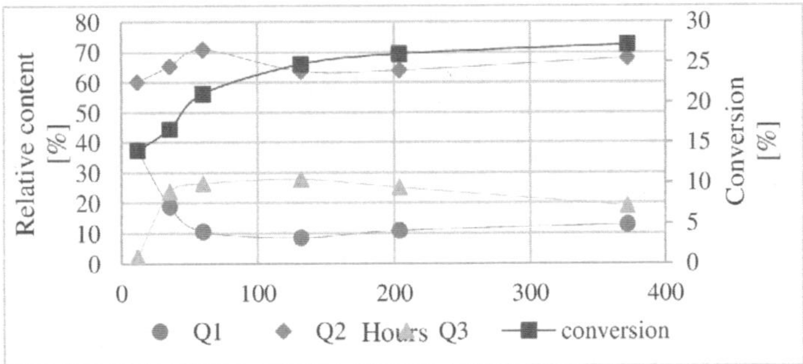

Figure 5. Relative content of Q motives in PVP modified B sol and relative conversion of TEOS to condensed species.

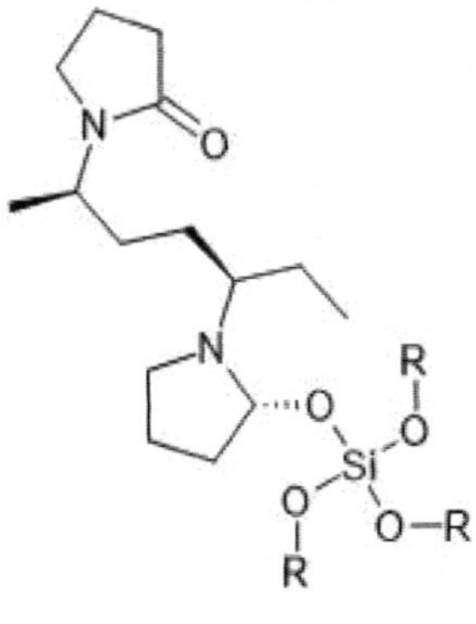

Figure 6. PVP – siloxane species interaction with low TEOS concentration.

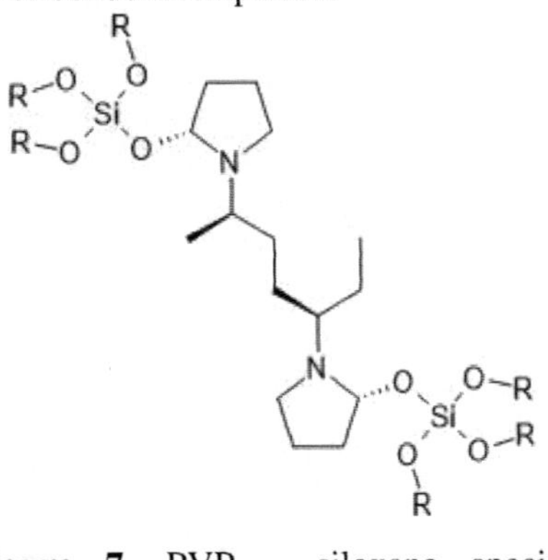

Figure 7. PVP – siloxane species interaction with high TEOS concentration.

Pure D sols have almost constant composition during measured interval. It has high TEOS to condensed species conversion. Relative content of Q2 motive structures is approximately 30% higher than Q^3 motive structures (figure 9). Relative content of ratio of cyclic to linear structures is 2.3:1 (figure 9). Higher Q^2 cyclic structures content should be cause by higher ethanol content in pure sol, resulting in more closely packed siloxane units. Secondary effect is slight increase in viscosity; however it is still necessary to modify the D sols with carrier polymer to achieve uniform continuous fibres.

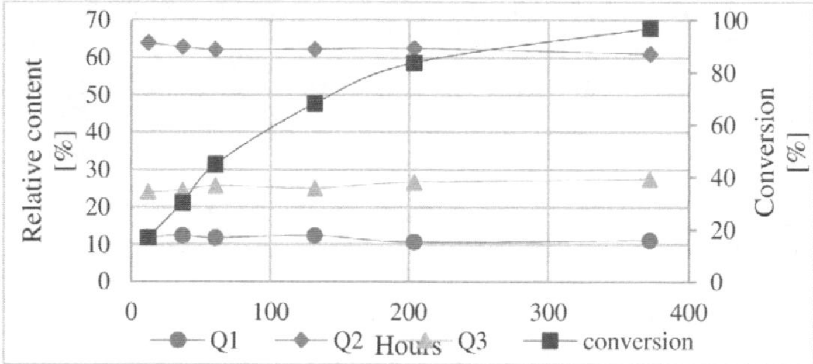

Figure 8. Relative content of Q motives in pure D sol and relative conversion of TEOS to condensed species.

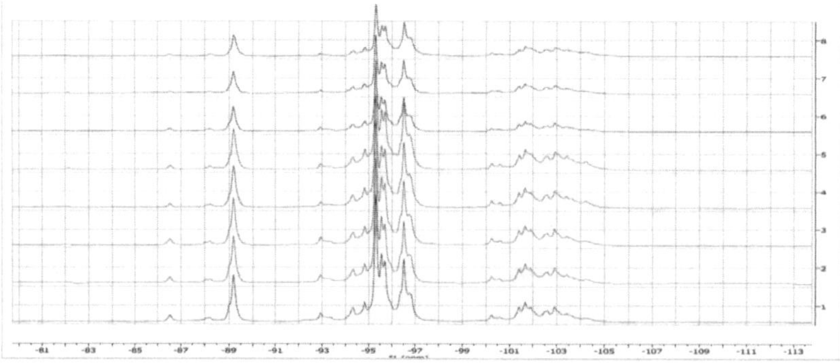

Figure 9. ^{29}Si NMR spectra of D sol

3.3. *Spinnability*

It was found that both B and D sols are spinnable. To achieve smooth uniform fibres is necessary to modify sols viscosity. During this study PVP was used for this purpose prior spinning process. B sol series samples without were workable for at least 2 months at ambient temperature. Samples stored at 5 °C stayed all workable over 6 month and some over 1 year. D sols have shown higher stability and were spinnable for at least 4 months at 25 °C and were workable for over 1 year. All tested additives reduced workability time significantly. Sols modified with PVP samples gelated after 1 moth at room temperature and after 2 months at 5 °C. Addition of $AlCl_3$ did not lead to gelation but formation of slowly sedimenting particles. After a week viscosity and surface tension of sol decreased to the point where electrospinning was no longer possible, and dispersion was electrosprayed instead. Precipitation of particles was could be caused by impurities in $AlCl_3$ or undissolved crystals of the same working as seed crystals. Introduction of Boehmite caused faster gelation rate of the sol. Speed of gelation was bound to the age of pure TEOS sol. Solution created from fresh TEOS sol was workable for 14 days, 6 months old TEOS sol gelated after 4 days. In this case storing sols at 5 °C and 25 °C did not cause any significant change in gelation speed.

3.4. *Fibre morphology*

All sols that did not age with PVP had to be modified with PVP to enhance viscosity prior spinning. Fibres electrospun from pure solutions B and D had similar results and diameter averaged at 840 nm and 838 nm respectively. As spun were straight with uniform thickness and no beads and with high porosity (figure 10). After calcination average diameter was 386 nm and 381 nm, fibres were no longer porous, and beads appeared (figure 11). B sols suffered with beads formation much more than fibres from D sols. D sol fibres had more uniform fibres (figure 12). Addition of $AlCl_3$ led to removal of beads, but difference in diameters was insignificant (figure 13).

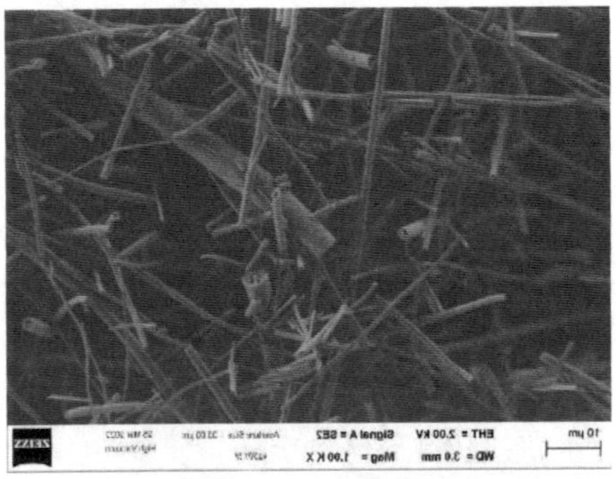

Figure 10. Electrospun fibres from B sol

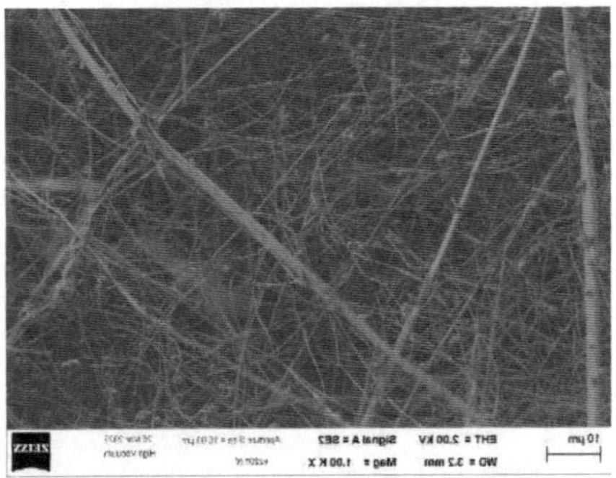

Figure 11. Fibres from B sol after calcination.

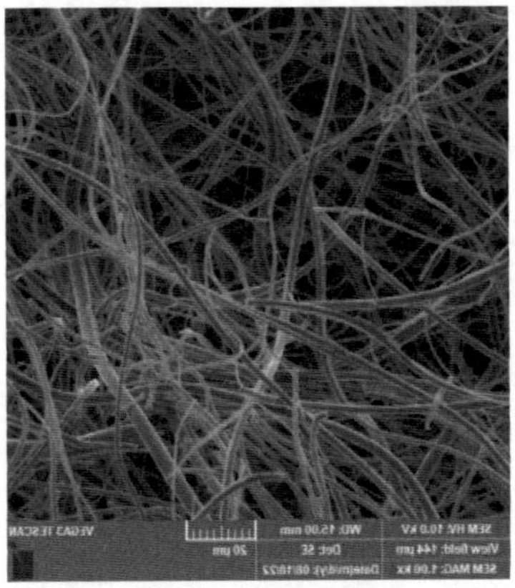

Figure 12. Electrospun fibres from D sol

Figure 13. Electrospun fibres from D sol doped with $AlCl_3$

4. Conclusion

Prepared pure sols have workability over 2 months in ambient temperatures and over 6 months at 5 °C. During preparation of Mullite nanofibers workability of sols is 4-14 days based on TEOS sol age. Sols with PVP had gellated within a month at ambient temperatures and 3 months at 5 °C. Samples with $AlCl_3$ did not form gel but solid particles precipitated after 7 to 10 days. Amorphous glass nanofibers were successfully prepared with average diameter 386 nm, higher ethanol content leads to lower relative content of linear Q^2 structures in contrary to PVP content during ageing.

[1] Yang X-h He P Xia Y-y 2009 Preparation of mesocellular carbon foam and its application for lithium/oxygen battery *Electrochem. Commun.* **11** 1127-30.

[2] Zhang B Kang F Tarascon J-M Kim J-K 2016 Recent advances in electrospun carbon nanofibers and their application in electrochemical energy storage *Prog. Mater. Sci.* **76** 319-80.

[3] Soltani I Macosko CW 2018 Influence of rheology and surface properties on morphology of

nanofibers derived from islands-in-the-sea meltblown nonwovens *Polym. J.* **145** 21-30.
[4] Zdraveva E Fang J Mijović B Lin T 2017 *Electrospun nanofibers (Structure and Properties of High-Performance Fibers)* ed G Bhat chapter 11 267-300.
[5] Yoon J Yang H-S Lee B-S Yu W-R 2018 Recent Progress in Coaxial Electrospinning: New Parameters, Various Structures, and Wide Applications *Advanced Materials* **30** 1704765.
[6] Liu Y Sagi S Chandrasekar R Zhang L Hedin NE Fong H 2008 Preparation and characterization of electrospun SiO2 nanofibers *Journal of nanoscience and nanotechnology* **83** 1528-36.
[7] Angammana CJ Jayaram SH 2011 Analysis of the Effects of Solution Conductivity on Electrospinning Process and Fiber Morphology *IEEE Transactions on Industry Applications* **47** 1109-17.
[8] Sill TJ von Recum HA 2008 Electrospinning: Applications in drug delivery and tissue engineering *Biomaterials* **29** 1989-2006.

Corrosion processes of sintered materials based on Fe

M Sedlaříková[1], M Zatloukal[1], J Kuchařík[1], P Čudek[1], G Fafilek[2], E Doleželová[1]

[1] Department of Electrical and Electronic Technology, FEEC, Brno University of Technology, Technická 10, 616 00 Brno, Czech Republic
[2] Technische Universität Wien, Forschungsgruppe Elektrochemische Methoden und Korrosion, Getreidemarkt 9, Vienna, Austria

sedlara@vutbr.cz

Abstract. All currently used metal-based implants have the disadvantage of often requiring secondary surgery to remove the implant. This is why there is a lot of research into biodegradable bone implants these days. Biodegradable iron-based alloys are potential candidates for use as materials for temporary implants. This work deals with the current topic of Fe-based biodegradable sintered materials and their corrosion. These materials are suitable for applications where gradual decomposition is required. Sintered materials based on Fe-Mg are investigated in this work. The manufactured samples were immersed in NaCl solution and simulated body fluid – SBF. Subsequently, changes in corrosion potential and current, pH, conductivity and observations under the microscope were regularly analysed.

1. Introduction

Bone injuries such as fractures often result in permanent bone damage such as weakness and structural or volumetric changes in the fused bone. Treatment for such injuries can be long and expensive. Biodegradable materials may offer an improvement to such treatment, as they could provide the necessary short-term fixation and support for the fractured bone to be fused together. Biodegradable bone support will be gradually absorbed by corrosion processes and safely removed from the patient's body. Various compositions of such materials and manufacturing processes are examined in different corrosion environments to evaluate a number of properties such as corrosion rate and corrosion potential. The key parameters for biodegradable materials are mechanical strength, flexibility, tissue compatibility, and fast degradability.

Corrosion is the spontaneous, gradual transformation of metals or non-metallic organic and inorganic materials. It arises as a result of a chemical or electrochemical reaction of the base material with the external environment. Corrosion occurs in liquids, gases, and various chemicals. The results of corrosion can be very different. The appearance of the material may change or it may completely disintegrate. Corrosion is caused by electrochemical processes. The main corrosive agents include the hydroxide group, which is labelled OH. In addition, anions that arise from acids can act as corrosive agents ($(CO_3)^{2-}$, $(Cl)^-$, $(NO_3)^-$, $(SO_4)^{2-}$).

Corrosion can be described by:

- Internal corrosion mechanism
- Type of corrosion environment
- External conditions
- Type of corrosion attack

Electrochemical corrosion is a case of metal corrosion in electrolytes (conductive medium). It consists of two partial reactions – the anode reaction and the cathode reaction. Both the anode and cathode reactions are interrelated and cannot occur without each other unless an external electric current passes through the corroding metal. The anode reaction represents the oxidation of the metal and is therefore its own type of corrosion. The depolarizing (cathode) reaction represents the reduction of oxygen dissolved in the electrolyte or the discharge of a hydrogen ion.

Electrochemical corrosion can be divided into atmospheric corrosion, corrosion in water, corrosion in soils, corrosion in aqueous solutions of acids, bases and salts.

Metals that are more susceptible to corrosion include manganese, iron, aluminium and lead. Metals that are more difficult to corrode include gold, silver and titanium [1].

2. Experimental part

2.1. Preparation and Description of Samples

Manufacturing of biodegradable bone implants samples can be described in the following steps. The first step was to prepare the material for sintering and the next step was sintering. In the first step, one of the batches was prepared by mixing the powdered metal material with an organic polymer carrier that formed the basic structure. The mixture was poured into a mold; the solvent was evaporated and the samples were dried. The carrier polymer was then removed by annealing. Powders of iron, magnesium and silver were used to prepare two sets of alloys: Fe-0.5Ag-1Mg (sample b) Fe-0.5Ag-3Mg (sample c) (all in wt%). Another batch used polyurethane (PUR) foam that was immersed in a mixture of powdered metals and demineralized water. Sample (e) is composed of pure iron. Another sample was prepared from powdered iron and magnesium Fe-3Mg (sample a) by using Polystyrene (PS). Pure iron powder was also used to make sample d). Other samples were prepared using the PUR method. Drying was used to remove all water and annealing was used to remove the polymer matrix carrier. The second step was sintering using a defined temperature and time. The temperature of 450 degrees Celsius was maintained for 2 hours. The temperature was then raised to 1120 degrees Celsius and held for one hour. This was done in a protective non-corrosive atmosphere of nitrogen or argon. The resulting material from both batches is hard and free of any polymer matrix carrier and is ready for examination and exposure to a corrosive environment. The samples were subjected to two different corrosion solutions. The first solution was sodium chloride and water, the second was simulated body fluid (SBF). Samples number b), c) and e) were soaked in salt solution, and sample a) was exposed to simulated body fluid (SBF). Samples number b) and c) were made by soaking polyurethane foam in a mixture of powdered metal and water. In the case of sample a), the organic polymer matrix carrier method was used. Figure 1 shows the SEM morphology of sintered alloys (a) Fe-3Mg, (b) Fe-1Mg-0.5Ag, (c) Fe-3Mg-0.5Ag, (d) Fe.

Figure 1. Environmental scanning electron microscope images of internal structure of samples prepared by a) sample with PS (Fe-3Mg), b) sample with PUR (Fe-1Mg-0.5Ag), c) sample with PUR (Fe-3Mg-0.5Ag), d) sample with PUR (Fe).

2.2. Solution preparation

The first solution that was prepared was NaCl. 100ml of distilled water and 0.9 g of NaCl were used. The second solution that was prepared was simulated body fluid (SBF) with ion concentrations almost identical to those of human blood plasma. This solution was approved in 2003 by Technical Committee ISO/TC150 of the International Organization for Standardization. To prepare 1000 ml of SBF, 700 ml of ion-exchanged and distilled water and the reagents listed in Table 1 were used. 1M-HCl was used to lower the pH. [4]

Table 1. Reagents used for preparing 1000 ml of SBF [4]

Order	Reagent	Amount	Purity (%)	Formula weight
1	NaCl	8.035 g	99.5	58.44
2	$NaHCO_3$	0.355 g	99.5	84.00
3	KCl	0.225 g	99.5	74.55
4	$K_2HPO_4 \cdot 3H_2O$	0.231 g	99.0	228.22
5	$MgCl_2 \cdot 6H_2O$	0.311 g	98.0	203.30
6	1.0M-HCl	39 ml	-	-
7	$CaCl_2$	0.292 g	95.0	110.98
8	Na2SO4	0.072 g	99.0	142.04
9	Tris	6.118 g	99.0	121.13

Nominal ion concentrations of SBF in comparison with those in human blood plasma is show in table 2.

Table 2. Comparison of SBF with human blood plasma [4]

	Ion concentrations ($mmol \cdot l^{-1}$)							
Ion	Na^+	K_+	Mg_{2+}	Ca_{2+}	Cl_-	HCO^-_3	PO_4^{3-}	SO_4^{2-}
Blood plasma	142.0	5.0	1.5	2.5	103.0	27.0	1.0	0.5
SBF	142.0	5.0	1.5	2.5	148.8	4.2	1.0	0.0

The NaCl solution contains only some of the ions found in the human body. In comparison, the SBF solution contains most of the ions represented in the human body and can thus much better simulate the internal environment of the human body. SBF contains both chloride and phosphate ions, which act differently on the passive layer that forms during corrosion. Phosphate ions support the formation of the passive layer, while chloride ions disrupt it. Depending on the activity of these ions, the value of the corrosion potential shifts to positive or negative values.

2.3. pH measurement

The pH of the solution was one of the measured values. This measurement shows how fast the corrosion processes are over time and how the corrosion rate changes – in some cases the rate is higher at first and then slows down, or it can be linear. From these measurements it is possible to approximately determine how much metal is being degraded and how much remains inside the body. It is also possible to estimate whether the values can be dangerous for the human organism or not.

The first pH measurement was made for a pure solution. The samples were then placed in a solution simulating the environment of the human body, and pH measurements were taken every other month. The changes in pH values are described in Table 3. The pH measurement in Figure 2 shows that the corrosion of sample a) was slower than the other samples.

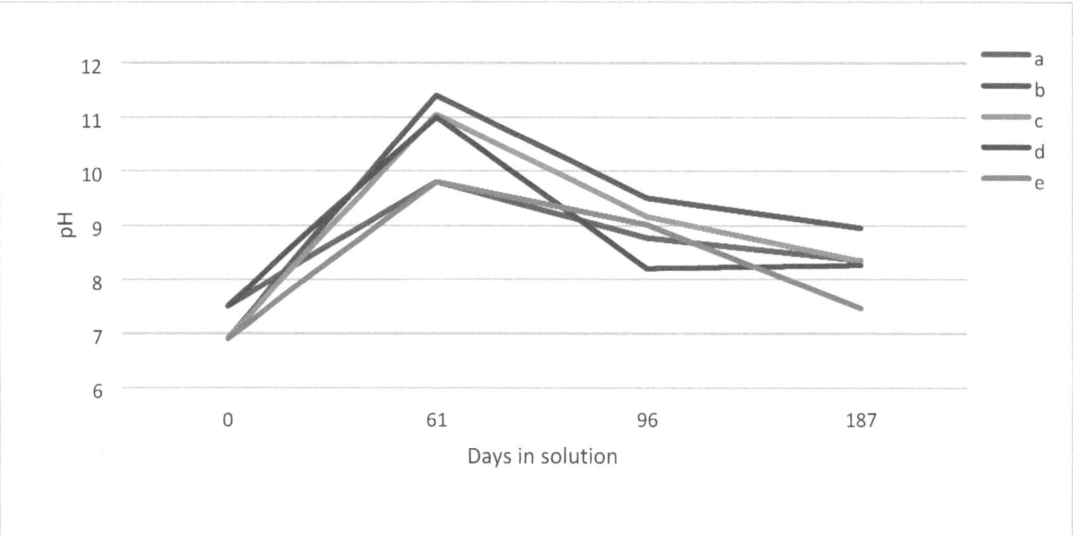

Figure 2. Changes of pH of the samples

Table 3. pH values of solutions and its changes in time

Solution	Sample	pH of solutions			
SBF	a) Fe-3Mg	7.50	9.80	8.77	8.35
NaCl	b) Fe-0.5Ag-1Mg	6.90	11.40	9.50	8.95
NaCl	c) Fe-0.5Ag-3Mg	6.90	11.05	9.16	8.34
SBF	d) Fe	7.50	11.00	8.20	8.26
NaCl	e) Fe	6.90	9.80	9.01	7.46
Number of days		0	32	61	88

Between day one and day 61, the pH was increasing. It was caused by an oxidation reaction in the release of electrons.

$$M \rightarrow M^{n+} + ne^- \quad (1)$$

$$H_2O + 2e^- \rightarrow H_2 + 2OH^- \quad (2)$$

$$2H_2O + 4e^- \rightarrow 4OH^- + 4e^- \quad (3)$$

The iron was dissolved in the solution during this reaction. After day 61, the pH began to decrease. It is caused by the reduction reaction of iron.

$$2Fe^{2+} + OH^- \rightarrow Fe(OH)_2 \quad (4)$$

The SBF solution is a solution containing chlorides, sulfates, and phosphates (Table 2), and the increase in pH value due to corrosion of pure iron promotes the precipitation and deposition of phosphates [5].

Reaction in SBF:

$$Fe(OH)_2 + Cl^- \rightarrow FeClOH + OH^- \quad (5)$$

$$FeClOH + H^+ \rightarrow Fe^{2+} + Cl^- + H_2O \quad (6)$$

$$Fe^{2+} + O_2 + 3OH^- \rightarrow Fe(OH)_3\downarrow + O^-_2 \quad (7)$$

$$Fe(OH)_3 + 2Cl^- \rightarrow FeCl_2OH + 2OH^- \quad (8)$$

$$FeCl_2OH + H^+ \rightarrow Fe^{3+} + 2Cl^- + H_2O \quad (9)$$

$$2PO^3{}_4^- + 3Ca^{2+} \rightarrow Ca_3(PO_4)_2\downarrow \quad (10)$$

$$2PO^3{}_4^- + 3Mg^{2+} \rightarrow Mg_3(PO_4)_2\downarrow \quad (11)$$

$$2PO^3{}_4^- + 3Fe^{2+} + 8H_2O \rightarrow Fe_3(PO_4)_2 \cdot 8H_2O \quad (12)$$

$$PO^3{}_4^- + Fe^{3+} \rightarrow FePO_4\downarrow \quad (13)$$

A serious problem is the alkalizing effect, for example, with biodegradable materials based on magnesium alloys [6]. A study by Song, G. L. mentioned that the corrosion of magnesium implant caused an increase in the pH value of the local body fluid near the magnesium alloy [5], which could adversely affect the balance of pH-dependent physiological reactions and lead to alkaline poisoning effect [7].

Another measurement was made to determine the corrosion potential and current. The measurements were performed on an AUTOLAB TYPE II potentiostat, using a three-electrode arrangement consisting of a working electrode (PIGE), an auxiliary electrode (platinum) and a reference electrode (fully saturated calomel). The measurement was carried out in a 9% NaCl solution and the potentiodynamic method was chosen, for which a step of 0.001 V and a scan rate of 0.01 V·s-1 were set. Before the measurement, the samples were properly cleaned using isopropyl alcohol and distilled water, then they were dried for 4 hours, and then a metal powder was obtained from the sample, which was then applied to the working electrode.

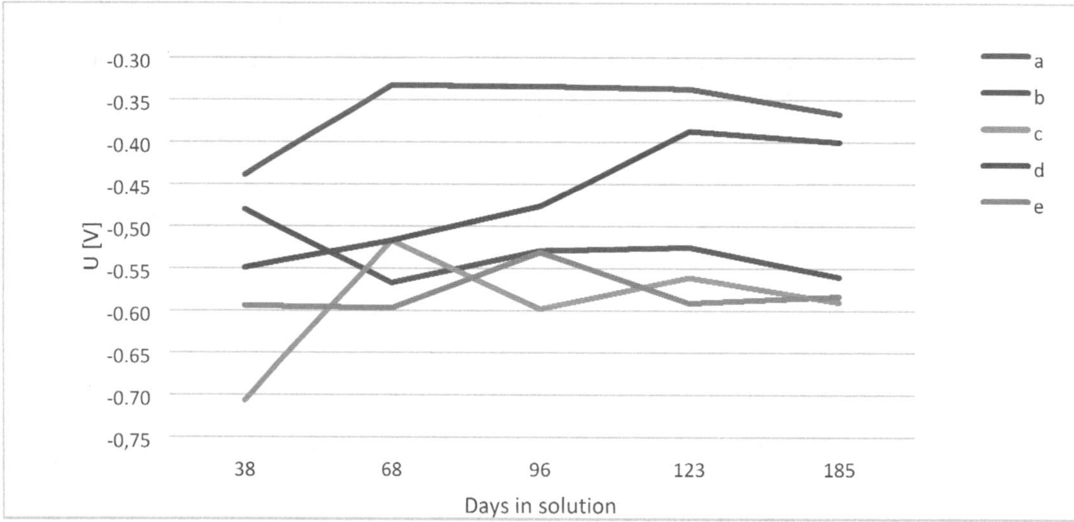

Figure 3. Difference in corrosion potential between samples a), b), c), d) and e) versus time (days)

Table 4. Changes in corrosion potential of samples in SBF and NaCl solutions as a function of time

Solution	Sample	Corrosion potential (V)				
SBF	a) Fe-3Mg	-0.439	-0.333	-0.334	-0.327	-0.367
NaCl	b) Fe-0.5Ag-1Mg	-0.480	-0.567	-0.529	-0.525	-0.560
NaCl	c) Fe-0.5Ag-3Mg	-0.707	-0.517	-0.598	-0.561	-0.590
SBF	d) Fe	-0.549	-0.517	-0.476	-0.387	-0.400
NaCl	e) Fe	-0.594	-0.597	-0.531	-0.591	-0.583
Number of days		38	68	96	123	185

The increase in corrosion potential in the first days is due to the formation of OH^- at the beginning of dissolution, then the corrosion potential decreases because the reaction of $2Fe^{2+} + OH^-$ forms a passivation layer on the surface of the sample. The possibility of the formation of passive layers was also supported by the measured pH (Figure 2), which was alkaline and thus indicated the formation of hydroxyl anions. The reduction in the corrosion potential of sample b is caused by the lower mass of magnesium in this sample compared to samples a and c. This measurement is shown in Figure 3 and the values are shown in Table 4.

The corrosion current is related to how many electrochemical reactions are taking place. If more reactions in the sample take place, the corrosion current is larger. Sample a) was placed in SBF solution. The SBF solution contains many ions that increase the corrosion current. There is a similarity between the current curves of samples b and c. The difference is due to the different concentration of magnesium in the samples. Figure 4 shows the corrosion currents of the samples and the values are shown in Table 5.

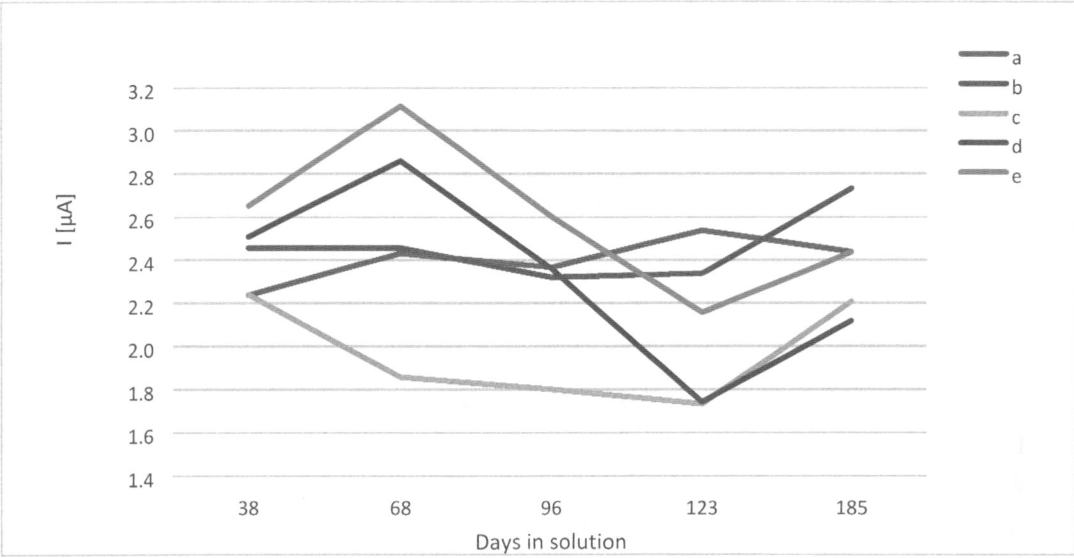

Figure 4. Difference in corrosion current between samples a), b), c), d) and e).

Table 5. Changes of corrosion current of the samples a), b), c), d) and e).

Solution	Sample	Corrosion current [µA]				
SBF	a) Fe-3Mg	2.240	1.858	1.802	1.734	2.207
NaCl	b) Fe-0.5Ag-1Mg	2.457	2.457	2.319	2.338	2.732
NaCl	c) Fe-0.5Ag-3Mg	2.237	2.431	2.367	2.537	2.438
SBF	d) Fe	2.508	2.859	2.359	1.743	2.118
NaCl	e) Fe	2.652	3.114	2.602	2.157	2.437
Number of days		38	68	96	123	185

The process of precipitation of ions in the solution caused a significant decrease in conductivity in sample a). On the other hand, for samples b) and c) there was a slower decrease in the conductivity value. After 123 days, the value increases because the chloride ions disrupt the passivation layer on the surface of the samples.

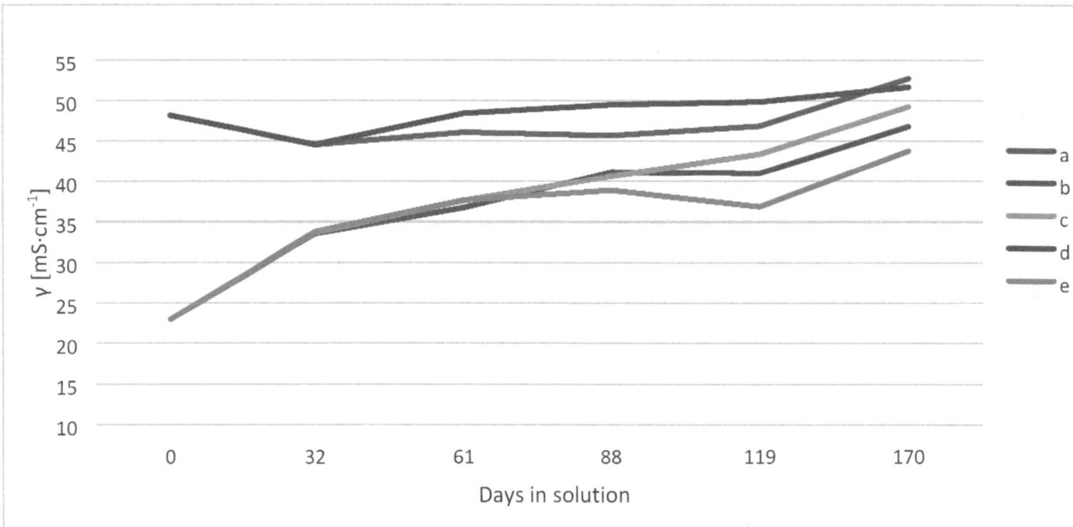

Figure 5. Difference in corrosion solution conductivity between samples a), b), c), d), e) depending on time (days).

Table 6. Changes of corrosion solution conductivity of the samples a), b) c), d) and e)

Solution	Sample	Corrosion solution conductivity [$mS \cdot cm^{-1}$]					
SBF	a) Fe-3Mg	48.165	44.535	46.080	45.683	46.863	52.723
NaCl	b) Fe-0.5Ag-1Mg	22.976	33.532	36.725	41.174	41.011	46.807
NaCl	c) Fe-0.5Ag-3Mg	22.976	33.687	37.624	40.647	43.401	49.244
SBF	d) Fe	48.165	44.573	48.404	49.479	49.850	51.658
NaCl	e) Fe	22.976	33.771	37.678	38.889	36.898	43.785
Number of days		**0**	**32**	61	88	119	170

3. Conclusion

Experiments have shown that corrosion occurs and changes its behaviour over time according to the amount of corrosion products that are generated in different samples. Corrosion of the samples occurs when they are submersed in approximately 15 millilitres of the solution. But human blood flows throughout the whole body and is purified in the kidneys. It is likely that the corrosion of the implant would not be slowed as much by the corrosion products in solution when implanted in the human body. Observations under the microscope (Figure 1) showed the structure of the samples and allowed a better observations of the porosity. These observations clearly showed that the porosity of the samples was highly dependent on the type of foam used and not dependent on the combination of metals used. Multiple reactions occur simultaneously in the sample during the degradation process. During the dissolution of metals - oxidation, metal cations are formed, which in combination with chloride anions can form metal chlorides. Hydroxyl ions are also produced. Hydrolysis breaks down metal chlorides, forming metal hydroxides, and increases the concentration of hydrogen ions, which can counteract the hydroxyl ions and lower the pH, making the solution more acidic.

Acknowledgments
This work was supported by the project of specific research at Brno University of Technology: Materials and Technologies for Electrical Engineering IV, reg. no. FEKT-S-20-6206.

References

[1] Doleželová E 2022 Determination of the corrosion potential of powdered materials, Second laboratory task. Brno [cit. 2022-07-11]. University of Technology Brno, Faculty of Electrical Engineering and Communication, Department of Electrical and Electronic Technology

[2] Tkáčová T 2020 Biodegradabilní kostní implantáty na bázi železa Brno [cit. 2020-06-02]. https://www.vutbr.cz/studenti/zav-prace/detail/127492. Diploma Thesis. University of Technology Brno, Faculty of Electrical Engineering and Communication, Department of Electrical and Electronic Technology

[3] Galanová Z 2017 Biodegradable bone implants. Brno. Diploma Thesis. University of Technology Brno, Faculty of Electrical Engineering and Communication, Department of Electrical and Electronic Technology

[4] Kokubo T and Takadama H 2006 How useful is SBF in predicting in vivo bone bioactivity? *Biomaterials* **27**, **15**, 2907-2915 ISSN 0142-9612

[5] Zhang E, Chen H and Shen F 2010 Biocorrosion properties and blood and cell compatibility of pure iron as a biodegradable biomaterial. *J Mater Sci: Mater Med* **21**, 2151–2163 https://doi.org/10.1007/s10856-010-4070-0

[6] Song G L, Song S Z 2007 A possible biodegradable magnesium implant material. *Adv Eng Mater.* **9(4)** 298–302

[7] Song G L 2007 Control of biodegradation of biocompatible magnesium alloys. *Corros Sci.* **49(4)** 1696–701

Cooling ability of smooth and dimpled surfaces

L Dobsakova[1] and T D Nguyen[1]

[1] Department of Mechanical Engineering , University of Defence, Brno, Czech Republic

lenka.dobsakova@unob.cz

Abstract. Reliable cooling of electronic components is an essential factor in electrical engineering. Every electronic device produces a certain amount of heat during its operation. This heat can cause overheating and a device can suffer damage. Therefore, the issue of cooling plays a significant role and is an important theme appropriate for continuously finding new improvements.

The paper describes a quantification of the heat transfer coefficients between a body and a surrounding environment. Experimental research of the heat transfer by cooling is focused on the plane smooth and structured surfaces. Surface structures improving the heat transfer rate are of dimples. The cooling process was measured by using a thermal imaging camera to obtain a cooling curve – the change in temperature time course during air cooling. The analysis of the cooling curve gradient allows us to identify convective and radiant components of the heat transfer. The presented results in the form of diagrams include different cases of air flow around smooth and dimpled surfaces.

1. Introduction

Many electrical devices produce a large amount of heat during their work. The calculation of those devices' cooling is very important because it can have a fundamental effect on their function. This paper deal with two united thoughts. For one thing, the paper considers increasing of cooling ability of heat sinks by using a structured surface. For another this paper studies a quantification of the heat transfer coefficients. Heat transfer is composed of two phenomena – the convective component and the thermal radiation component. A heat transfer calculation is often calculated only with the convective component because its amount is a dominant part of the heat transfer, while the radiant component is neglected. This paper describes the identification and quantification of convective and radiant components of heat transfer for two types of surfaces – smooth and structured surfaces.

The surface structure is made of simple dimples. The data needed for calculation were obtained from measurement by using a thermal imaging camera for different cases of airflow conditions. Samples material is copper, often used for heatsinks manufacturing because of its high thermal conductivity.

The dimpled surface is often studied in research of drag reduction [1], [2]. Some papers are describing the heat transfer of structured surface in comparison to smooth surface by convective heat transfer inside a cooling passage [3]. In this paper is described heat transfer in open area, where the heat exchange by radiation between body and the environment comprises a important account for total heat transfer. In this paper, a beginning of considerations is described, which will continue by CFD simulations with place emphasis on determination of heat transfer coefficients quantification.

2. Theoretical background of heat transfer

One of the fundamental equations of the heat transfer theory is the equation of heat transfer rate $\dot{Q}$ (see Equation 1). This equation expresses how much heat Q some object absorbs or transfers to its surrounding during a time unit τ. Where the heat Q represents the amount of thermal energy which the object can bear inside it. The heat is given by the object's mass m, its specific heat coefficient c, and current temperature in degrees Celsius, marked by the small letter t (in this paper, the temperature used in Kelvin degree is marked by capital letter T for better orientation in equations).

$$\dot{Q} = \frac{dQ}{d\tau} = m\,c\frac{dt}{d\tau} \tag{1}$$

The heat is transferred simultaneously by two physically different ways – by convection and radiation. So, the total heat transfer rate is the sum of the convective heat transfer rate Q_c and the radiant heat flux Q_r:

$$\dot{Q} = \dot{Q}_c + \dot{Q}_r \tag{2}$$

2.1. Convective heat transfer rate

Convective heat flux is in a simplified way a heat transfer between solid and surrounding fluid as a consequence of fluid flow and can be expressed by Newton's law of cooling:

$$\dot{Q}_c = \alpha\, A(t_w - t) \tag{3}$$

where α is the convective heat transfer coefficient, A is the heat transfer area of the wall, and $(t_w - t)$ is the temperature difference between the wall and fluid.

Convection is a complicated and diverse process. The heat transfer ratio by conduction and fluid flowing depends on flowing character (laminar or turbulent), thermo-kinetic quantities, and hydro-dynamic effects. The convective heat transfer coefficient is not simple to determine, because it depends on fluid flow velocity and character, fluid properties, wall shape and size, and the like. So, the coefficient value is determined by using the similarity theory.

2.2. Radiant heat transfer rate

Radiant flux is the thermal energy radiated from an object's surface in form of electromagnetic waves. The case of radiation between an object and the surrounding environment can be expressed by the simplified equation:

$$\dot{Q}_r = A\,\varepsilon_1 C_0\left[\left(\frac{T_1}{100}\right)^4 - \left(\frac{T_2}{100}\right)^4\right] \tag{4}$$

where C_0 is the blackbody radiant coefficient, ε_1 is the emissivity of an object, and temperatures in Kelvin degrees T_1 and T_2 express temperatures of an object's surface and surrounding environment. [4]

3. Measured objects

Two types of surfaces were measured. Both samples were 10 mm thick copper square-shaped plates with a side 150 mm long. One sample was with a smooth surface, the second was with a dimpled surface. Both plate surfaces were supplied by a special coat with precisely defined emissivity $\varepsilon = 0.96$. The dimpled copper plate is shown in Figure 1 and both plates' parameters are written in Table 1.

Figure 1. The dimpled cooper plate

Table 1. Parameters of copper plates.

Quantity	Smooth plate	Dimpled plate
Mass m [kg]	2.037	1.828
Specific heat capacity c [$J.kg^{-1}.K^{-1}$]	384.6	384.6
Surface Area A [m^2]	0.0225	0.0232

The dimpled plate has 182 dimples. Dimples have approximately the same size. A size difference was given by the accuracy of a manufacturing machine. Dimples have a spherical cap shape and their surface was analyzed by the surface profilometer TALYSURF CLI 1000. A depth radius and a surface shape diameter were measured and the results of four examined dimples are shown in Figures 2 and 3.

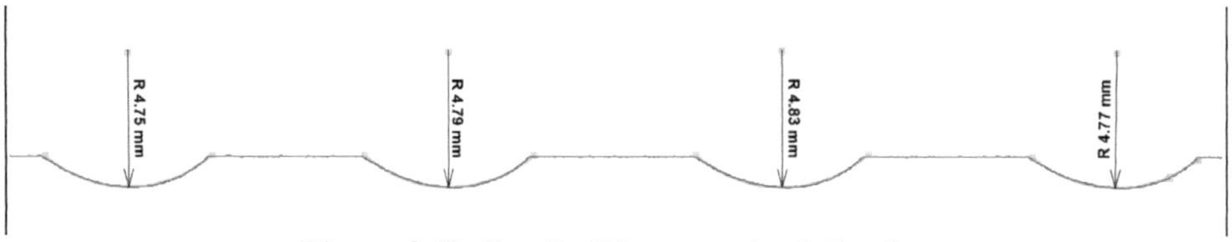

Figure 2. Radius R of four examined dimples

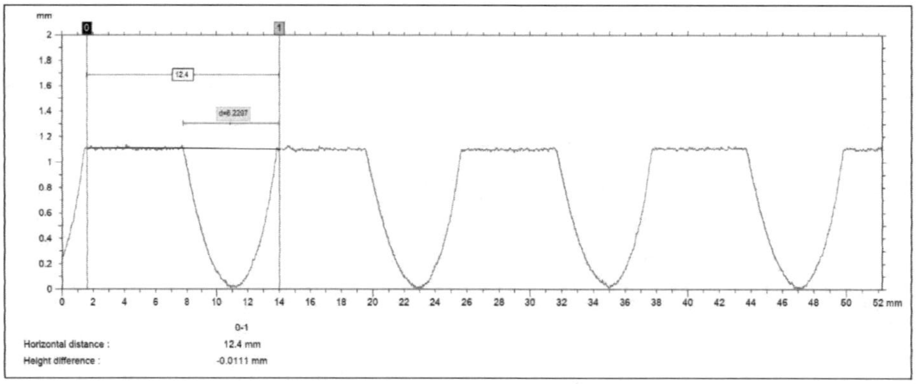

Figure 3. The measurement result of the one dimple cap base diameter.

Average values calculated from measured data were considered as the standard size of dimples, which was used for surface calculation of the dimpled plate. The dimpled surface structure increased by 3 % in comparison to the smooth plate.

4. Measurement

The experiment was designed to measure the change in the temperature time course of the plate surfaces during air cooling. Both copper plates were heated in a furnace to a specified temperature, about 130 °C. They were then potted in a heat-insulating material bed from fiberglass, ensuring that five plate surfaces were covered and just only one of them had contact with the surrounding ambient air. The physical phenomenon in the cooling process of the two samples has no difference, so the experiment of them was performed in the same way for both convection cases. For the natural convection case, the plates were kept in normal atmospheric conditions while the experiment for forced convection was performed by creating an airflow with the speed v = 3 m/s, v = 6 m/s, and incidence angle between the direction of airflow and plate surface φ = 0°, φ = 30°. In the Table 2 is the listing of executed measurement cases. Plates were positioned for all cases in a vertical position.
The thermal imaging camera FLIR T640 then allowed to time record the temperature measurement of the plate surfaces, which were provided with a special coating of precisely defined emissivity (ε = 0.96). It took a thermal photo every 30 seconds and the measured values showed a cooling curve corresponding to every case. Those values also were used for the following calculations and analysis. The thermal imaging camera is calibrated with accuracy ± 2 °C or ± 2 % of reading [5].
The measuring setup schema is in Figure 4 and examples of images from the thermal imaging camera are in Figure 5 for both measured plates.

Table 2. The list of measured cases.

	No.	Used symbol	v [m.s^{-1}]	φ [°]	$t_{arf} = t_{atm}$ [°C]
	1.	$v_0φ_0$	0	0	22.4
	2.	$v_3φ_0$	3	0	21.9
Smooth plate	3.	$v_3φ_{30}$	3	30	22.0
	4.	$v_6φ_0$	6	0	22.0
	5.	$v_6φ_{30}$	6	30	22.1
	1.	$v_0φ_0$	0	0	22.4
	2.	$v_3φ_0$	3	0	21.9
Dimpled plate	3.	$v_3φ_{30}$	3	30	22.0
	4.	$v_6φ_0$	6	0	22.0
	5.	$v_6φ_{30}$	6	30	22.1

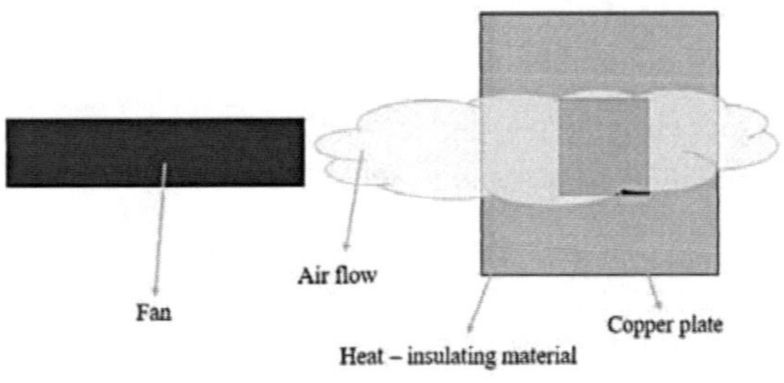

Figure 4. Scheme of heat transfer experimental setup for plate

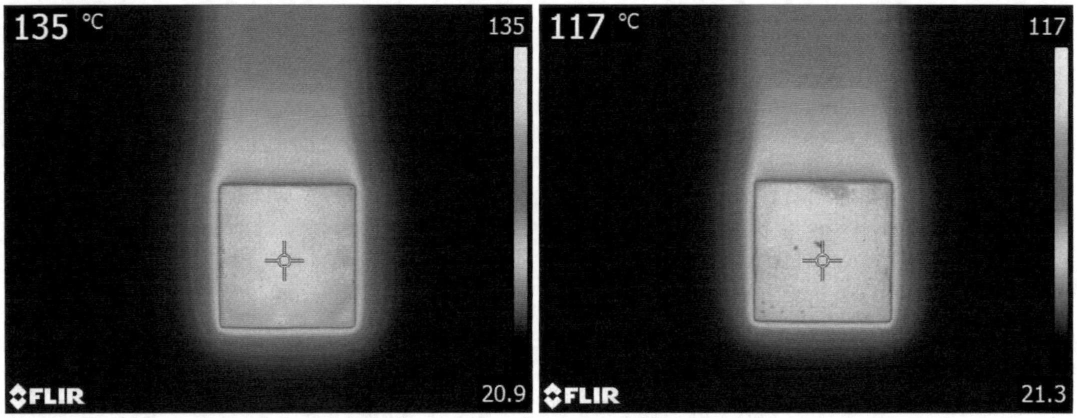

Figure 5. Images from thermal imaging camera for smooth (on the left) and dimpled plates (on the left) during the measurement of natural convection case (v = 0 m/s)

5. Data analysis

Besides the cooling curve measured by the thermo-imaging camera, another important result of the study is heat transfer rate Q and its components (convective and radiant). These equations below were used for calculating heat transfer rate values.

The total heat transfer was calculated numerically for the time interval $\Delta\tau = 30$ s as follows:

$$\dot{Q} = m\,c\frac{dt}{d\tau} = \frac{mc}{\Delta\tau}\Delta t = \frac{mc}{\Delta\tau}(t_\tau - t_{\tau+30}) \tag{5}$$

where t_τ is the approximation plate temperature at the time τ and $t_{\tau+30}$ is the approximation plate temperature at a time (τ +30).

The radiant flux calculation uses the adapted equation (4)

$$\dot{Q}_r = A\,\varepsilon_1 C_0\left[\left(\frac{T_{aver}}{100}\right)^4 - \left(\frac{T_{atm}}{100}\right)^4\right] = A\,\varepsilon_1 C_0\left[\left(\frac{\frac{T_\tau + T_{\tau+30}}{2}}{100}\right)^4 - \left(\frac{T_{atm}}{100}\right)^4\right] \tag{6}$$

where T_{atm} is the surrounding environment temperature, T_{aver} is the average plate temperature for every interval of 30 seconds, and T_τ and $T_{\tau+30}$ are approximated temperatures at the time τ and (τ + 30).

The convective heat flux is calculated from the adapted equation (2):

$$\dot{Q}_c = \dot{Q} - \dot{Q}_r \tag{7}$$

Subsequently, the convective heat transfer coefficient α and total heat transfer coefficient α_{total} were calculated from the adapted Equation (3):

$$\alpha = \frac{\dot{Q}_c}{A(t_{aver} - t_{arf})} \qquad and \qquad \alpha_{total} = \frac{\dot{Q}}{A(t_{aver} - t_{arf})} \tag{8}$$

where t_{arf} is the airflow temperature.

 2382 (2022) 012020 doi:10.1088/1742-6596/2382/1/012020

6. Experimental results

6.1. Cooling curves

Measured cooling curves by using a thermal imaging camera are shown in Figure 6 for all cases mentioned above. The cooling curves of the smooth plate are on the left figure, and the measured cooling curves of the dimpled plate are on the right figure.
By comparing graphs, the cooling curves of the dimpled plate for all cases are steeper than the cooling curves of the smooth plate. So, the cooling process is faster at the beginning. A difference increases with a faster airflow v and a greater angle of incidence φ.

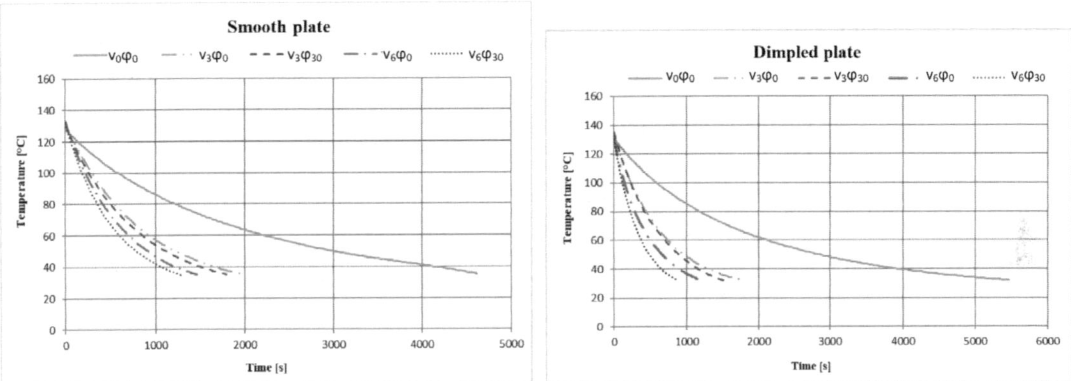

Figure 6. Cooling curves of both plates for all examined cases of convection

6.2. Heat transfer rates

The heat transfer rate Q and its components (convective and radiation) depending on time are shown in Figures 7 – 9 for both plates. In figures are just cases with the incidence angle between a direction of airflow and a plate surface $\varphi = 0°$ because the incidence angle change makes minimal differences in heat transfers.
For the natural convection case (see Figure 7), the difference between convective and radiant heat transfer rates is small. Both plates were heated to temperature 130°C and then cooled by natural convection. The time of cooling on the temperature 41 °C is 4035 second for smooth plate and 3795 second for dimpled plate. (The measurement of the dimpled plate continued further to lower temperature, but for good comparison was used temperature 41 °C.) The dimpled plate was cooling faster by 240 second.
The convective heat transfer of the smooth plate has proportion 55 % on total heat transfer at beginning and has almost constant course during the cooling (on average 53.9 %). The radiate heat transfer rate of the smooth plate has the proportion 45 % at beginning (during the cooling on average 46.1 %). Heat transfer proportions of the dimpled plate has the same course. The convective heat transfer has proportion 50.2 % at beginning has almost constant course during the cooling (on average 51.7 %). The radiate heat transfer rate the proportion 49.8 % at beginning (during the cooling on average 48.6 %).
From the comparing of both graphs, the total and convective heat transfer rates are higher for smooth plate. So, the smooth plate should be cooled faster. But, the radiation caused the important part of cooling. The radiant flux is higher for dimpled plate and caused faster cooling. From the beginning, the difference between radiant flux of dimpled and smooth plate is 5.88 %, better for dimpled plate. The convective heat flux is bigger for smooth plate all time of measurement. From the beginning, the difference between smooth and dimpled plate is 7.93 % and it change during the measurement. The duration of the radiant flux difference is important for cooling of dimpled plate by natural convection and will be studied and verified in further research.

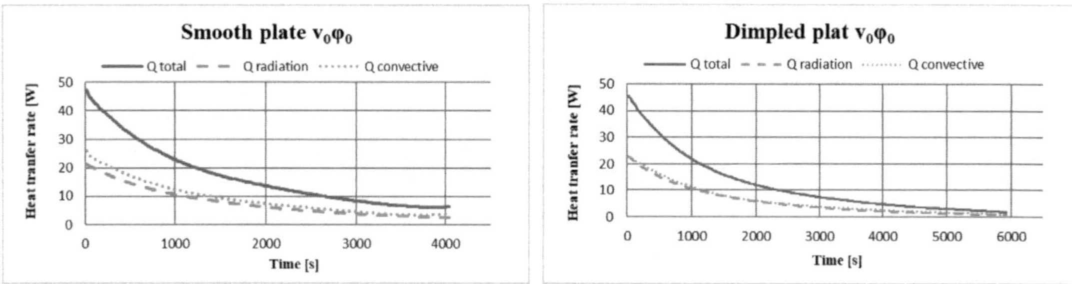

Figure 7. Heat transfer rates for the smooth and dimpled plates for the natural convection case.

For the forced convection case with v = 3 m/s, the convective heat transfer is dominant, but the radiate component is still high. The dimpled plate have both heat transfers components higher than the smooth plate. The time of cooling on the temperature 35 °C is 1995 second for smooth plate and 1545 second for dimpled plate. The dimpled plate was cooling faster by 450 second.

The convective heat transfer of the smooth plate has proportion 76 % on total heat transfer at beginning and increases during the plate cooling (on average 81.4 %). The radiate heat transfer rate of the smooth plate has the proportion 26 % at beginning and decreases during the cooling (on average 18.6 %). Heat transfer proportions of the dimpled plate has the same course. The convective heat transfer has proportion 80 % at beginning and increases during the plate cooling (on average 84.1 %). The radiate heat transfer rate the proportion 20 % at beginning and decreases during the cooling (on average 15.9 %).

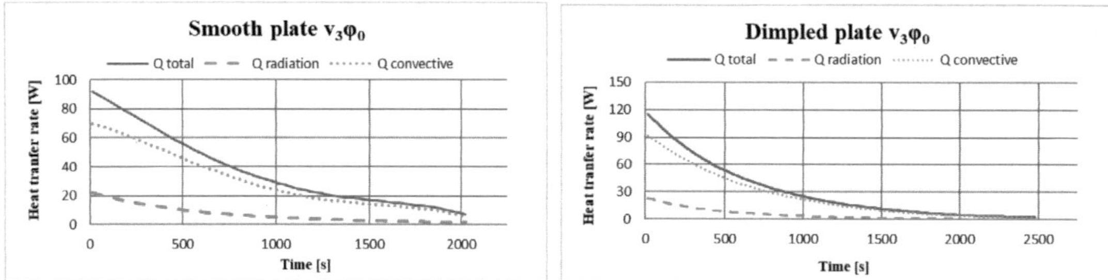

Figure 8. Heat transfer rates for smooth and dimpled plates for the forced convection case v = 3 m/s.

For the forced convection case with v = 6 m/s, the convective heat transfer is dominant and the radiant component has almost neglecting size. The course of heat transfer rate for case v = 6 m/s is same as for case v = 3 m/s. The both heat transfer components have higher size for the dimpled plate. The time of cooling on the temperature 33 °C is 1635 second for smooth plate and 1185 second for dimpled plate. The dimpled plate was cooling faster by 450 second, same as in case with v = 3 m/s.

The convective heat transfer of the smooth plate has proportion 83.9 % on total heat transfer at beginning and increases during the plate cooling (on average 86.7 %). The radiate heat transfer rate of the smooth plate has the proportion 16.1 % at beginning and decreases during the cooling (on average 13.3 %). Heat transfer proportions of the dimpled plate has the same course. The convective heat transfer has proportion 87.4 % at beginning and increases during the plate cooling (on average 88.3 %). The radiate heat transfer rate the proportion 12.6 % at beginning and decreases during the cooling (on average 11.7 %).

The cooling of both plates have expected course.

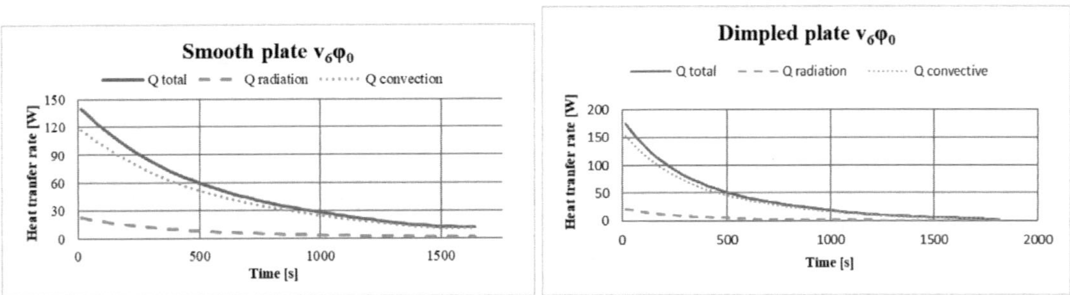

Figure 9. Heat transfer rates for smooth and dimpled plates for the forced convection case v = 6 m/s.

6.3. Heat transfer coefficients

Heat transfer coefficients computed from Equations (8) depending on temperatures are shown in Figures 10 and 11. The measured results were approximated by a linear trend line. In case of heat transfer coefficients with the incidence angle $\varphi = 0°$, trend lines of smooth and dimpled plates are similar in increasing ways. Trend lines of heat transfer coefficients of dimpled plate with the incidence angle $\varphi = 30°$ in comparison to lines of $\varphi = 0°$ have the inverse inclination.

For the natural convection case, the convective heat transfer coefficient of smooth plate has higher value then convective coefficient of dimpled plate. In comparison to forced convection with v = 3 m/s and v = 6 m/s, convective coefficients of smooth plate have lower value than dimpled plate. It verifies the importance of radiation heat flux during the cooling process with natural convection.

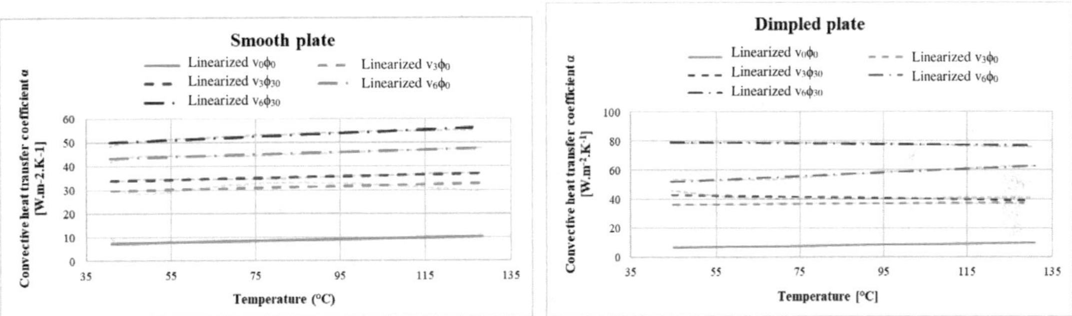

Figure 10. Convective heat transfer coefficient of smooth and dimpled plates

In the total heat transfer coefficients cases, only the coefficients of velocity $\varphi = 0°$ measurements are shown in Figure 11. Total heat transfer coefficients have higher values in comparison to convective coefficients because they are computed from total heat transfer including the radiation component (see Equation 8).

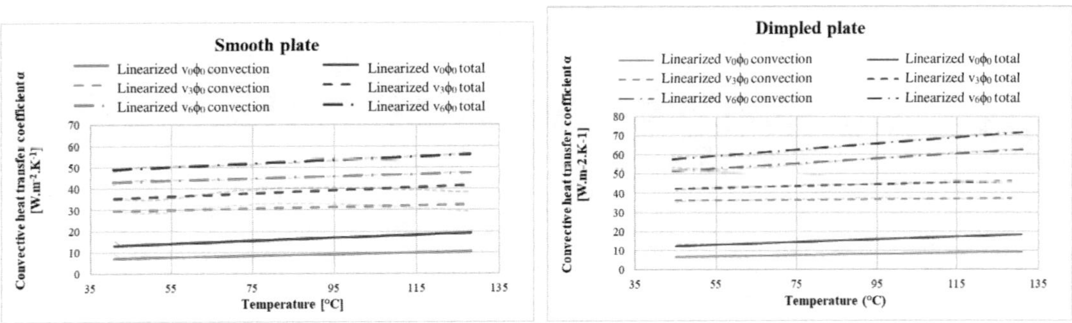

Figure 11. Convective and total heat transfer coefficient of smooth and dimpled plates

7. Conclusion

The presented results show that the heat transfer measurement by using a thermal imagining camera is a suitable method for a cooling process evaluation.

The cooling process measurement of the smooth and the dimpled plate proved that the dimpled plate has steeper cooling curves in all measured cases. So the dimpled plate cooling process is faster, especially in higher temperatures. The analysis of measured data pointed out that the radiant component of the heat transfer during the natural convection cooling process occupies a significant part of the total heat transfer rate. Neglecting the radiant component in a heat transfer calculation could bring out a large deviation. In case of natural convection, radiant heat flux cause faster cooling process for dimpled plate even if total and convection heat transfer have higher value. The dimples on the surface cause increased heat transfer rate in lower temperatures with inclined airflow.

The dimpled surface impact on the heat transfer process will continue by using CFD simulation. A special interest will be done to study and verify of the radiant flux during a natural convection.

Acknowledgments

The paper has been prepared thanks to the support of the project "VAROPS (DZRO FVT 3) Military autonomous and robotic systems" and Surface technology in applications special techniques SV20-216.

References

[1] Spálenský V and Rozehnal D 2017 CFD simulation of dimpled sphere and its wind tunnel verification *MATEC Web of Conferences*. **107** 00077.

[2] Tay C, Khoo B and Chew Y 2015 Mechanics of drag reduction by shallow dimples in channel flow *Physics of Fluids* **27** 035109.

[3] Ashif P and Rakesh K 2019 Heat transfer performance investigation of the spherical dimple heat sink and inclined teardrop dimple heat sink *Numerical Heat Transfer, Part A: Applications*, **76** 73-86.

[4] Horák V and Kulish V 2011 *Thermodynamics* (Brno: University of Defence)

[5] User's manual: Flir T6xx series [online]. Wilsonville USA: Flir Systems, Inc. [cited 7.7.2022]. Available from https://www.globaltestsupply.com/product/flir-t640-infrared-camera

Analysis of gas thermal conductivity at low pressures using a mathematical-physical model

P Šabacká[1]

1 Department of Electrical and Electronic Technology, Brno University of Technology, Technická 10, 616 00 Brno, Czech Republic

E-mail: xhlava44@vutbr.cz

Abstract. As part of the research in the field of pumping vacuum chambers in the Environmental Electron Microscope, research on supersonic flow through apertures is being carried out at the Department of Electrical and Electronic Technology of the Brno University of Technology in cooperation with the Institute of Scientific Instruments of the CAS. This paper deals with the effect of low pressures on the thermal conductivity of gases that occurs when observing samples in the Environmental Electron Microscope.

1. Introduction

At the Department of Electrical and Electronic Technology of Brno University of Technology, in cooperation with ISI Brno, research on supersonic flow at low pressures is underway [1, 2, 3].

Due to the presence of a gas pressure in the order of units to thousands of Pa in the sample chamber, the environmental scanning electron microscope (ESEM) allows to observe electrically non-conductive or semiconductive samples without charge artifacts (Danilatos), or to observe sensitive biological [4] and polymeric [5] samples in their native state without any damage with special methods [6], or to study these samples in dynamic in-situ experiments [5]. Signal electrons are detected in the ESEM using special ionization [7, 8] or scintillation detectors [9, 10, 11], whose images can be correlated with light microscope images [12].

The results are used in the design of vacuum chambers of the Environmental Electron Microscope [13, 14]. In planned future research, the shape and distribution of shock waves during the formation of a supersonic flow behind a small aperture separating two chambers with different pressures will be investigated [15, 16].

2. Physics of low prsessures

Like diffusion or viscosity, heat conduction in gases can be explained using the concepts of kinetic theory [17].

In agreement with equation (1), the heat flux density q (W m^{-2}) can be assumed to be proportional to the temperature gradient, since temperature is a measure of the internal energy of ideal gases:

$$q = \frac{Q}{S\mathrm{d}t} = -\lambda \frac{\mathrm{d}T}{\mathrm{d}x} \tag{1}$$

In this equation, Q (J m^{-2} s^{-1}) is the heat transferred through a cross-section S (mm), perpendicular to the flow axis, in time dt. The variable λ (W m^{-1} K^{-1}) is the heat conductivity coefficient.
The heat flux density can be determined similarly to the momentum flux density:

$$q = \frac{(n_{+}+n_{-})}{S\mathrm{d}t}\bar{l}\frac{\mathrm{d}u}{\mathrm{d}x} = \bar{l}\frac{(n_{+}+n_{-})}{S\mathrm{d}t}\frac{\mathrm{d}u}{\mathrm{d}t}\frac{\mathrm{d}T}{\mathrm{d}x} \tag{2}$$

Where n_{+}/n_{-} indicates the number of molecules passing through the cross-section of the content S (mm) in the positive or negative direction of the coordinate system in time dt. The variable u (U m^{-1}) in equation (2) denotes the mean value of energy per molecule and can be expressed as:

$$u = \frac{U_m}{N_A} \tag{3}$$

Where U_{m} (J mol^{-1}) represents the molar internal energy and N_{A} (mol^{-1}) the Avogadro's constant.
The mean energy of a single molecule can also be expressed using the specific heat capacity c_{v} (J kg^{-1} K^{-1}) and the mass m_0 (g mol^{-1}) of the molecule:

$$u = m_0 c_v T \tag{4}$$

This statement implies:

$$\frac{\mathrm{d}u}{\mathrm{d}T} = m_0 c_v \tag{5}$$

and we can rewrite relation (2) in the form:

$$q = -\frac{1}{2}N_V m_0 \bar{v}\bar{l}c_v\frac{\mathrm{d}T}{\mathrm{d}x} \tag{6}$$

where N_{V} (mol m^{-3}) is the density of molecules. Comparing with (1), we find that:

$$\lambda = \frac{1}{2}\rho\bar{v}\bar{l}c_v = \eta c_v \tag{7}$$

Since specific heat capacity is not a function of gas pressure, the dependence of the thermal conductivity coefficient on pressure and temperature is the same as for dynamic viscosity. The coefficient is proportional to the square root of the temperature and independent of the gas pressure. However, the pressure independence is only valid if the pressure is not too low, as is the case in electron microscopy. At very low gas pressures, thermal conductivity, like viscosity, becomes pressure dependent. As the pressure decreases, the thermal conductivity coefficient decreases.

The reason for this phenomenon is the size of the mean free path of the molecules. If the gas pressure in the container gradually decreases, the mean free path of the molecules increases $\bar{l}$. If the mean free path is much smaller than the size of the space, the molecules make a disordered motion accompanied by a large number of collisions and transfer kinetic energy from the higher temperature places in the gas to the colder places. However, once the mean free path of the gas molecules becomes comparable to or even larger than the dimension of space, the heat conduction mechanism changes. In this case, the molecules move between the walls of space with virtually no collisions between them, transferring energy from the warmer wall to the colder wall. In this case, the total energy transferred from one wall to the other depends mainly on the density of molecules N_V in the space. However, the density of molecules N_V is directly proportional to the pressure p of the gas at constant temperature. It follows that if we reduce the pressure of the gas from the value at which the mean free path of the molecule approximately corresponds to the dimension of the space, the thermal conductivity coefficient λ decreases.

The direct proportionality between the thermal conductivity coefficient λ (W m $^{-1}$ K $^{-1}$) and the pressure p (Pa) of the gas in the low-pressure region, where instead of mutual collisions of gas molecules there are mainly collisions of gas molecules with wall molecules, can be derived from equation (7) if we put in it instead of the mean free path $\bar{l}$ (m) of molecules the dimension of space d (m). If we express the density ρ (g m^{-3}):

$$\rho = \frac{pm_0}{kT} \tag{8}$$

and using the relationship for mean speed:

$$\bar{v} = \sqrt{\frac{8kT}{\pi m_0}} \tag{9}$$

we get a relationship after modification:

$$\lambda = \frac{8}{3\pi}\frac{c_v}{v}pd \tag{10}$$

The relationship shows that the thermal conductivity coefficient is directly proportional to the pressure p (Pa) of the gas if the mean free path of the molecules is equal to or greater than the dimension of the space between which the heat is conducted through the gas. At the same time, it follows from (10) that the thermal conductivity coefficient depends not only on the properties of the gas itself, but also on the dimension of the internal space d (m).

The mean free path of the molecules is therefore inversely proportional to the gas pressure. If we reduce the pressure of the gas in a confined space from atmospheric pressure, the moment the pressure falls below the value at which the mean free path begins to exceed the system dimension d (m), significant changes in the gas properties occur. If the pressure drops so that the so-called high vacuum condition is satisfied:

$$\bar{l} \gg d, \tag{11}$$

molecules and charged particles will move without colliding with each other. Only at the walls of the system there will be so-called external collisions. This will affect the transport processes. The thermal conductivity coefficient of gases does not depend on the pressure until equation (11) becomes valid. The same applies to viscosity. The practical consequence is the thermal insulating properties of the vacuum and the free, unhindered motion of the bodies in the vacuum system by the gas molecules. The gas flow under condition (11), which is a molecular gas flow, has quite different properties from the flow at atmospheric pressure, where it is determined by the mutual friction of the individual gas layers of thickness corresponding to the mean path $\bar{l}$.

The course of the thermal conductivity of air as a function of gas pressure can then be expressed by the following graph (Figure 1).

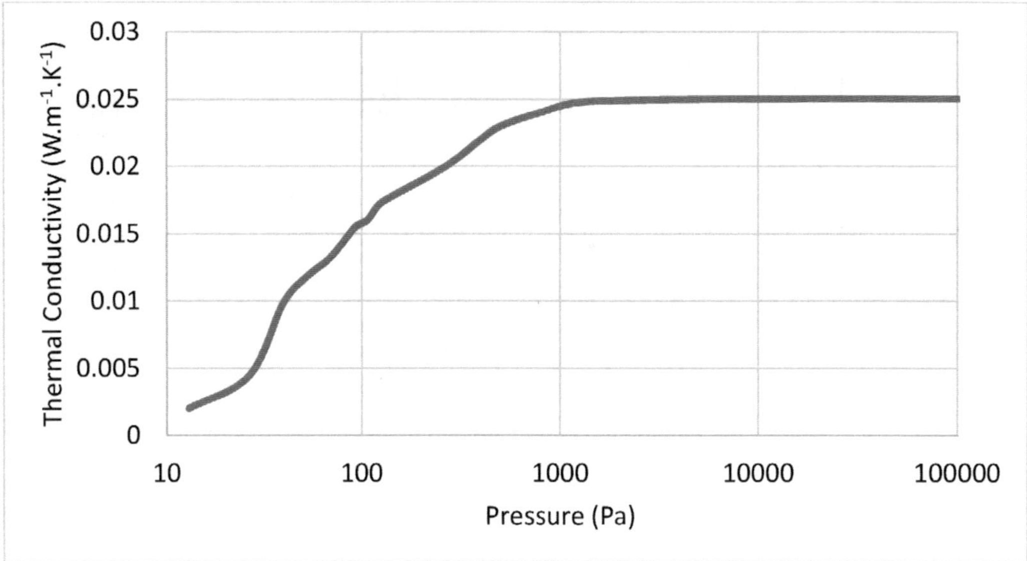

Figure 1. Dependence of thermal conductivity on pressure.

3. Experimental chamber

At the Institute of Scientific Instruments of the Academy of Sciences of the Czech Republic in cooperation with Department of Electrical and Electronic Engineering of the Brno University of Technology, an experimental chamber has been developed for research of supersonic flow at low pressures (Fig. 2) [18, 19]. It consists of two chambers separated by a small aperture in the form of a Laval nozzle, on which supersonic flow is generated beyond the critical diameter. However, for gas thermal conductivity studies, this aperture is closed and only one of the chambers is used, which is pumped to the planned operating pressure and a sample is placed in its centre of mass.

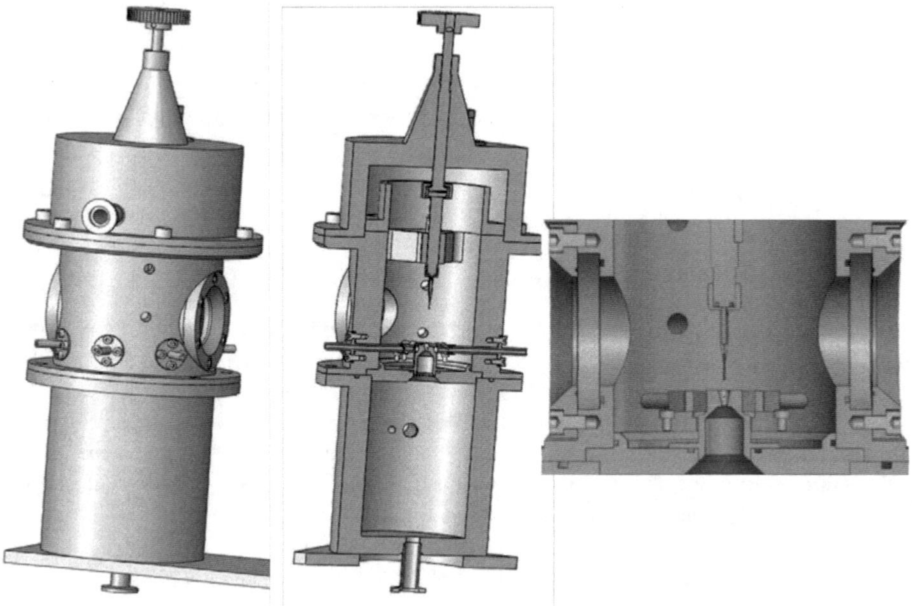

Figure 2. Experimental chamber.

4. Mathematical-physical model

For mathematical-physical analyses to map the expected thermal conductivity conditions in the experimental chamber for the planned experiments, a mathematical-physical model of the chamber was created in the form of a simplified 2D volumetric model. The chamber in the configuration for measuring the thermal conductivity of gases is cylindrical with a base diameter of 80 mm and a height of 100 mm. A sample, also cylindrical in shape, with a base diameter of 2 mm and a height of 30 mm, is inserted into the chamber and placed exactly in the centre of the chamber (Fig. 3). In the Fig. 3 the Boundary Conditions are marked, where orange color points to rotation axis of 2D volume model, green color Input, blue color Output, where the pressure values are set according to examined variant according to Table 1. Purple color points to experimental chamber body with the heat transfer values from the chamber to the surroundings. This transfer is setusing Heat transfer coefficient with value of 100 $W.m^{-2}.K^{-1}$ I thickness of the wall of 25 mm and Free Stream Temperature of 24°C.

Figure 3. Boundary Conditions.

The thermal conductivity of a gas is relatively independent of pressure when the pressure is close to 1 atm. The thermal conductivity becomes dependent on the magnitude of the pressure only at higher or, on the other hand, at very low pressures.

Three analyses were performed where the operating pressure in the chamber was set to the values shown in Table 1,

Table 1. Boundary conditions.

P_A (Pa)	P_{EU} (Pa)	P_{EL} (Pa)
101 325	5 000	267

Where P_A is the atmospheric pressure, P_{EU} is typical upper pressure value in the sample chamber in ESEM and P_{EL} is typical lower pressure value in the sample chamber in ESEM.

The sample was heated to 100 °C and a time-varying mathematical and physical analysis of the cooling of the sample under the given pressure conditions was performed.

5. Results and discussions

In these simulations, the thermal conductivity of an aluminum sample placed in a chamber with atmospheric pressure was compared with a chamber depleted to a pressure of 5000 Pa, which is the upper limit of the working pressure in the Environmental Electron Microscope for observing samples, and a pressure of 267 Pa, which is in the contrary the lower limit. The results are shown in Fig. 4.

Journal of Physics: Conference Series **2382** (2022) 012021 doi:10.1088/1742-6596/2382/1/012021

The results show that at pressures above 5000 Pa, there is only a slight decrease in the thermal conductivity of the gas in the chamber compared to the conductivity at atmospheric pressure, but at pressures in the lower range of the ESEM operating conditions, a significant decrease should already be expected.

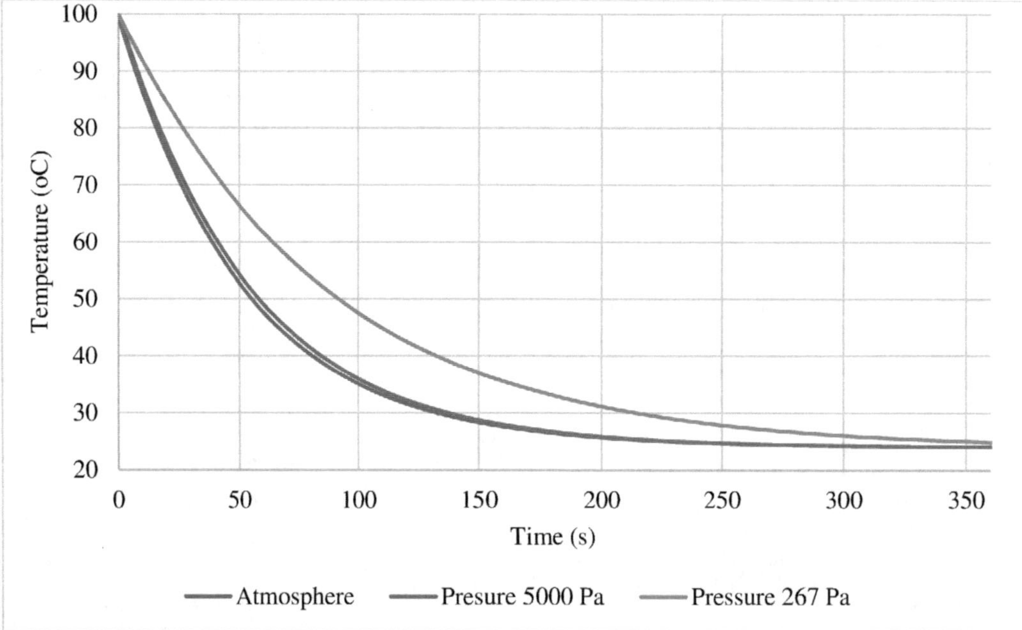

Figure 4. Dependence of thermal conductivity over time at specified pressure levels.

Figure 5 shows the temperature distribution in the sample and the ambient gas at a cooling time of 100 s. A slight difference between the Atmospheric Pressure variant and the 5000 Pa pressure can be seen, which according to Figure 1 is already slightly in the region where the thermal conductivity drops. However, there is already a significant difference between the Atmospheric pressure or 5000 Pa pressure and the 367 Pa pressure, where a much slower heat transfer is evident. In the figures it can be seen that the hotter gas rises upwards in the X-axis.

The results are the first basis for upcoming experiments in the experimental chamber, and the overall results will inform experiments looking at moist samples heated or cooled by a Peltier cell in the ESEM.

The results of given simulations, which will be followed by further studies, are important basis for mapping temperature condition on cooled or heated samples in the Environmental electron microscope. The quality of the resulting image obtained from the microscope is determined, among others, by the environment in which the sample is placed. Due to hydration and increased pressure up to 2000 Pa, the specimen chamber is adapted so a sample containing water dries out more slowly and lasts in its natural state for the duration of its monitoring, but the time is limited. One of the factors influencing the lifetime of the sample and also the quality of results in whether the sample holds the required temperature at which it is to be monitored during the pumping periods of the chamber and the monitoring time at working pressure. The problem is that the monitored surface of the sample is very susceptible to heat exchange with the environment, and yet the temperature conditions are supplied from the Peltier element on the specimen table – that is, through its entire width. Thus, the thermal conductivity of the surrounding gas has great impact on this problem. The true is, that at a lower working pressure, the sample can dry out faster and the observation time is shorter, but it is possible to obtain a sharper image, and due to the low thermal conductivity, it is possible to keep the sample eve on its surface at the desire working temperature and heat is not removed or supplied from its surface so quickly. Given mathematical – physics analyzes will help to find desire compromise in between the variables: thermal conductivity of surrounding gas, the working pressure in the specimen chamber, which oppositely affects the rate of drying out and scattering of the primary electron beam.

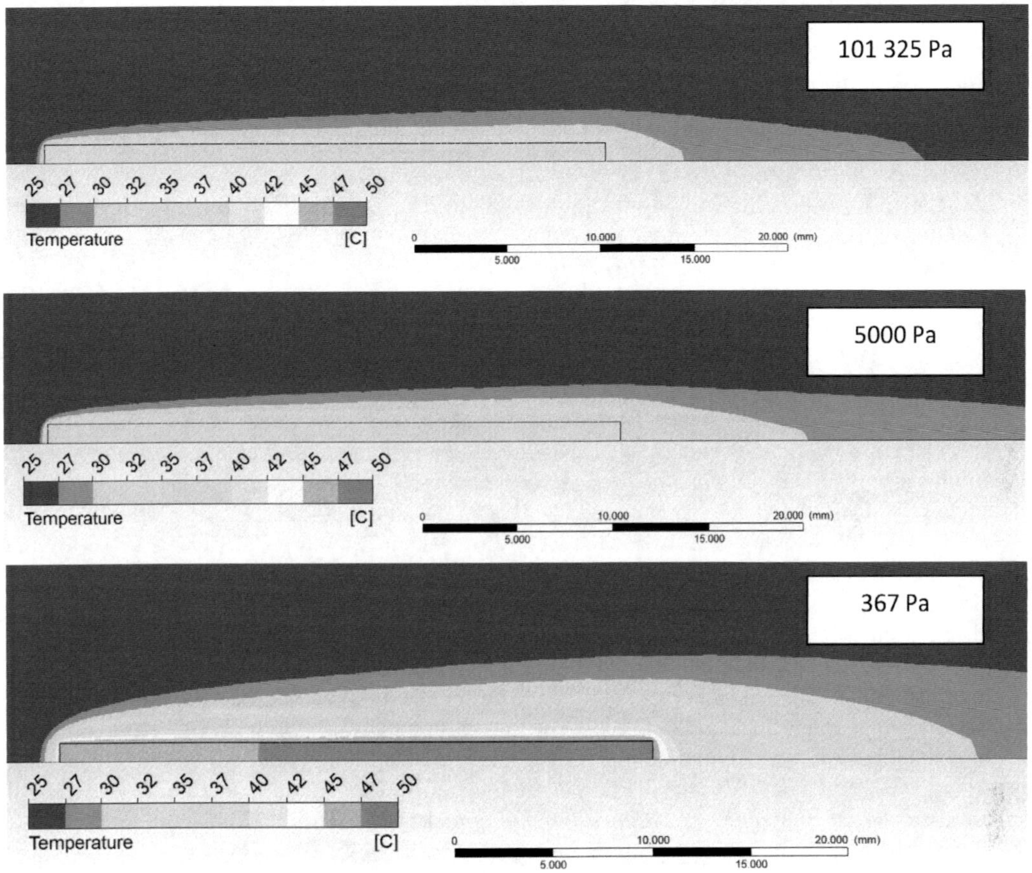

Figure 5. Comparison of results in 100 s.

Acknowledgments

This work was supported by the project of specific research at Brno University of Technology: Materials and Technologies for Electrical Engineering IV, reg. no. FEKT-S-20-6206.

6. References

[1] Maxa J, Neděla V, Jirák J, Vyroubal P and Hladká K 2012 Analysis of gas flow in a secondary electron scintillation detector for ESEM with a new system of pressure limiting apertures *Advances in Military Technology* **7 (2)** 111-116

[2] Maxa J, Bílek M, Hlavatá P, Vyroubal P and Lepltová K 2016 Comparisons Using Methods of Continuum Mechanics and Monte Carlo at Differentially Pumped Chamber *Advances in Military Technology* **11 (2)** 143-150

[3] Bílek M, Maxa J, Hlavatá P and Bayer R 2017 Modelling and simulation of a velocity field within supersonic flows in low-pressure areas *ECS Transactions* **81(1)** 311–316

[4] Neděla V, Hřib J, Havel L, Hudec J and Runštuk J 2016 Imaging of Norway spruce early somatic embryos with the ESEM, Cryo-SEM and laser scanning microscope *Micron* **84(May)** 67-71

[5] Neděla V, Tihlaříková E, Maxa J, Imrichová K, Bučko M and Gemeiner P 2020 Simulation-based optimisation of thermodynamic conditions in the ESEM for dynamical in-situ study of spherical polyelectrolyte complex particles in their native state *Ultramicroscopy* **211**

[6] Tihlaříková E, Neděla V, Dordevic B 2019 In-situ preparation of plant samples in ESEM for energy dispersive x-ray microanalysis and repetitive observation in SEM and ESEM *Scientific Reports* **9(FEB)** 2300
[7] Neděla V, Konvalina I, Lencová B and Zlámal J 2011 Comparison of calculated, simulated and measured signal amplification in variable pressure SEM *Nuclear Instruments & Methods in Physics Research Section A* **645 (1)** 79-83
[8] Neděla V, Konvalina I, Oral M and Hudec J 2015 The Simulation of Energy Distribution of Electrons Detected by Segmental Ionization Detector in High Pressure Conditions of ESEM. *Microscopy and Microanalysis* **21(4)** 264-269
[9] Jirák J, Neděla V, Černoch P, Čudek P and Runštuk J 2010 Scintillation SE detector for variable pressure scanning electron microscopes *Journal of Microscopy* **239 (3)** 233-238
[10] Neděla V, Tihlaříková E, Runštuk J and Hudec J 2018 High-efficiency detector of secondary and backscattered electrons for low-dose imaging in the ESEM *Ultramicroscopy* **184A(JAN)** 1-11
[11] Vyroubal P, Maxa J, Neděla V, Jirák J and Hladká K 2013 Apertures with Laval Nozzle and Circular Orifice in Secondary Electron Detector for Environmental Scanning Electron Microscope *Advances in Military Technology* **8 (1)** 59-69
[12] Stelate A, Tihlaříková E, Schwarzerová K, Neděla V and Petrášek J 2021 Correlative Light-Environmental Scanning Electron Microscopy of Plasma Membrane Efflux Carriers of Plant Hormone Auxin *Biomolecules* **11(10)** 1407
[13] Hlavatá P, Maxa J, Bílek M and Neděla V 2017 Impact of the shape of the differentially pumped chamber on critical flow character *ECS Transactions* **81(1)** 317–322
[14] Hlavatá P, Maxa J, Bílek M, Vyroubal P and Bayer R 2017 Influence of critical flow in the differentially pumped chamber AQUASEM *Advances in Military Technology* **12(2)** 301–310
[15] Maxová A, Maxa J and Šabacká P 2020 The impact of pumping velocity on temperature running in chambers of experimental system *ECS Transactions* **99(1)** 317–323
[16] Maxa J, Hlavatá P and Vyroubal P 2019 Analysis of impact of conic aperture in differentially pumped chamber *Advances in Military Technology* **14(1)** 151–161
[17] Charles University *Scripts for Physics* (source: https://physics.mff.cuni.cz/kfpp/skripta/kurz_fyziky_pro_DS/display.php/molekul/6_2)
[18] Hlavatá P, Maxa J and Vyroubal P 2018 Analysis of pitot tube static probe angle in the experimental chamber conditions *ECS Transactions* **87(1)** 369–375
[19] Maxa J, Hlavatá P and Vyroubal P 2018 Using the ideal and real gas model for the mathematical - Physics analysis of the experimental chambre *ECS Transactions* **87(1)** 377–387

Journal of Physics: Conference Series **2382** (2022) 012022 doi:10.1088/1742-6596/2382/1/012022

Mathematical and physical analysis of air compression waste heat dissipation used as energy storage

J Maxa[1]

[1] Department of Electrical and Electronic Technology, Brno University of Technology, Technická 10, 616 00 Brno, Czech Republic

E-mail: maxa@vutbr.cz

Abstract. As part of the research in the field of pumping vacuum chambers in the Environmental Electron Microscope, research on supersonic flow through apertures is being carried out at the Department of Electrical and Electronic Technology of the Brno University of Technology in cooperation with the Institute of Scientific Instruments of the CAS. This research, which deals with gas compression and expansion, utilizes this experience in the paper that deals with the issue of gas heating during compression when filling a storage tank as an energy storage device. This heating of the gas is a loss relative to the stored energy and there is an interest in limiting this heating or using this heat in other ways. This paper deals with limiting the heating of compressed air when filling a storage tank as energy storage.

1. Introduction

At the Department of Electrical and Electronic Technology of Brno University of Technology, in cooperation with ISI Brno, research on supersonic flow at low pressures is underway [1, 2, 3, 4, 5]. The results are used in the design of vacuum chambers of the Environmental Electron Microscope [6, 7, 8, 9, 10]. The research on vacuum chamber pumping on the experimental chamber allows experiments on gas compression in the container and heat transfer during compression [11, 12, 13,14].

There are many ways to compare energy storage. The most common comparison is by system efficiency. Figure 1 shows the relation of efficiency to the lifetime of energy storage technologies [15].

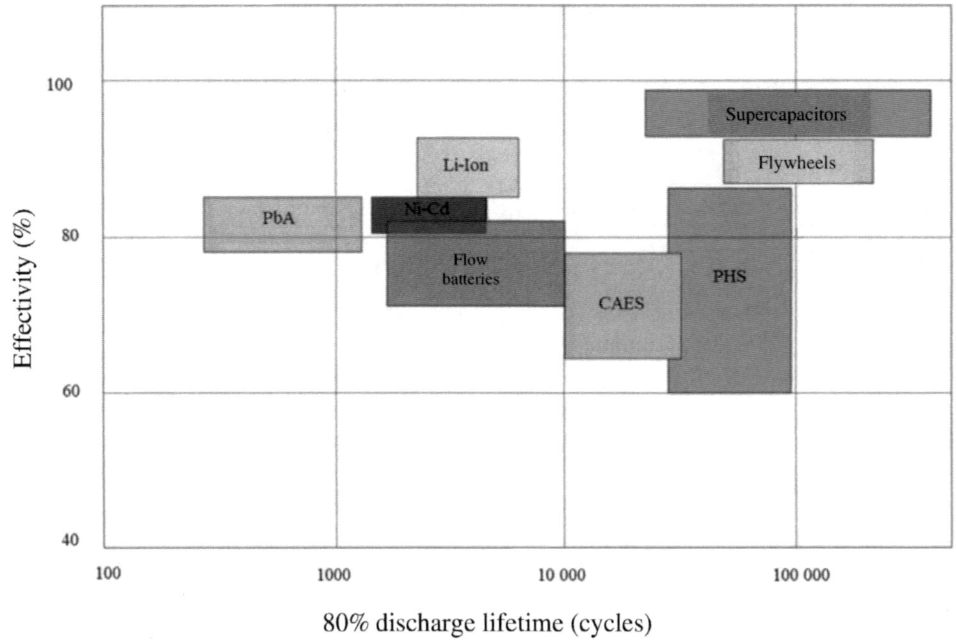

Figure 1. Relationship of efficiency to lifetime for different types of energy storage.

where PbA stands for Lead-acid accumulators, Ni-Cd stands for Ni-Cd accumulators, Li-Ion stands for Li-Ion accumulators, CAES stands for Compressed Air Energy Storage and PHS stands for Pumped Hydro Storage.

However, the disadvantage is that the described systems only store energy but do not produce it. Their price must therefore be covered by the difference between the price of the energy taken from the grid and the price of the energy supplied to the grid. Or the difference between the price invested in energy storage and the price invested in production [16].

Therefore Charles Barnhart and Sally Benson from Stanford University made a comparison of different technologies based on the so-called ESOI, which stands for Energy Stored on Invested, the ratio of energy stored to energy invested in production. The ESOI value is calculated from the ratio of the number of cycles of a given technology, its efficiency and depth of discharge to the so-called grey energy, which is the energy required for production, see equation (1). The comparison is shown in Figure 2. [17] The higher the ESOI value the better, because it means that the stored energy greatly exceeds the energy required for production.

$$ESOI = \frac{Stored\ energy}{Grey\ energy} = \frac{\lambda \eta D}{\varepsilon_g} \tag{1}$$

Where λ is the number of cycles, η is the effectivity D is the depth of discharge (discharge to 80%) ε_g is the energy invested in production (grey energy).

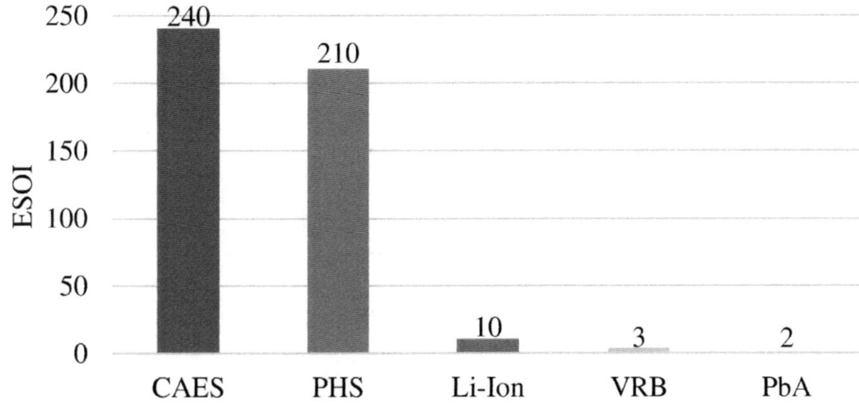

Figure 2. ESOI – the ratio of energy stored on energy invested in production.

where VRB stands for Vanadium redox flow batteries.

This quoted text was presented without any major discussion of the different types of energy storage and only shows the advantage of using compressed air as an energy storage device where it is appropriate, for example as a stationary storage device like a pumped storage plant. Therefore, it will not replace the battery system for mobile applications.

2. Mathematical and physical model

In this part of the research, the simulation of air pumping in ANSYS was carried out on a model example in which a bottle with the dimensions shown in Figure 3 was used as a reservoir instead of an underground storage tank. The dimensions were chosen according to a commonly used compressor tank.

This is a 2D model of a bottle where the dimension b is the radius of the base. ANSYS Fluent simulation software was used to simulate the process of air compression into the container and subsequent cooling.

The mathematical-physical analysis was carried out using Ansys Fluent, which applies the continuum method and uses the Stock Navier equations and the finite volume method to solve the problems. Since the following calculations do not involve large gradients, such as critical flow, the Pressure Based coupled calculation scheme with the second upwind discretization with the Advection Upstream Splitting Method (AUSM) was used in the calculation.

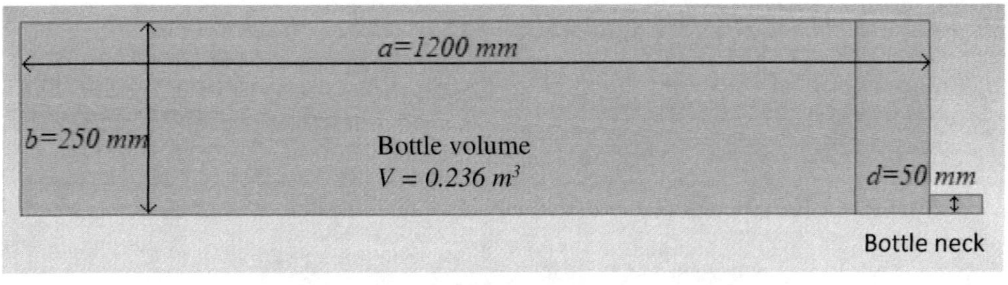

Figure 3. 2D axisymmetric model of the bottle.

Air from the compressor was pumped into the neck of the bottle with a mass flow rate inlet of $Q_m = 0.007$ kg/s. The value was taken from the compressors used. Furthermore, the heat transfer coefficient H = 25 W/m2·K [18] was set on the walls of the bottle. This value was chosen according to the heat transfer to the surroundings in a space with a still air without flow. The temperature of the air pumped into the bottle is the compressor outlet temperature and was set to 40 °C according to the given equations (2) [18].

$$T_2 = T_1 \left[1 + \frac{\left(\frac{P_2}{P_1}\right)^{\left(\frac{k-1}{k}\right)} - 1}{\eta} \right] \tag{2}$$

Where η is isentropic efficiency, P_1 is suction pressure, P_2 is discharge pressure, T_1 and T_2 are suction and discharge temperatures and k is Poisson's constant. The values were taken with respect to the gas - air used.

The pumping process was carried out for 209 s, increasing the pressure to 709 kPa and the temperature to 61.49 °C. The pumping was then stopped and the cylinder cooled freely until a time of 600 s, at which time it cooled to a temperature of 27,75 °C and the pressure dropped to 638 kPa.

This model example was used to investigate the behaviour of the air during pumping into the store, the heating of the air during pumping and the pressure drop during subsequent cooling.

Subsequently, several cases in which the air heating during pumping is minimised in different ways were designed and simulated. The bottle model was modified by adding rings (Figure 4). The rings, based on the technology used, reduced the heating of the air during bottle pressurization:

Variant A - Only embedded heat absorbing rings.
This was chosen because it is the least economically demanding type of solution

Variant B - Heat absorbing rings acting as a thermocouple
This option was used because of the low cost option of waste heat recovery.

Variant C - Refrigerator rings cooled to -10 °C for 72 s.
This variant - as well as Variant D - is more demanding due to the need of a pump for the heat exchanger. Variant C is highly pre-cooled once, but has no pumping power during pumping.

Variant D - Rings pre-cooled to 10 °C before pumping begins. They were no longer cooled during pumping.

This variant is chosen to compare the previous pre-cooled variant with the variant continuously cooled to a lower temperature. The two actively cooled variants are thus compared, with passive cooling and energy consumption using a thermocouple.

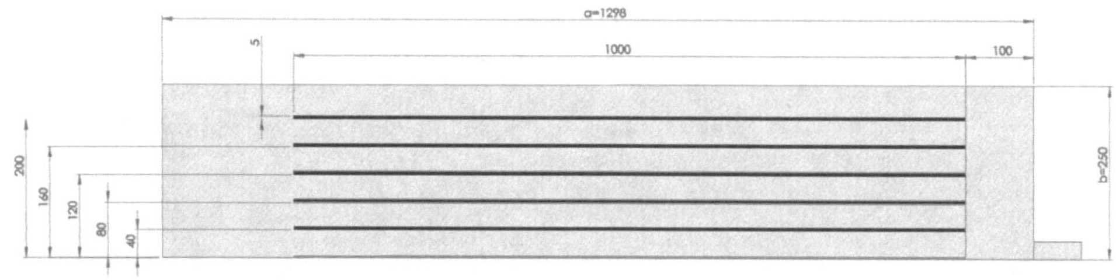

Figure 4. 2D axisymmetric model of a bottle with embedded cooling elements.

2.1. Variant A

In the first case, the rings served only as inserts without fulfilling the function of any of the technologies. The idea was therefore to see if just the added volume of material could limit the heating of the air during compression.

Pumping was terminated at a time of 223 s, after reaching a comparable maximum pressure value to the model example, namely 707 382 Pa. A comparison of the temperature and pressure evolution during pumping relative to the model example (pumping only) is shown in Figure 4.

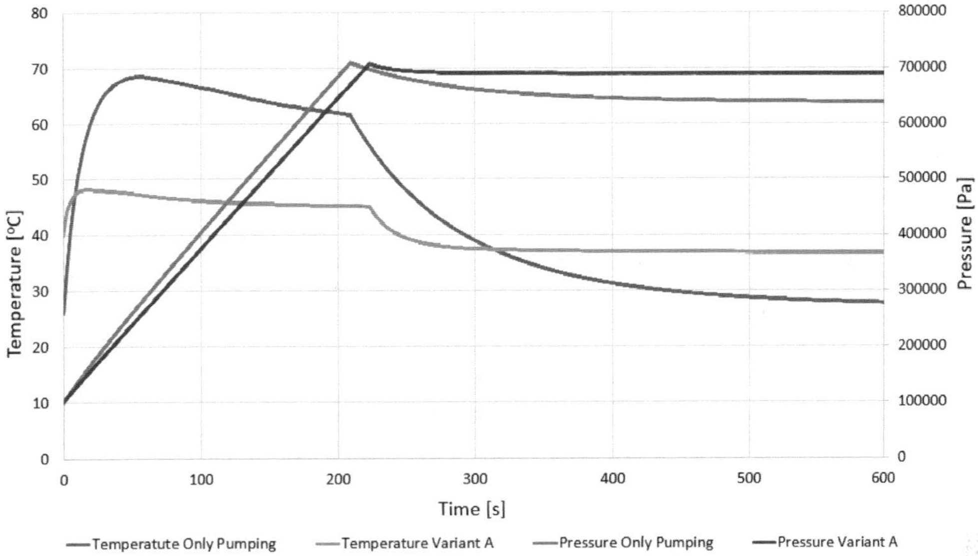

Figure 5. Temperature and pressure time dependence compared to the model example (Variant A).

2.2. Variant B

In another case, the rings were used as the warm end of a thermocouple. The cold end was the ambient air outside the bottle. The rings were set to negative power based on the temperature difference. The power value was obtained from the graph in Figure 4.

Pumping was terminated at 224 s, after reaching a comparable maximum pressure value to the model example of 710 041 Pa. A comparison of temperature and pressure during pumping relative to the model example is shown in Figure 6.

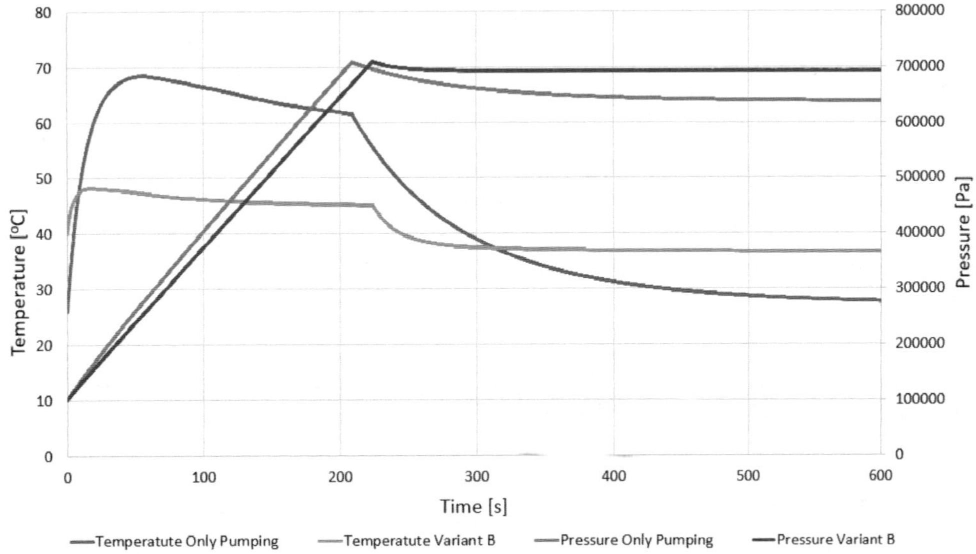

Figure 6. Temperature and pressure time dependence compared to the model example (Variant B).

2.3. Variant C

In this case, the rings were used as a cooling circuit.

They were cooled to -10 °C for 72 s. The pumping was stopped at 254 s, after reaching a comparable maximum pressure value to the model example of 709 540 Pa.

A comparison of the temperature and pressure evolution during pumping relative to the model example is shown in Figure 7.

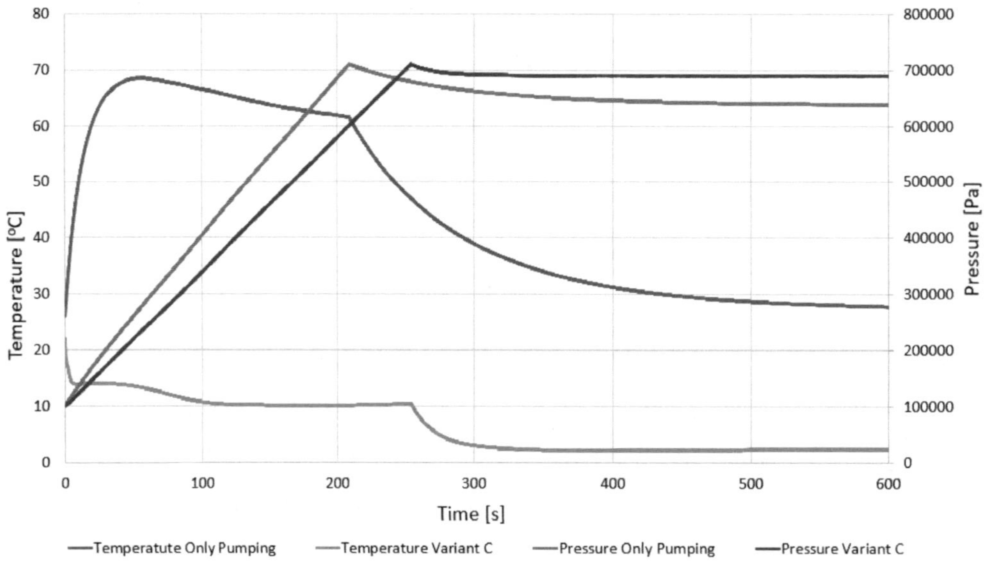

Figure 7. Temperature and pressure time dependence compared to the model example (Variant C).

2.4. Variant D

In the last case, the rings were also used as cooling elements, but they were pre-cooled to 10 °C before the pumping started and they were no longer cooled during pumping.

The pumping was terminated at 240 s, after reaching a comparable maximum pressure value to the model example of 710 560 Pa.

A comparison of temperature and pressure during pumping relative to the model example is shown in Figure 8.

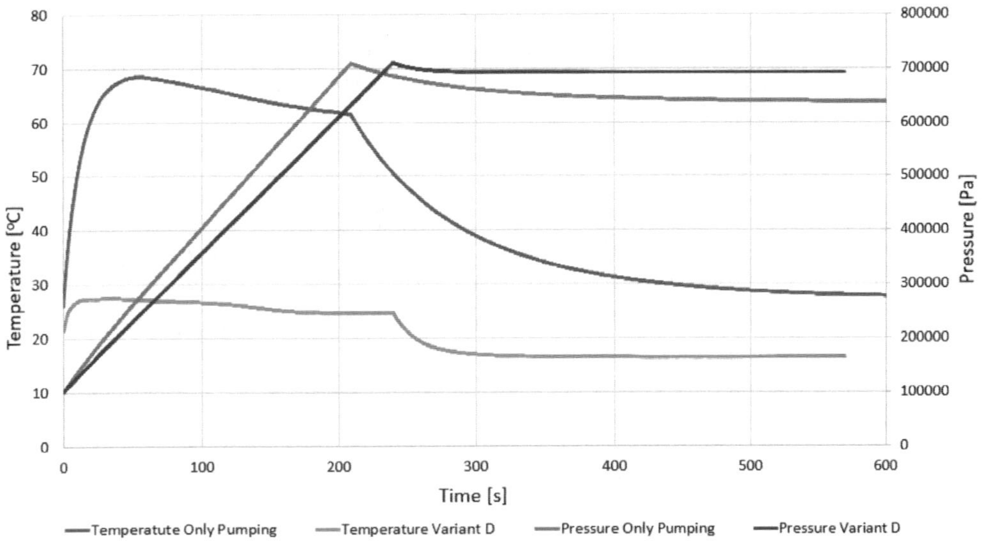

Figure 8. Temperature and pressure time dependence compared to the model example (Variant D).

3. Results and discussions

The results show that heat removal by any of the tested variants has an effect of reducing the pressure in the cylinder by up to 50 kPa (Figure 9). The rings themselves prevent rapid heating of the gas in the cylinder by their large surface area and mass, which removes the heat from the pressurized gas by convection and does not require any external power such as a refrigeration compressor. The thermocouple option demonstrates that there is a combination of the effect of the rings (added mass), and the energy extraction by the thermocouples, but only with a small difference. The option with cooled rings at -10 °C proves to be unnecessarily expensive. The pre-cooled to 10 °C option, which no longer requires external power to be supplied during pumping, shows better results than the previous options, due to the fact that it can keep the compressed gas at approximately room temperature. The pressure level of the compressed gas is not affected by temperature changes which cause it to expand further and ultimately shrink.

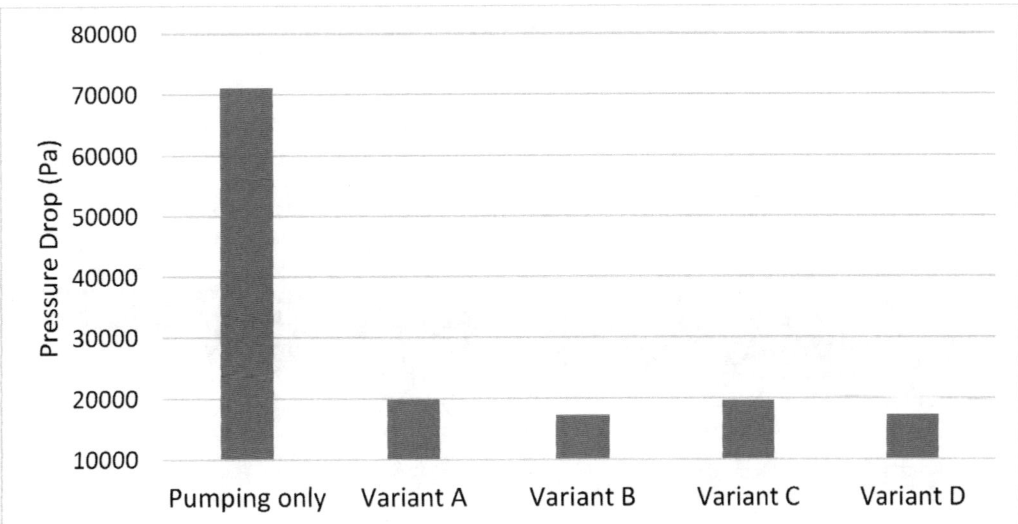

Figure 9. Comparison of pressure drop when using supplementary rings on the container.

3.1. Comparison of the temperature profiles from the simulation at the hottest moment

The following figures show the tank temperature for each variant at the time of maximum temperature, that is at 50 s. The Pumping Only variant is the hottest option at first look and is tens of degrees warmer than the other variants (Figure 10). The next result that differs is Variant C, which in contrast is significantly cooler than the other three approaches, where the profile is not so different (Figure 13).

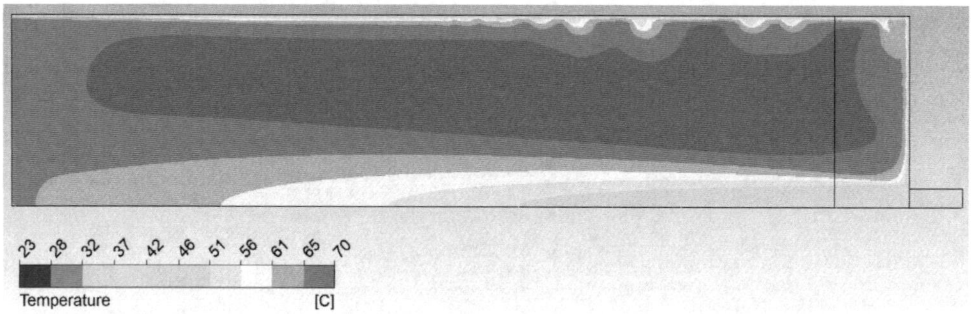

Figure 10. Pumping only.

Figure 11. Variant A.

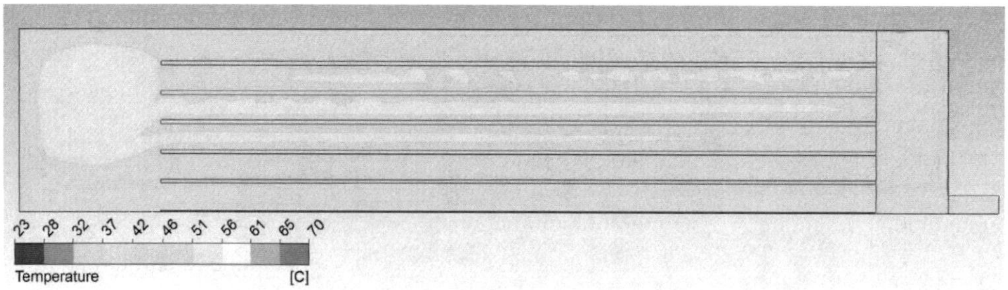

Figure 12. Variant B.

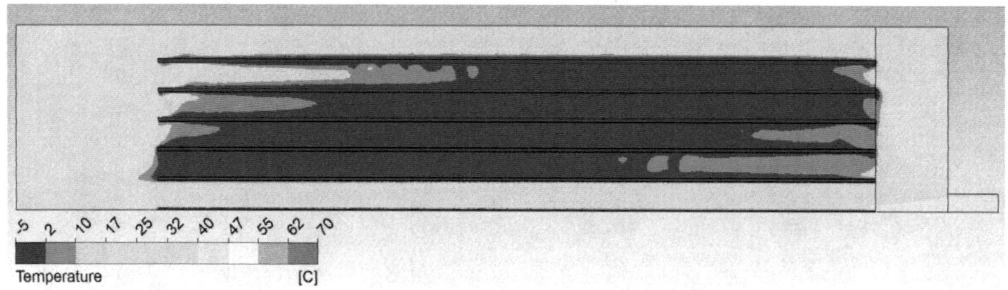

Figure 13. Variant C.

Figure 14. Variant D.

3.2. Comparison of temperature profiles from simulation at the time of end of pumping

The following figures (Figures 15 - 19) show the temperature in the tank at the time of shutting off the pumping, that is at the time of 209-250 s. In all cases, the temperature dropped at the time of shutting off the pumping or at the time of maximum pressure compared to the start of pumping, as seen in the previous comparison.

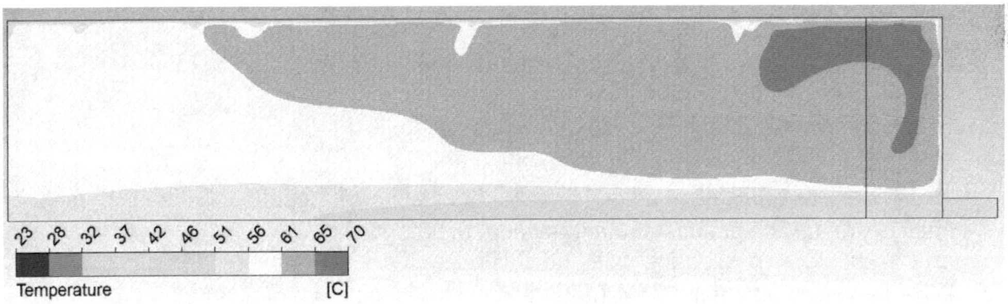

Figure 15. Pumping only.

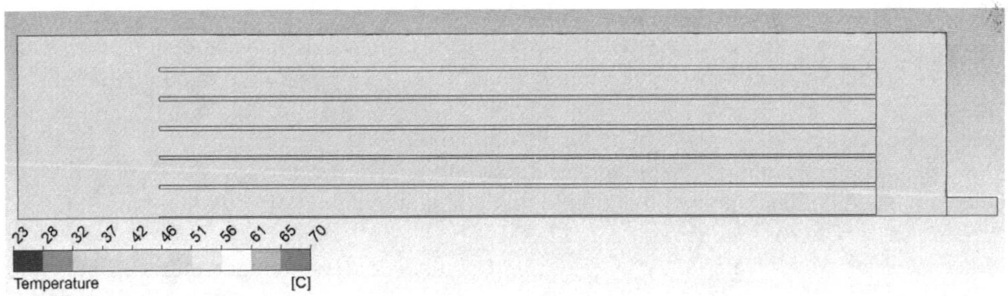

Figure 16. Variant A.

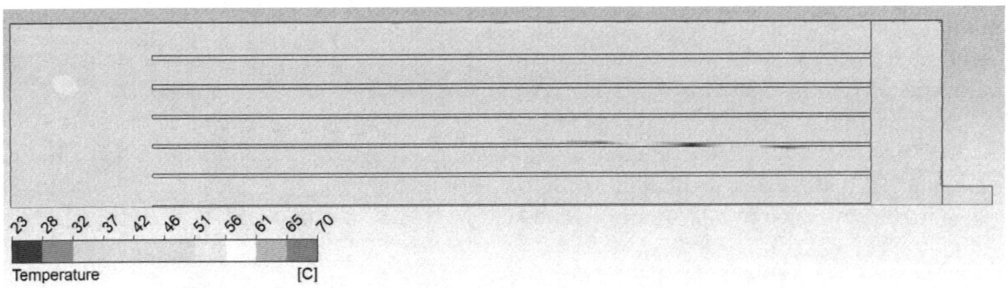

Figure 17. Variant B.

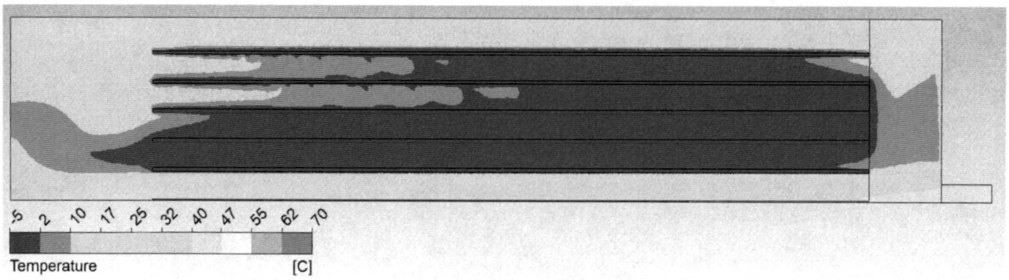

Figure 18. Variant C.

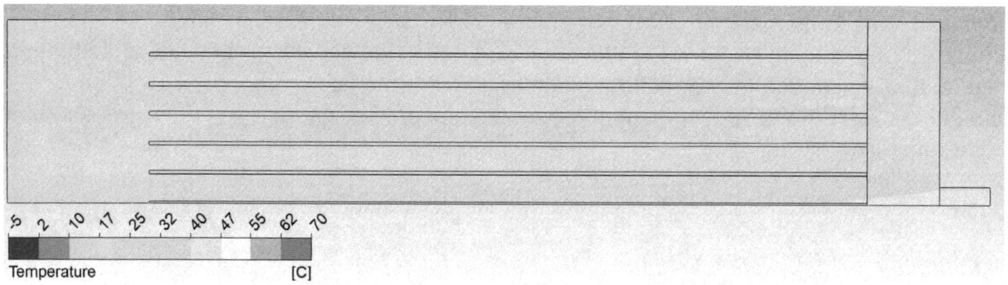

Figure 19. Variant D.

The analyses showed the suitability of the chosen options. It turns out that for smaller devices, the passive Option A or, in combination with thermocouples, Option B is sufficient. For larger devices, it is already worth to choose active cooling using an exchanger. Depending on the size of the storage tank and thus the compression time during the filling of the tank, the pre-cooling and the power of the heat exchanger are selected.

Acknowledgments

This work was supported by the project of specific research at Brno University of Technology: Materials and Technologies for Electrical Engineering IV, reg. no. FEKT-S-20-6206.

4. References

[1] Maxa J, Neděla V, Jirák J, Vyroubal P and Hladká K 2012 Analysis of gas flow in a secondary electron scintillation detector for ESEM with a new system of pressure limiting apertures *Advances in Military Technology* **7 (2)** 111-116

[2] Maxa J, Bílek M, Hlavatá P, Vyroubal P and Lepltová K 2016 Comparisons Using Methods of Continuum Mechanics and Monte Carlo at Differentially Pumped Chamber *Advances in Military Technology* **11 (2)** 143-150

[3] Bílek M, Maxa J, Hlavatá P and Bayer R 2017 Modelling and simulation of a velocity field within supersonic flows in low-pressure areas *ECS Transactions* **81(1)** 311–316

[4] Hlavatá P, Maxa J, Bílek M and Neděla V 2017 Impact of the shape of the differentially pumped chamber on critical flow character *ECS Transactions* **81(1)** 317–322

[5] Hlavatá P, Maxa J, Bílek M, Vyroubal P and Bayer R 2017 Influence of critical flow in the differentially pumped chamber AQUASEM *Advances in Military Technology* **12(2)** 301–310

[6] Tihlaříková E, Neděla V and Dordevic B 2019 In-situ preparation of plant samples in ESEM for energy dispersive x-ray microanalysis and repetitive observation in SEM and ESEM *Scientific Reports* **9(FEB)** 2300

[7] Neděla V, Hřib J, Havel L, Hudec J and Runštuk J 2016 Imaging of Norway spruce early somatic embryos with the ESEM, Cryo-SEM and laser scanning microscope *Micron* **84(May)** 67-71

[8] Vyroubal P, Maxa J, Neděla V, Jirák J and Hladká K 2013 Apertures with Laval Nozzle and Circular Orifice in Secondary Electron Detector for Environmental Scanning Electron Microscope *Advances in Military Technology* **8 (1)** 59-69

[9] Neděla V, Konvalina I, Lencová B and Zlámal J 2011 Comparison of calculated, simulated and measured signal amplification in variable pressure SEM *Nuclear Instruments & Methods in Physics Research Section A* **645 (1)** 79-83

[10] Neděla V, Konvalina I, Oral M and Hudec J 2015 The Simulation of Energy Distribution of Electrons Detected by Segmental Ionization Detector in High Pressure Conditions of ESEM. *Microscopy and Microanalysis* **21(4)** 264-269

[11] Jirák J, Neděla V, Černoch P, Čudek P and Runštuk J 2010 Scintillation SE detector for variable pressure scanning electron microscopes *Journal of Microscopy* **239 (3)** 233-238

[12] Neděla V, Tihlaříková E, Runštuk J and Hudec J 2018 High-efficiency detector of secondary and backscattered electrons for low-dose imaging in the ESEM *Ultramicroscopy* **184A(JAN)** 1-11

[13] Neděla V, Tihlaříková E, Maxa J, Imrichová K, Bučko M and Gemeiner P 2020 Simulation-based optimisation of thermodynamic conditions in the ESEM for dynamical in-situ study of spherical polyelectrolyte complex particles in their native state *Ultramicroscopy* **211**

[14] Stelate A, Tihlaříková E, Schwarzerová K, Neděla V and Petrášek J 2021 Correlative Light-Environmental Scanning Electron Microscopy of Plasma Membrane Efflux Carriers of Plant Hormone Auxin *Biomolecules* **11(10)** 1407

[15] Libich J 2020 Superkondezátor vs. baterie: parametry a použití (source: https://oze.tzb-info.cz/akumulace-elektriny/21462-superkondezator-vs-baterie-parametry-a-pouziti)

[16] Lázňovský M 2013 Kam se solární elektřinou? Do stlačeného vzduchu nebo na kopec (source: https://www.idnes.cz/technet/veda/skladovani-energie-pro-sitove-sluzby.A130822_141253_veda_mla)

[17] Barnhart Ch J and Benson S M 2013 On the importance of reducing the energetic and material demands of electrical energy storage *Energy & Environmental Science* **6** 1083-1092

[18] Moshfeghian M 2011 Compressor Calculations: Rigorous Using Equation of State vs Shortcut Method (source: http://www.jmcampbell.com/tip-of-the-month/2011/11/compressor-calculations-rigorous-using-equation-of-state-vs-shortcut-method/)

Mathematical and physical analysis of the character and position of shock waves during pumping of vacuum chambers

R Bayer[1]

1 Department of Electrical and Electronic Technology, Brno University of Technology, Technická 10, 616 00 Brno, Czech Republic

E-mail: Robert.Bayer@vut.cz

Abstract. As part of the research in the field of pumping vacuum chambers in the Environmental Electron Microscope, research on supersonic flow through apertures is being carried out at the Department of Electrical and Electronic Technology of the Brno University of Technology in cooperation with the Institute of Scientific Instruments of the CAS. This paper deals with the character and location of shock waves during pumping of vacuum chambers of the experimental chamber as a basis for the forthcoming experiment of mapping the given shock waves using the Schlieren optical method.

1. Introduction

At the Department of Electrical and Electronic Technology of Brno University of Technology, in cooperation with ISI Brno, research on supersonic flow at low pressures is underway [1, 2, 3].

Due to the presence of gas pressures in the order of units to thousands of Pa in the sample chamber, the environmental scanning electron microscope (ESEM) allows to observe electrically non-conductive or semiconductive samples without charge artifacts (Danilatos), or to observe sensitive biological [4] and polymeric [5] samples without damaging them in their native state using special methods [6], or to study these samples in dynamic in-situ experiments [5]. Signal electrons are detected in the ESEM using special ionization [4, 7] or scintillation detectors [8, 9, 10], whose images can be correlated with light microscope images [11].

The results are used in the design of vacuum chambers of the Environmental Electron Microscope [12, 13, 14]. During the forthcoming research, the shape and distribution of shock waves will be studied when supersonic flow occurs behind a small aperture separating two chambers with different pressures.

The aim of the study will be to evaluate the state and behaviour of shock waves in dilute gas at the continuum boundary. Shock waves as a region of densified gas have a major influence on the passage of the primary electron beam and its scattering through a differentially pumped chamber. A given region of the condensed gas forms a scattering enhancing barrier. The study and subsequent experiment will demonstrate the behaviour of shock waves at the boundary of the continuum mechanics and will allow the subsequent modification of the shock wave conditions by nozzle shaping, in particular by limiting the perpendicular shock wave at the expense of oblique shock waves, which do not form such a strong barrier to the electron beam.

The problem is mainly addressed by Dr. Danilatos, who has published several papers dealing with gas flow under given differential pumping conditions.

In [15], the author analyses the character of the supersonic flow and the location of the Mach disk with respect to the open expansion space or the insertion of supersonic flow rectifiers.

Journal of Physics: Conference Series **2382** (2022) 012023 doi:10.1088/1742-6596/2382/1/012023

In publication [16], the author analyses the gas flow through the aperture as a function of its size and the controlled backpressure, using the example of the FEI electron microscope (now ThermoFisher). There he describes the evolution of densities and pressures when gas flows through a small aperture separating two chambers with a large pressure gradient.

In [17], the author discusses the effect of the aperture force on the electron beam losses during the passage of differential pumping in relation to the ability to better pump a given space to the required pressure. This is followed up in publication [18] with a study of the effect of nozzle angle on the problem.

2. Experimental chamber

The current Experimental Chamber consists of two chambers separated by a small opening that simulates the condition created by differential pumping in an electron microscope.

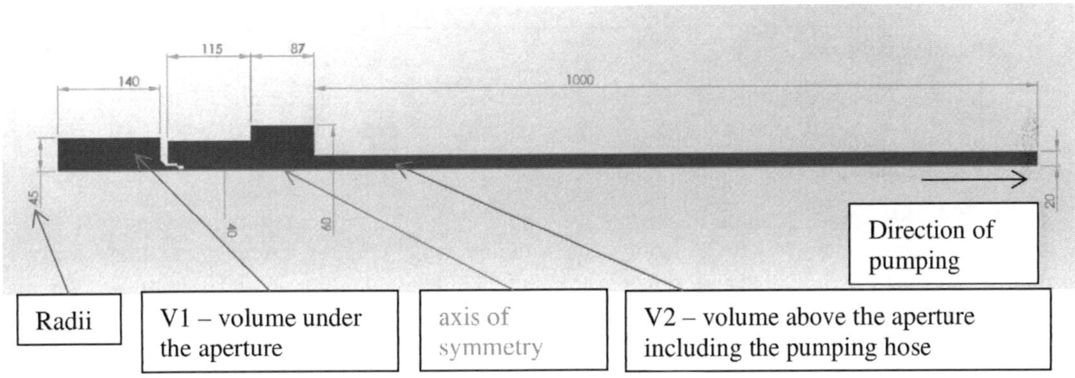

Figure 1. 2D axisymmetric model.

Figure 1 shows a schematic of the given chambers P1 and P2 separated by the nozzle as it will be used for the calculation in Ansys Fluent. This calculation was performed as 2D axisymmetric.

- Chamber volume $V1 = 2,813,743.82$ mm^3
- Chamber volume $V2 = 896,023.41$ mm^3
- Diameter of the aperture separating the two chambers $d1 = 2$ mm,
- Initial temperature $T_0 = 297.15$ K
- Initial pressure $P_0 = 101,325$ Pa
- Wall heat transfer convention = 25 W/m^{2}K
- The parameters of the rotary pump PFEIFFER DUO 65 M were used for pumping - Discharging velocity - $V_p = 62$ m^3/hr.

Time-varying mathematical and physical tasks were performed to determine the pressure ratios in both chambers during the actual pumping.

For the actual analysis, a 2D axisymmetric model of the pumping of the experimental chamber was created.

For the mathematical and physical analyses, Ansys Fluent system was used, which uses continuum mechanics and uses the Stock Navier equations for the calculations. These equations are valid in the region of Knudsen number up to 10^{-3} as seen in Figure 2.

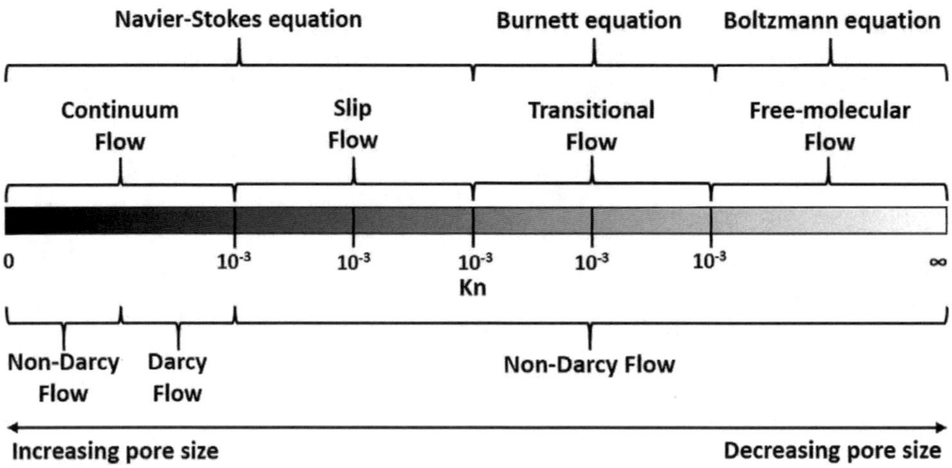

Figure 2. Classification of different flow regimes based on Knudsen number (Kn), its correlation with pore size in shale, and the equations used for solving these flow models [19].

Previous analyses have been performed confirming that in the simulations we are dealing with we are still in the region where the Stock Navier equations are valid [20, 21]. Comparative analyses have also been performed with published results of Dr. Danilatos, who used the Monte Carlo method for the case of differentially pumped chamber analysis, proving the agreement [3].

The analysis was solved using the mechanism of the continuum method using the Ansys Fluent system, which uses the finite volume method [3]. Due to the critical flow present in the computed region, a Density Based coupled calculation scheme with a second upwind discretization with the Advection Upstream Splitting Method (AUSM) [14, 2] was used in the calculation.

3. Results analysis

The results of Dr. Danilatos [15] were used to analyse the location of the Mach disk under the given low-pressure conditions:

$$z_M = 0.67 D_{kr} \sqrt{\frac{P_0}{P_1}} \quad (1)$$

Distance of the Mach disk is approximately 7.1 mm from the critical cross-section (see position of critical cross-section in Figure 3 and Mach disk locations in Table 1). Subsequent analysis of the ANSYS Fluent results showed the same Mach disk distance (Figure 3). The Mach disk is located in front of the perpendicular shock wave.

Table 1. Pressure ratios and Mach Disk locations.

Time (s)	Pressure in V1 (Pa)	Pressure in V2 (Pa)	P_v (Pa)	P_v/P_0 (-)	M_v (-)	Zm (mm)
1	49,496	1,782	1,083	0.022	3.14	9.1
2	27,210	939	639	0.023	3.11	8.8
3	15,878	561	410	0.026	3.03	8.3
4	9,538	342	273	0.029	2.96	7.1
5	5,950	210	190	0.032	2.89	7.5
6	3,433	121	127	0.037	2.8	7
7	1,843	65	83	0.045	2.67	6.3

The calculation of the ratios of the state variables on the shock wave is based on the physics of isentropic one-dimensional flow. This leads to relations that determine the relationship of pressures, densities, velocities and Mach numbers before and after the shock wave [23].

For isentropic flow, the following applies:

$$\frac{v_v}{v_{kr}} = \left[\frac{(\chi+1)M^2}{2+(\chi-1)M^2}\right]^{\frac{1}{2}} \quad (2)$$

$$\frac{v_v}{v_o} = \left[\frac{2}{2+(\chi-1)M^2}\right]^{\frac{1}{2}} \quad (3)$$

$$\frac{T_v}{T_o} = \frac{2}{2+(\chi-1)M^2} \quad (4)$$

$$\frac{p_v}{p_o} = \left[\frac{2}{2+(\chi-1)M^2}\right]^{\frac{\chi}{\chi-1}} \quad (5)$$

$$\frac{\rho_v}{\rho_o} = \left[\frac{2}{2+(\chi-1)M^2}\right]^{\frac{1}{\chi-1}} \quad (6)$$

$$\frac{\rho_v}{\rho_{kr}} = \frac{A_{kr}}{A} = M\left[\frac{\chi+1}{2+(\chi-1)M^2}\right]^{\frac{1\chi+1}{2\chi-1}} \quad (7)$$

where p_0 is the inlet pressure, p_v is the outlet pressure, T_0 is the inlet temperature, T_v is the outlet temperature, v_0 is the inlet velocity, v_v is the outlet velocity, v_{kr} is the critical velocity, ρ_0 is the inlet density, ρ_v is the outlet density, M is the Mach number, χ is the gas constant = 1.14, A is the calculated cross section and A_{kr} is the critical cross section.

In the nozzle that separates the two chambers V1 and V2, a critical flow is generated with a subsequent supersonic flow ending in a shock wave. This shock wave reflects the physical principle of the change in the ratio of state variables before and after the shock wave, as will be shown below in comparison with the isentropic flow theory describing the change in state variables when passing through the shock wave. This theory is represented by equations (8) to (11). In the equations, the index 2 denotes the value behind the shock wave and the values with index 1 represent the values at the Nozzle End location. For Mach number, it is the ratio of the Mach number value at the point of highest expansion and behind the shock wave.

This phenomenon can be seen in the graphs (Figures 4 and 5) plotting the Mach number values and the static pressure profile along the path behind the nozzle (Figure 3). Figure 5 shows the gas expansion behind the nozzle. From the above relations, the phenomenon seen in (Figure 4) is that the magnitude of the expansion and Mach number depends on the ratio of the pressures in chamber V1, denoted as P_0, and the pressure at the Nozzle End position P_v. By gradually reducing the pumping pressures, their ratio changes, and the magnitude of expansion decreases.

$$M_2^2 = \frac{2+(\varkappa-1)M_1^2}{2\varkappa M_1^2-(\varkappa-1)} \quad (8)$$

$$\frac{T_2}{T_1} = 1 + \frac{2(\varkappa-1)}{(\varkappa+1)^2}\cdot\frac{1+\varkappa M_1^2}{M_1^2}\cdot(M_1^2-1) \quad (9)$$

$$\frac{p_2}{p_1} = 1 + \frac{2\varkappa}{\varkappa+1}(M_1^2-1) \quad (10)$$

$$\frac{\rho_2}{\rho_1} = \frac{V_1}{V_2} = \frac{(\varkappa+1)M_1^2}{2+(\varkappa-1)M_1^2} \quad (11)$$

Table 2 contains the ratios of the variables according to equations (8) and (9). Comparing this theory with the results obtained from Ansys (Figure 4 to 8), it can be seen that they match.

Table 2. Pressure ratios and Mach Disk locations in time.

Time (s)	M2 (-)	ϱ_2/ϱ_1 (-)	T_2/T_1 (-)	p_2/p_1 (-)	M_1(-)
1	0.468	3.97	2.835	11.26	3.14
2	0.469	3.96	2.81	11.11	3.11
3	0.473	3.88	2.715	10.544	3.03
4	0.478	3.82	2.632	10.06	2.96
5	0.482	3.763	2.55	9.578	2.89
6	0.488	3.663	2.451	8.98	2.8
7	0.498	3.527	2.311	8.15	2.67

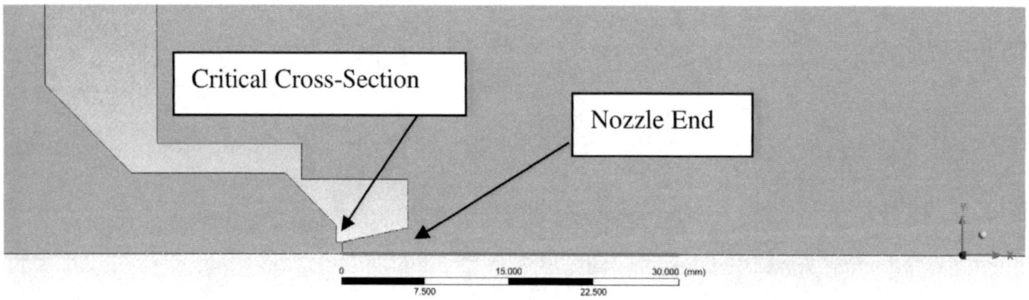

Figure 3. Axis of symmetry - denoted Line 1 in calculations.

The correspondence with the isentropic flow theory can be seen in Figure 4 and Table 1, where the Mach number magnitude at the Nozzle End in Table 1 from relation (5) and (7) from the cross-sections of the surfaces in the aperture with a diameter of 2 mm and at the Nozzle End with a diameter of 4.52 mm matches the results obtained from the mathematical and physical analysis in Ansys Fluent in Figure 4.

Similarly, as shown in Table 1 for the P_v/P_0 ratio for each time, its result corresponds to equation (5).

Table 3 shows the ratios of the velocity values before the aperture, at the critical region of the aperture and at the gas expansion point, and the temperature and density before the aperture and at the expansion point.

Table 3. Velocity, temperature, density ratios and Mach numbers in time.

Time (s)	v_v/v_{kr} (-)	v_v/v_o (-)	T_v/T_o (-)	ϱ_v/ϱ_o (-)	ϱ_v/ϱ_{kr} (-)	M_v (-)
1	1.995	0.58	0.337	0.066	0.207	3.14
2	1.99	0.584	0.341	0.68	0.213	3.11
3	1.971	0.594	0.353	0.074	0.232	3.03
4	1.955	0.603	0.363	0.08	0.245	2.96
5	1.937	0.612	0.375	0.086	0.262	2.89
6	1.914	0.624	0.389	0.095	0.286	2.8
7	1.878	0.642	0.412	0.109	0.323	2.67

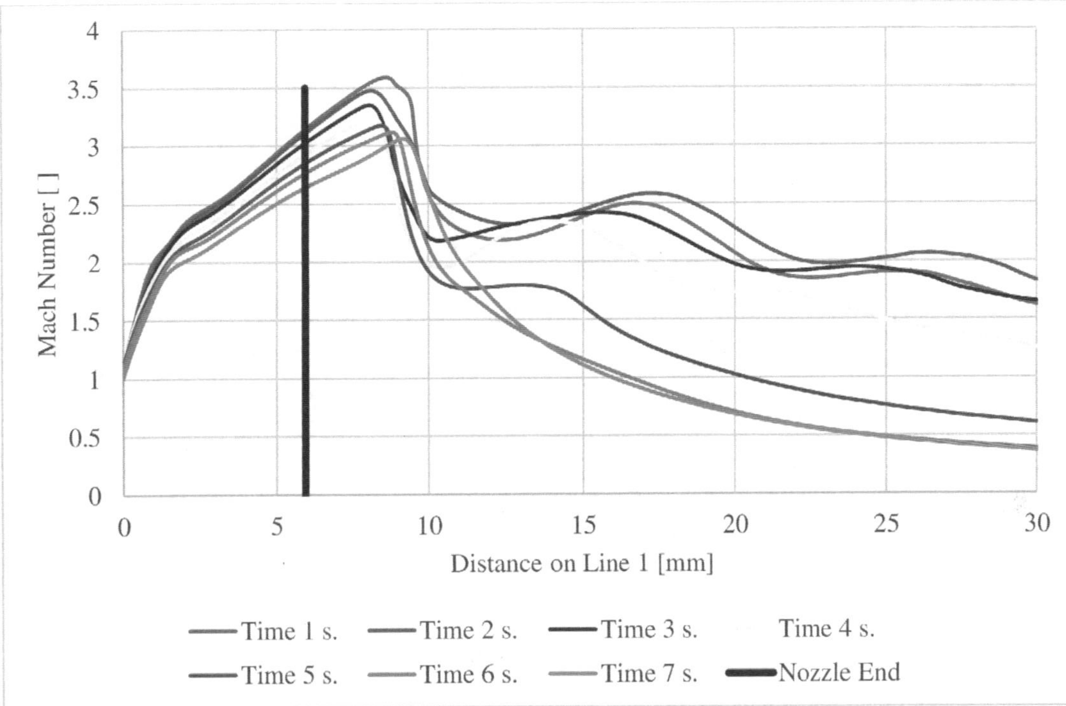

Figure 4. Evolution of Mach Number on Line 1.

The magnitude of the expansion and its progression is directly dependent on the flow pressure behind the nozzle (Figure 5). It is obvious that higher expansion causes the formation of a series of Mach disks.

Figure 6 shows the velocity evolution behind the aperture. The results in Figure 6 are consistent with the one-dimensional flow theory given in equations (2) and (3). In this case, the velocity ratios at the expansion peak and at the critical nozzle cross section are v_v/v_{kr} and the velocity ratios at the expansion peak and the velocity before entering the nozzle are v_v/v_o. The given ratios from equations (2) and (3) are provided for the respective Mach number in Table 3.

Similarly, Figure 7 shows a temperature profile behind the aperture. Again, the results in this figure are consistent with the one-dimensional flow theory given in equation (4), and the given ratio of Temperature at the expansion peak and Temperature before entering the aperture T_v/T_o for the corresponding Mach number is given in Table 3.

Similarly, Figure 8 shows the gas density behind the aperture. Again, the results in this plot are consistent with the one-dimensional flow theory given in equations (6) and (7). In this case, the ratios of the densities at the expansion peak and at the critical cross section of the aperture ϱ_v/ϱ_{kr} and the ratio of the densities at the expansion peak and the densities before entering the aperture ϱ_v/ϱ_o. The given ratios from equations (6) and (7) are given for the corresponding Mach number in Table 3.

It can be concluded that the results of the above one-dimensional flow theory based on equations (2) to (7) match the results obtained from the Ansys Fluent system shown in Figures 4 to 8.

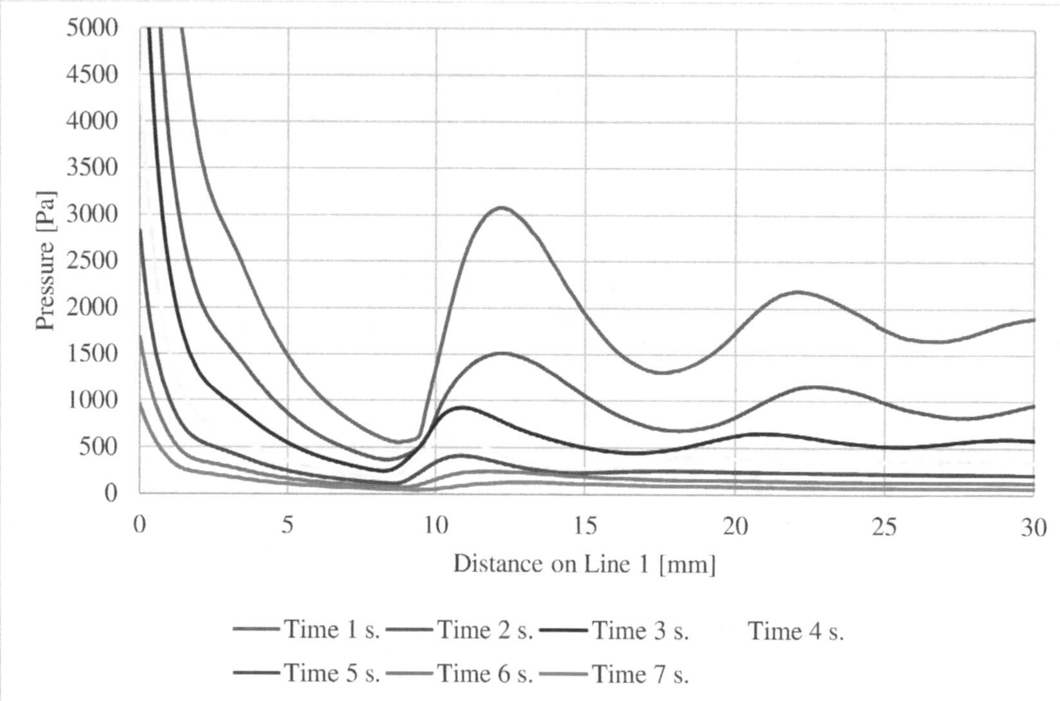

Figure 5. Pressure evolution on Line 1.

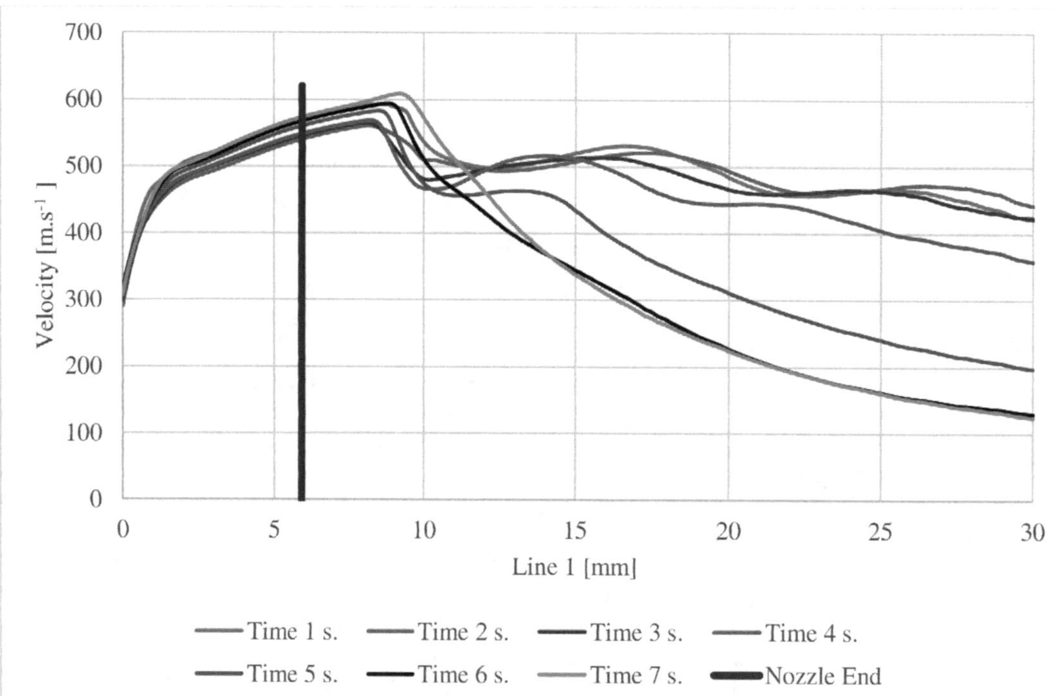

Figure 6. Velocity evolution on Line 1.

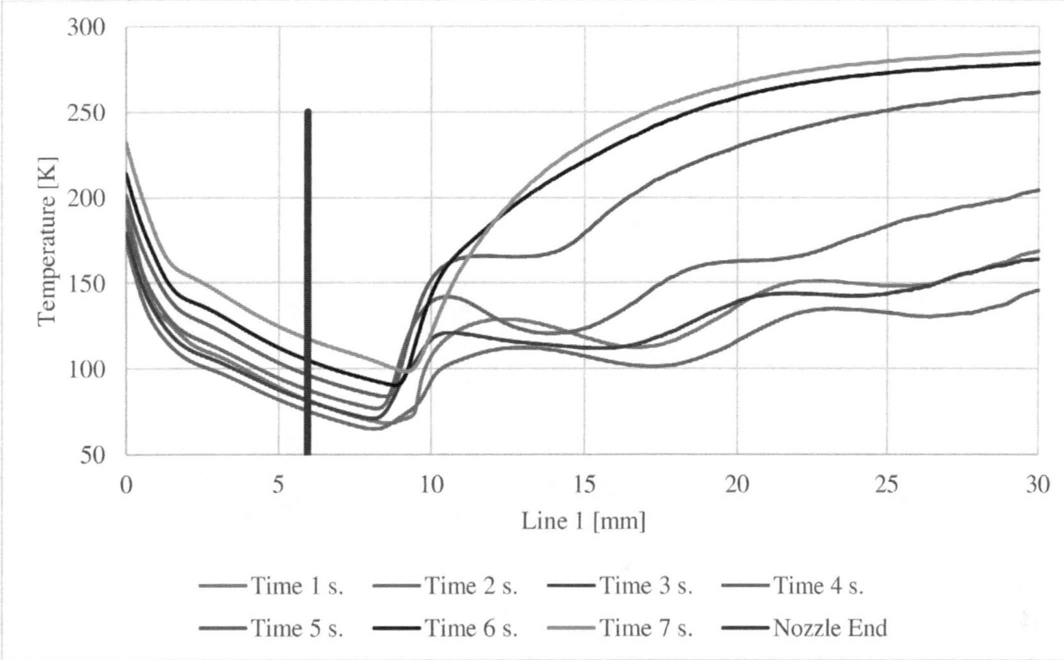

Figure 7. Temperature evolution on Line 1.

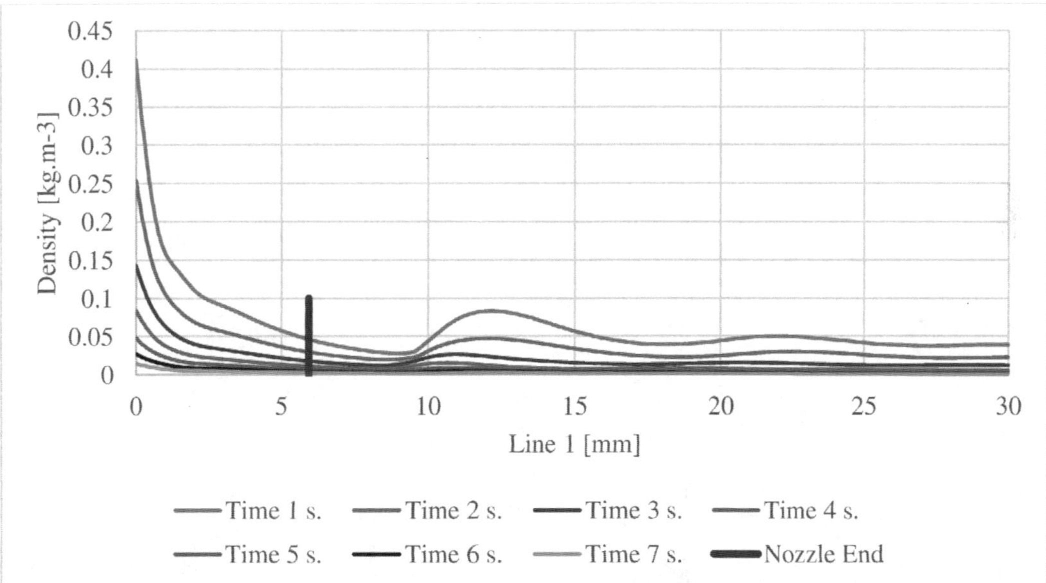

Figure 8. Density evolution on Line 1.

As the ratio of the chamber V1 P_0 and pressure at the point of maximum expansion P_v decreases and hence the expansion and Mach number decrease, the size and intensity of Mach disks decrease (see Figures 9 to 15).

The next figures show the compression during pumping. As the pressure ratios gradually decrease, the strong compression with additional compression behind the nozzle changes from strong compression with additional pressure This fact is depicted in Figures 9 to 15 and affects the manner and character of the shock waves described in Figure 13. Figures 9 to 16 are the basis for planned experiments where the location and shape of shock waves at the continuum mechanics boundary will be evaluated using Schlieren method.

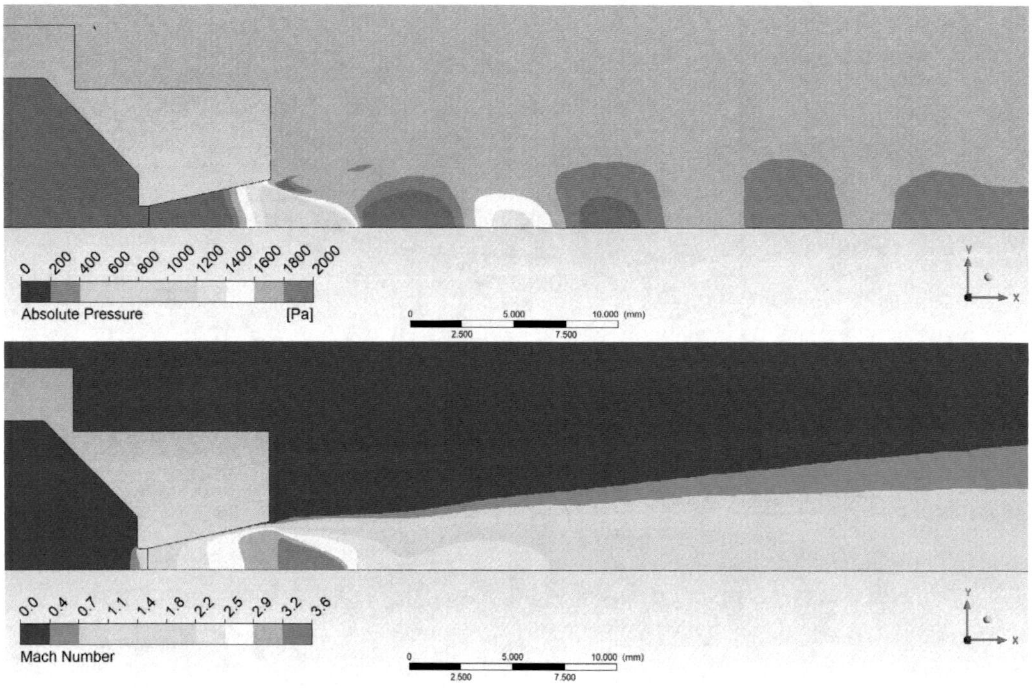

Figure 9. Pressure and Mach Number distribution in time of 1 s.

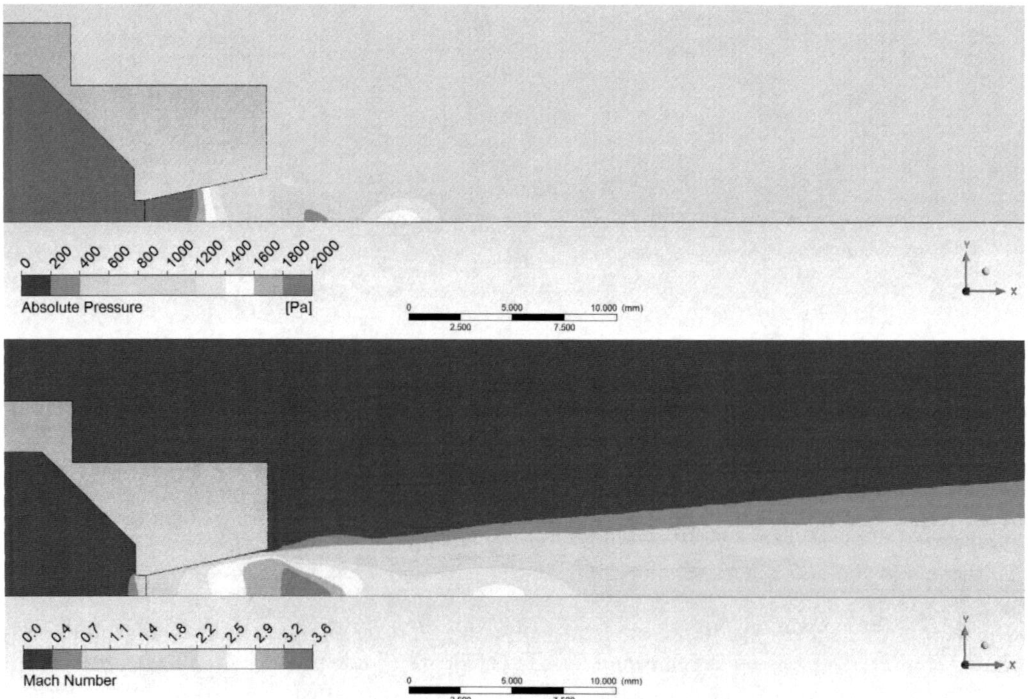

Figure 10. Pressure and Mach Number distribution in time of 2 s.

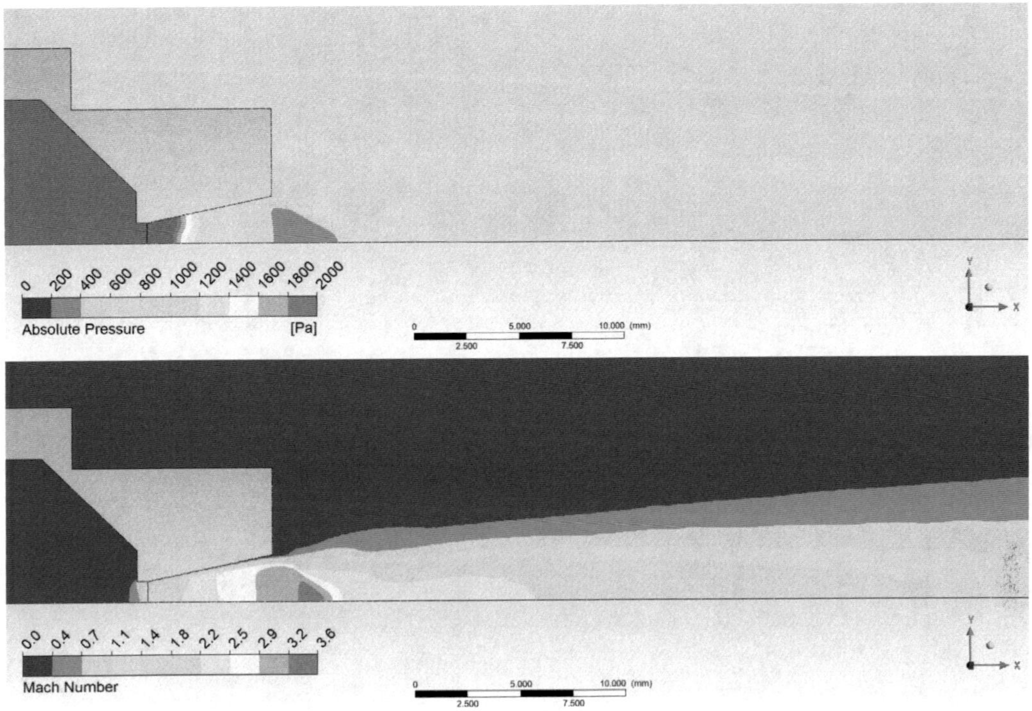

Figure 11. Pressure and Mach Number distribution in time of 3 s.

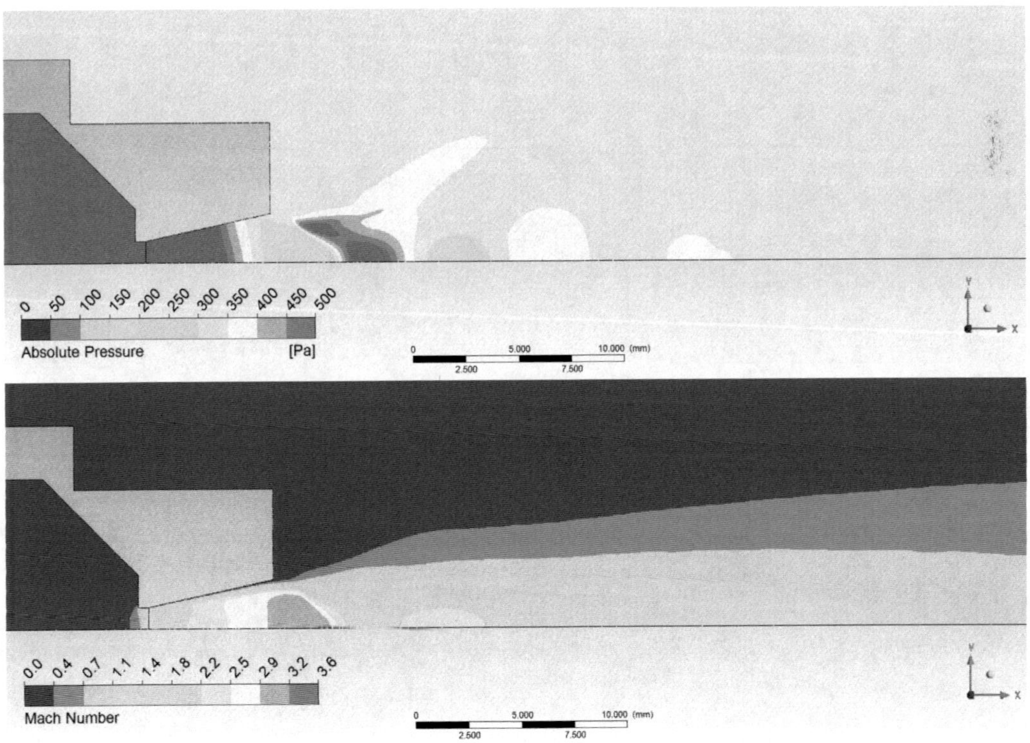

Figure 12. Pressure and Mach Number distribution in time of 4 s.

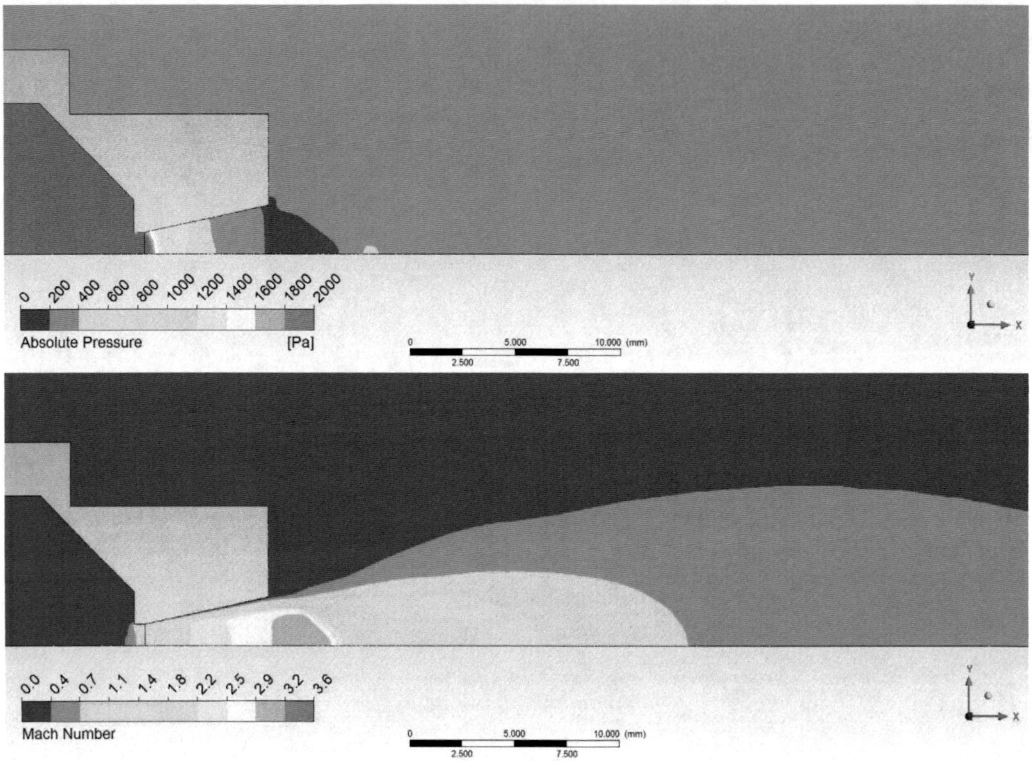

Figure 13. Pressure and Mach Number distribution in time of 5 s.

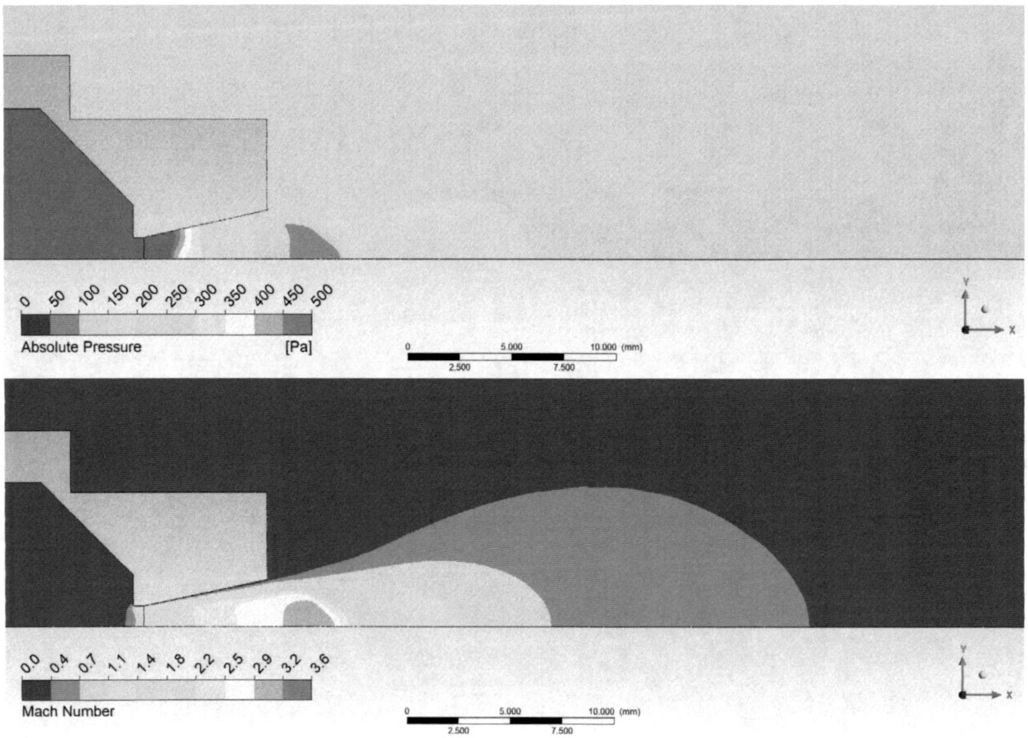

Figure 14. Pressure and Mach Number distribution in time of 6 s.

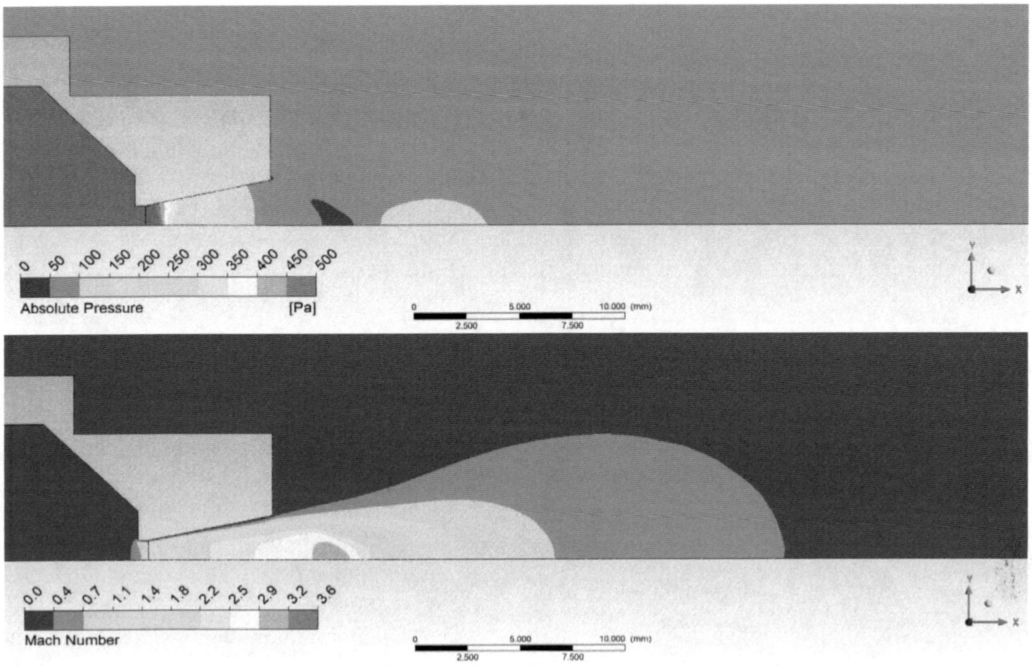

Figure 15. Pressure and Mach Number distribution in time of 7 s.

The magnitude of the expansion and the associated ratio of the chamber pressure V1 P_0 to the pressure at the Nozzle End P_v is related to the character of the shock waves that will be monitored in the experimental chamber using the Schlieren method. The higher pressure/expansion ratio produces a set of oblique shock waves with a perpendicular shock wave separating each period (Figure 16 Time 1 s), which move steadily at supersonic velocity (see Figure 4). By gradually decreasing the pressure ratio, especially at times between 5th and 7th second, the expansion reaches a subsonic state very quickly. Here, the series of shock waves ceases and only one weaker set of oblique shock waves ending in a perpendicular shock wave is produced.

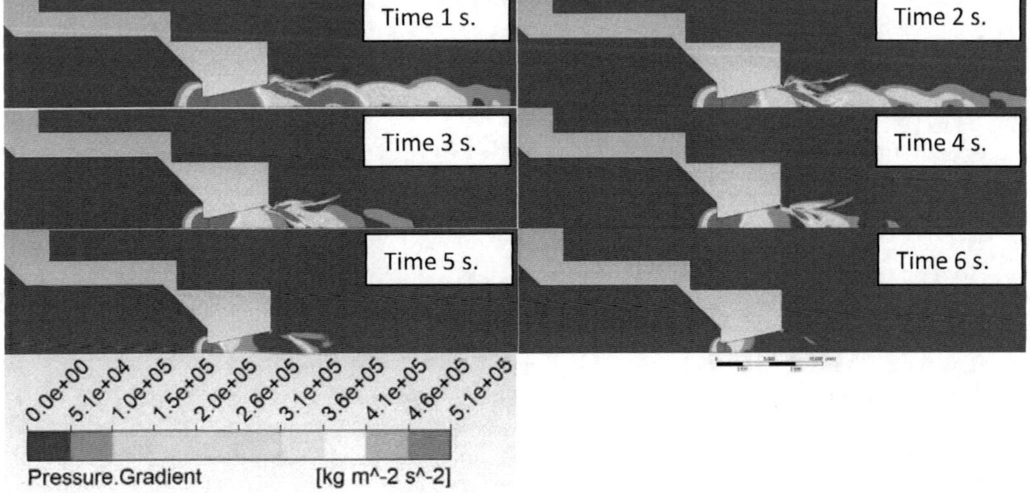

Figure 16. Location and intensity of shock waves during pumping of the experimental chamber.

4. Conclusions

The resulting shock waves in the supersonic flow during differential pumping have a large effect on the passage of the primary electron beam in the microscope. Figure 16 shows that within 3 s, in addition to the oblique shock waves, a perpendicular shock wave is formed, which has the biggest effect on the scattering of the electron beam. Presented results will be used as a basis for a forthcoming experiment to map shock waves at low pressures using the Schlieren method, which will be used to display and compare the measured values with those obtained by the mathematical-physical analysis shown in Figure 16 in order to further tune the computational model of supersonic flow at low pressures.

References

[1] Tihlaříková E, Neděla V and Dordevic B 2019 In-situ preparation of plant samples in ESEM for energy dispersive x-ray microanalysis and repetitive observation in SEM and ESEM *Scientific Reports* **9(FEB)** 2300
[2] Maxa J, Neděla V, Jirák J, Vyroubal P and Hladká K 2012 Analysis of gas flow in a secondary electron scintillation detector for ESEM with a new system of pressure limiting apertures *Advances in Military Technology* **7 (2)** 111-116
[3] Maxa J, Bílek M, Hlavatá P, Vyroubal P and Lepltová K 2016 Comparisons Using Methods of Continuum Mechanics and Monte Carlo at Differentially Pumped Chamber *Advances in Military Technology* **11 (2)** 143-150
[4] Vyroubal P, Maxa J, Neděla V, Jirák J and Hladká K 2013 Apertures with Laval Nozzle and Circular Orifice in Secondary Electron Detector for Environmental Scanning Electron Microscope *Advances in Military Technology* **8 (1)** 59-69
[5] Neděla V, Tihlaříková E, Runštuk J and Hudec J 2018 High-efficiency detector of secondary and backscattered electrons for low-dose imaging in the ESEM *Ultramicroscopy* **184A(JAN)** 1-11
[6] Bílek M, Maxa J, Hlavatá P and Bayer R 2017 Modelling and simulation of a velocity field within supersonic flows in low-pressure areas *ECS Transactions* **81(1)** 311–316
[7] Neděla V, Konvalina I, Lencová B and Zlámal J 2011 Comparison of calculated, simulated and measured signal amplification in variable pressure SEM *Nuclear Instruments & Methods in Physics Research Section A* **645 (1)** 79-83
[8] Neděla V, Hřib J, Havel L, Hudec J and Runštuk J 2016 Imaging of Norway spruce early somatic embryos with the ESEM, Cryo-SEM and laser scanning microscope *Micron* **84(May)** 67-71
[9] Neděla V, Konvalina I, Oral M and Hudec J 2015 The Simulation of Energy Distribution of Electrons Detected by Segmental Ionization Detector in High Pressure Conditions of ESEM. *Microscopy and Microanalysis* **21(4)** 264-269
[10] Jirák J, Neděla V, Černoch P, Čudek P and Runštuk J 2010 Scintillation SE detector for variable pressure scanning electron microscopes *Journal of Microscopy* **239 (3)** 233-238
[11] Neděla V, Tihlaříková E, Maxa J, Imrichová K, Bučko M and Gemeiner P 2020 Simulation-based optimisation of thermodynamic conditions in the ESEM for dynamical in-situ study of spherical polyelectrolyte complex particles in their native state *Ultramicroscopy* **211**
[12] Stelate A, Tihlaříková E, Schwarzerová K, Neděla V and Petrášek J 2021 Correlative Light-Environmental Scanning Electron Microscopy of Plasma Membrane Efflux Carriers of Plant Hormone Auxin *Biomolecules* **11(10)** 1407
[13] Hlavatá P, Maxa J, Bílek M and Neděla V 2017 Impact of the shape of the differentially pumped chamber on critical flow character *ECS Transactions* **81(1)** 317–322
[14] Hlavatá P, Maxa J, Bílek M, Vyroubal P and Bayer R 2017 Influence of critical flow in the differentially pumped chamber AQUASEM *Advances in Military Technology* **12(2)** 301–310
[15] Danilatos G D 2012 Velocity and ejector-jet assisted differential pumping: Novel design stages for environmental SEM *Micron* **43** 600-611
[16] Danilatos G D, Rattenberger J and Dracopoulos V 2011 Beam transfer characteristics of a commercial environmental SEM and a low vacuum SEM *Journal of Microscopy* **242** 166–180

[17] Danilatos G D 2009 Optimum beam transfer in the environmental scanning electron microscope *Journal of Microscopy* **234** 26–37
[18] Danilatos G D 2011 Figure of merit for environmental SEM and its implications *Journal of Microscopy* **244** 159–169
[19] Debanjan Ch and Vikram V 2021 A critical review on pore to continuum scale imaging techniques for enhanced shale gas recovery *Earth-Science Reviews* **217**
[20] Bayer R, Maxa J and Šabacká P 2021 Energy harvesting using thermocouple and compressed air *Sensors* **21**
[21] Šabacká P, Neděla V, Maxa J and Bayer R 2021 Application of Prandtl's theory in the design of an experimental chamber for static pressure measurements *Sensors* **21**
[22] Maxa J and Neděla V 2011 The Impact of Critical Flow on the Primary Electron Beam Passage through Differentially Pumped Chamber *Advances in Military Technology* **6(1)** 39–46
[23] Salga J and Hoření B 1997 *Tabulky Proudění Plynu* UNOB: Brno, Czech Republic

Acknowledgments
This work was supported by the project of specific research at Brno University of Technology: Materials and Technologies for Electrical Engineering IV, reg. no. FEKT-S-20-6206.

Evaluation of ballistic resistance of thermoplastic and thermoset composite panels

J Viliš[1], Z Pokorný[1], J Zouhar[2], R Vítek[1] and J Procházka[1]

[1] Department of Mechanical Engineering, Faculty of Military Technology, University of Defence, Brno 662 10, Czech Republic
[2] Faculty of Mechanical Engineering, Brno University of Technology, Institute of Engineering Technology, 616 69 Brno, Czech Republic

jindrich.vilis@unob.cz

Abstract. In this study, the ballistic resistance of composite panels made of solid fabrics was investigated. In the experimental part, the materials Twaron CT 747 TH110 impregnated with thermoplastic resin and Twaron CT 747 impregnated with thermoset resin were examined. The ballistic resistance of the composite panels was tested in the reduced version according to the NATO AEP-55 STANAG 4569 standard, protection level 1. The manufactured composite panels were tested by the rifle cartridges 7.62 x 51 mm FMJ M80. The high-speed camera was used to observe the impact of the missiles into the panel. As part of the evaluation of the experiment, the damage ranges of these panels were identified. The composite panel made of Twaron CT 747 TH110 impregnated with thermoplastic resin absorbed more energy of the impacting missiles than the panel made of Twaron CT 747 impregnated with thermoset resin. The results show that the panel made of Twaron CT 747 TH110 impregnated with thermoplastic resin, which was composed of 90 layers was not penetrated by the missile 7.62 mm M80 and complies with NATO AEP-55 STANAG 4569, protection level 1.

1. Introduction

The technological development of sophisticated weapons is still in progress and puts great emphasis on the development of new ballistic protection means for individuals and military equipments. These are the key elements that influence the results of combat operations around the world. The use of a ballistic means in a combat conflict must ensure the highest possible protection of the individuals during the operation. This means must provide protection against missiles, fragments and the effects of improvised explosive devices (IEDs). It should also allow for higher resistance to multi-hit. As part of research and development, the aim is to reduce the weight of ballistic protective means as much as possible while maintaining the required level of protection. The most widespread way to increase the ballistic resistance of these means, in addition to the choice of materials, is to increase its thickness. Increasing the thickness of the material, the ballistic resistance will increase, but at the same time, the total weight will increase. This will cause a decrease in the mobility of the individual and military equipment [1, 2, 3, 4, 5].

In order to increase the ballistic resistance and reduce the weight of the ballistic means, the materials are combined into panel systems to create multilayer armor with specific properties. These armors are designed to protect against small-caliber armor-piercing missiles from short and long firearms [5, 6]. Multi-layer armor combines the properties of hard materials (steel, ceramics) and malleable materials (polymers). When the missile hits the strike face of the hard plate, which is mostly made of ceramics (Al_2O_3, SiC, B_4C, or their combination), the missile deforms and loses a large part of its penetrating energy. Ceramic materials are characterized by high hardness, but with hardness come high brittleness. When the bullet hits the strike face plate, cracks in the ceramic layer are initiated and spread. To solve this problem, ceramics are applied to the malleable plate, which is the second most important component of the multilayer armor. This plate made of ballistic resistant fabrics must be as tough as possible to capture missile fragments and fragments of the strike face layer [6, 7, 8, 9].

The most significant breakthrough to increase the ballistic resistance of protective equipments while ensuring the required mobility was the discovery of solid textile fibres. Ballistic textiles are composed of high-performance textile fibres that overcome conventional textiles in terms of their mechanical properties. From the point of view of the production of ballistic protection means, polymer fibers are coming to the fore. Para-aramids (aromatic polyamide) and ultra-high molecular weight polyethylene (UHMWPE) are the most widely used materials for modern ballistic protection. Para-aramid fibers are characterized by low density, high tensile modulus, high toughness and low elongation at break. They are also characterized by resistance to higher temperatures and resistance to many chemicals except some acids and bases [9, 10, 11].

The main disadvantage of these fibers is degradation due to moisture and UV radiation. To reduce these negative properties, the fibers in the protective insert are inserted into a waterproof pocket and then they are stitched. Resistance to UV radiation can be increased by coating fibers. Based on their mechanical properties and ability to resist external influences, they provide a suitable combination for independent use in soft ballistic protection [10, 11, 12].

Currently, thermoset and thermoplastic resins are used to impregnate composite materials. The most commonly used thermoset resins include unsaturated polyesters (UP-R), vinyl esters (VE-R), epoxies (EP-R) and phenolics. Polypropylene (PP), polyamides (PA) or polyethylene (PE) are most commonly used for thermoplastic resins. Both thermosets and thermoplastics have their advantages and disadvantages. In comparison, thermosets, unlike thermoplastics, retain their strength and shape even when heated. They are easier to process. The fibers are easier to saturate and easier to wet. Components manufactured by using thermosets have excellent strength properties, but they are characterized by higher brittleness and lower ductility. The disadvantage is that after curing, thermosets cannot be reheated and shaped, which makes them difficult to recycle. The main advantages of thermoplastics include their mechanical properties. They retain their original shape even when the temperature changes. They are also characterized by higher toughness, the ability to withstand higher processing temperatures and the ability to tolerate damage. Repeated melting makes the recycling and processing of thermoplastics easier. However, it is undesirable for ballistic protection that they are softer at higher temperatures and more expensive compared to thermosets [12, 13, 14, 15].

In terms of ballistic protection design, it is necessary to realize for what specific situation we are designing a ballistic device.

This article is focused on the research of selected composite materials that are used for ballistic protection. The main goal was to determine the ballistic resistance of the selected materials and to determine their limit thicknesses. The composite panels were tested by the missiles 7.62 mm M80. The high-speed camera was used to record the impact of the missile and part of the fragmenting missile. The extent of damage to the composite panels was also assessed. The results from the shooting experiment provide a more detailed understanding of the penetration mechanism by the missile 7.62 mm M80 into the para-aramid fibre composite panels and provide valid data in the design of the multilayer armour.

2. Material and Methods

For the production of composite panels, the fabric made of para-aramid fibers Twaron CT 747 with the areal density of 410 g/m^2 was used in the experimental part. The properties of para-aramid fabric Twaron CT 747 are shown in Table 1.

Table 1. Properties of para-aramid fabric Twaron CT 747

Material	Thickness [mm]	Areal density [g/m^2]	Linear density [dtex]	Tensile strength module [GPa]	Composite density [kg/m^3]
Twaron CT 747	0.62	410	3360	115	1450

For the ballistic purposes, the Twaron CT 747 material was impregnated with thermoplastic resin TH110 and thermoset resin ER 68. The properties of thermoplastic resin TH110 and thermoset resin ER 68, as specified by the manufacturer, are given in Table 2.

Table 2. Properties of thermoplastic resin TH110 and thermoset resin ER 68

Resin	Density [g/m^3]	Colour
TH110	0.95	transparent
ER 68	1.2	transparent

2.1. Preparation and production of composite panels

In order to evaluate the ballistic resistance of the composite panels, the panels were manufactured in different thicknesses and with different number of layers. The panels that were made for the shooting experiment are shown in Table 3.

Table 3. Parameters of manufactured composite panels

Materials	Dimension [mm]	Number of layers	Thickness [cm]	Weight [kg]	Areal density [kg/m^2]
Twaron CT 747 TH110 (thermoplastic)	300 x 300	60	2.79	2.73	30.33
		30	1.33	1.41	15.67
Twaron CT 747 ER 68 (thermoset)	300 x 300	30	1.28	1.35	15.00
		30	1.27	1.34	14.89
		30	1.28	1.35	15.00

2.1.1. Twaron CT 747 TH110 (thermoplastic)

Material Twaron CT 747 TH110 impregnated with thermoplastic resin was manually cut to the size of 300 x 300 mm and layered to achieve the required thickness. The vacuum-packed composite material, which is ready to be pressed, is shown in Figure 2. The ZD 40 testing machine (400 kN), located at the Faculty of Mechanical Engineering in Brno, was used to fabricate the composite panels (Figure 1). The machine acted on the materials with the pressure of 10 bar. The pressing process (Figure 3) was carried out at 140 °C. The curing time was 15 minutes. Two composite panels were produced by the hot pressing method.

To test the ballistic resistance, the composite panel of 60 layers with the thickness of 2.79 cm, the weight of 2.73 kg and the areal density of 30.33 kg/m^2 was produced. The second produced panel was the panel of 30 layers, with the thickness of 1.33 cm, the weight of 1.41 kg and the area density of 15.67 g/m^2.

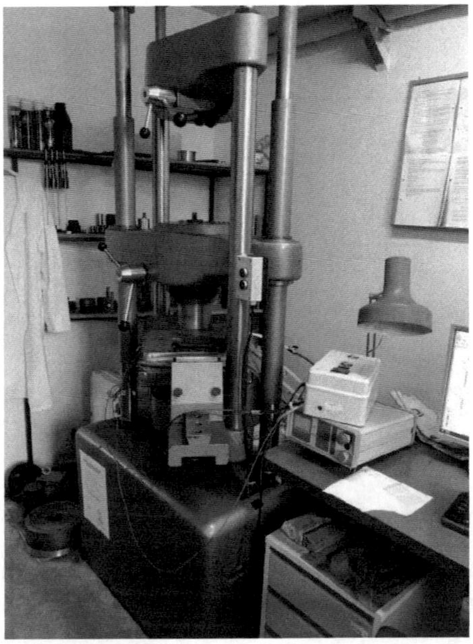

Figure 1. Testing machine ZD 40 (400 kN)

Figure 2. Vacuum-packed composite material

Figure 3. Pressing process

2.1.2. Twaron CT 747 ER 68 (thermoset)

Furthermore, to compare ballistic resistance, panels were made from Twaron CT 747 material, which was impregnated with thermoset resin ER 68. Autoclave technology was used to produce these panels. In the autoclave, the pressure of 10 bar was applied to the material. Curing took place at the temperature of 120 °C for 1 hour. The curing of the composite panels took place according to the curing cycle, which is shown in Figure 4. Three composite panels of 30 layers were produced by using autoclave technology, see Table 3.

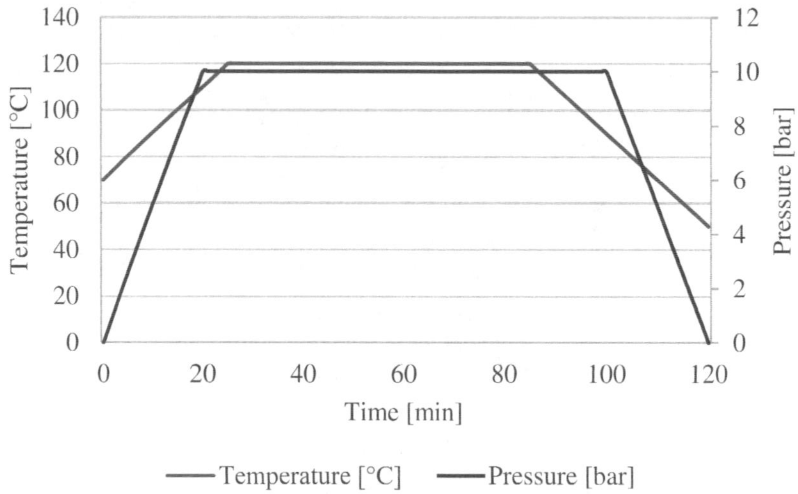

Figure 4. Curing cycle in the autoclave

2.2. Ballistic resistance testing

Ballistic resistance testing of the manufactured composite panels was carried out according to standard NATO AEP-55 STANAG 4569. This standard deals with the classification of the ballistic protection of lightly armored vehicles. It takes into account damage caused by missiles, artillery fragments and ground explosive charges (mines, improvised explosive devices) [16].

The description of the shooting and measuring station is evident from the block diagram, which is shown in Figure 5. The ballistic barrel was connected to the universal breech, which was attached to the shooting stool. The ballistic radar was placed at the level of the muzzle of the ballistic barrel, which measured the muzzle velocity of the missiles by using the Doppler effect. At the distance of 2.5 m in front of the ballistic barrel, laser gate designed to measure the velocity of the missiles was placed. Composite panels with dimensions of 300 x 300 mm were inserted into the frame for shooting tests. The distance between the ballistic barrel and the tested panel was 25 m. The measuring station included the optoelectronic target for measuring the velocities of the missiles and the high-speed camera for recording the impact and output velocities of the missiles.

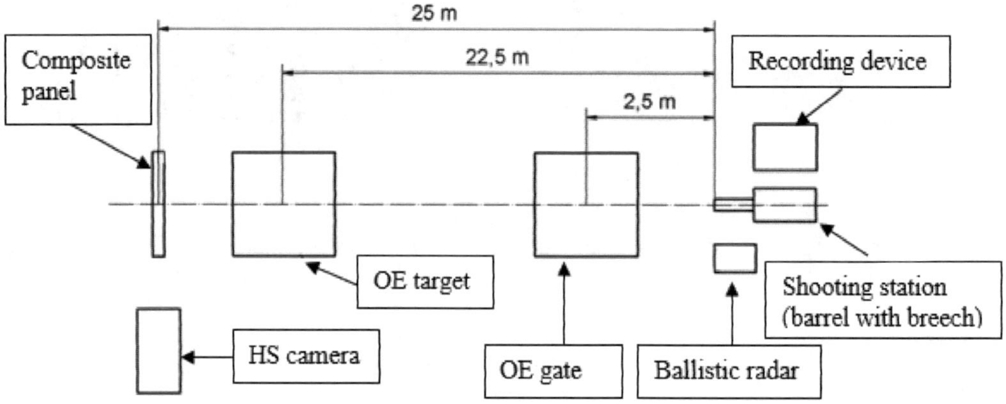

Figure 5. Block diagram of the shooting station

The ballistic resistance of the composite panels was tested by the rifle cartridges 7.62 x 51 mm NATO FMJ M80. The parameters of the rifle cartridge are listed in Table 4.

Table 4. Parameters of the rifle cartridge 7.62 x 51 mm NATO FMJ M80

Cartridge type	Missile type	Missile weight [g]	Manufacturer	Missile velocity [±15 m/s]
7.62 x 51 mm NATO	FMJ M80	9.55	S&B	833.00

3. Results and discussion

In this chapter, the results of the shooting experiment are presented. The composite panels, which were prepared and manufactured according to the above-described experimental methods, were tested by the rifle cartridges 7.62 x 51 mm NATO FMJ M80 with the parameters that are shown in Table 4.

3.1. Composite panel Twaron CT 747 TH110 (thermoplastic)

The first tested panel was the Twaron CT 747 TH110 impregnated with thermoplastic resin. The composition of the tested panel was consisted of composite panels of 60 and 30 layers, which were arranged in the row and fixed in the double frame with bolts and metal nuts. The thickness of this panel with the number of 90 layers was 4.12 cm and its weight was 4.14 kg. The panel was tested by the missile 7.62 mm FMJ M80. One missile was shot into the panel. The plastic deformation can be observed on the back side of the composite panel, which indicates that this is the limit thickness for the given rifle cartridge, see Figure 6.

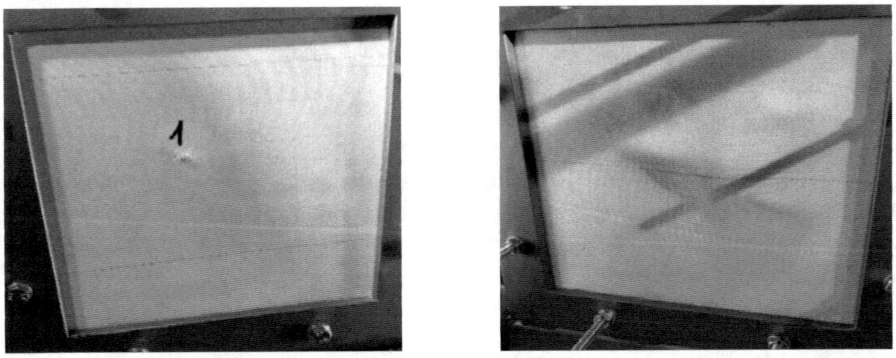

Figure 6. Strike and back face of the Twaron CT 747 TH110 (thermoplastic), 90 layers

In order to assess the ballistic resistance, this composite panel was also tested in the reduced version with 60 layers. One missile 7.62 mm NATO FMJ M80 was shot into the panel. This composite panel has been perforated, as shown in Figure 7. Detail of the panel penetration and missile fragmentation are shown in Figure 8.

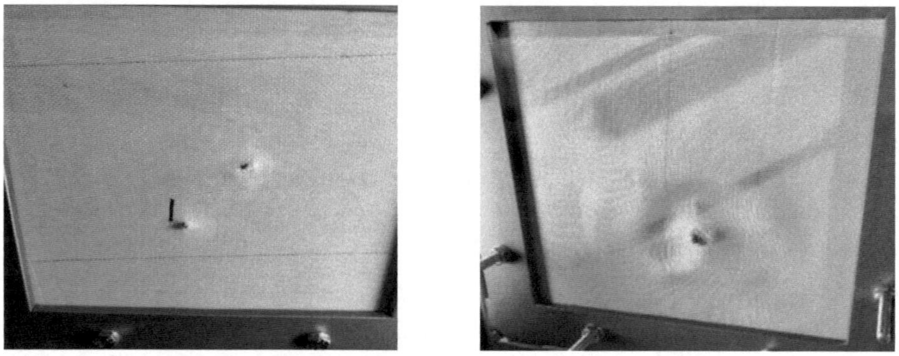

Figure 7. Strike and back face of the Twaron CT 747 TH110 (thermoplastic), 60 layers

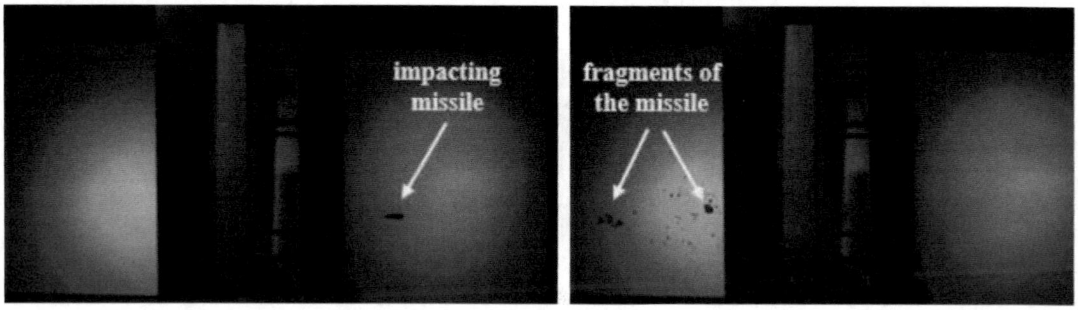

Figure 8. Detail of the panel penetration Twaron CT 747 TH110 (thermoplastic), 60 layers

3.2. Composite panel Twaron CT 747 ER 68 (thermoset)

The last tested panel was Twaron CT 747 ER 68 impregnated with thermoset resin. The tested panel was consisted of composite panels of 30, 30 and 30 layers, which were also arranged in the row and fixed in the double frame using bolts and metal nuts. The thickness of this panel with the number of 90 layers was 3.83 cm and its weight was 4.04 kg. The panel was tested by one missile 7.62 mm NATO FMJ M80. This missile completely penetrated the composite panel, see Figure 9. After perforation of the panel, the missile was also fragmented. Figure 10 shows the detail of the penetration of the Twaron CT 747 ER 68 (thermoset) by the high-speed camera.

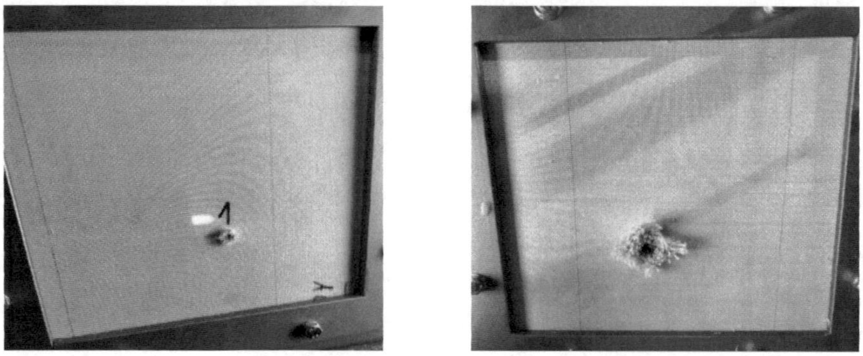

Figure 9. Strike and back face of the Twaron CT 747 ER 68 (thermoset), 90 layers

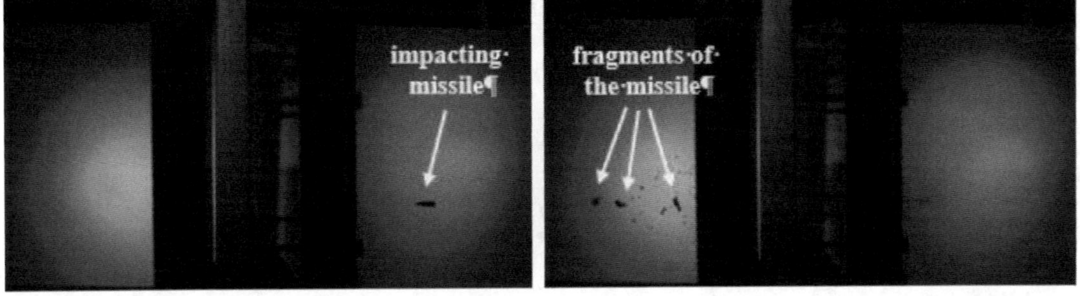

Figure 10. Detail of the panel penetration Twaron CT 747 ER 68 (thermoset), 90 layers

Table 5 shows the impact velocities of missiles at the distance of 25 m, output velocities of the missiles and their differences. Their kinetic energies are also listed here. The results were obtained from the ballistic radar (Prototypa-ZM, DRS-01) and from the high-speed camera (Photron, SA-Z).

Table 5. Outputs from the high-speed camera (Photron, SA-Z)

S. n.	Composite panel	Missile	v_{25} [m/s]	v_{out} [m/s]	Δ v [m/s]	E_{k25} [J]	E_{kout} [J]	Δ E_k [J]	Penetration
1.	Twaron CT 747 TH110, 90 layers	7.62 mm M80	834.0	0.0	834.0	3321.1	0.0	3321.1	No
2.	Twaron CT 747 TH110, 60 layers	7.62 mm M80	832.7	680.9	151.8	3310.9	2213.8	1097.1	Yes
3.	Twaron CT 747 ER 68, 90 layers	7.62 mm M80	830.5	233.4	597.1	3293.1	260.1	3033.0	Yes

From Table 5, it is clear that for the Twaron CT 747 TH 110 panel with the number of 60 layers, there was the decrease in missile velocity by 18.2 % (the decrease in kinetic energy by 33.1 %). In the case of impregnation with ER 68 thermosetting resin, the Twaron CT 747 ER 68 panel with 90 layers decreased in missile velocity by 71.9 % (the decrease in the kinetic energy of the missile by 92.1 %).

4. Conclusion

The study deals with the production and testing of composite panels designed for the ballistic protection of individuals and military equipment. Twaron CT 747 para-aramid fabric was used in the shooting experiment. This material was impregnated with thermoplastic and thermoset resin. Standard rifle cartridges 7.62 x 51 mm FMJ M80 were used to verify ballistic resistance. Subsequently, the ballistic resistance of the manufactured panels was examined.

Experiments show that the use of thermoplastic or thermosetting resin has the significant effect on ballistic resistance. The conclusions of the shooting experiment showed. After the impact by the missile 7.62 mm M80, the plastic deformation appeared on the back side of the Twaron CT 747 TH110 (thermoplastic) with the number of 90 layers (Figure 6), but there was no perforation. This panel absorbed all the kinetic energy of the missile. The panel Twaron CT 747 ER 68 with the number of 90 layers was perforated by the missile 7.62 mm M80 and this panel absorbed 92.1 % of the missile's kinetic energy from the high-speed camera results. In terms of visual evaluation of the damage mechanism, there were delamination, breakage and pull-out of the fibers due to the absorption of kinetic energy. The perforation of the panel was caused by the fact that thermoset resin has higher strength values compared to thermoplastic, but is more brittle and has lower ductility, which reduce the ballistic resistance of the used panel.

The results from the shooting experiment clearly show that the panel Twaron CT 747 TH110 impregnated with thermoplastic resin complies with the standard NATO AEP-55 STANAG 4569, protection level 1 and has potential in the development of new ballistic protective means.

Acknowledgements
The work presented in this paper was supported by the specific research project 2020 "SV20-216" at the Department of Mechanical Engineering, University of Defence in Brno and was supported by the Project for the Development of the Organization "VAROPS (DZRO VAROPS) Military autonomous and robotic assets" by the Ministry of Defence of Czech Republic.

References

[1] Bui XS Komenda J and Vítek R 2017 Frangibility of frangible bullet upon impact on a hard target *2017 International Conference on Military Technologies (ICMT)* pp 7-11 doi: 10.1109/MILTECHS.2017.7988721
[2] Skalický P and Komenda J 2019 Possibilities of Soft Ballistic Resistant Materials to Mitigate Dominant Effects of a Explosive Ordnance Detonation *2019 International Conference on Military Technologies (ICMT)* pp 1-5 doi: 10.1109/MILTECHS. 2019.8870127
[3] Rubio I Rodríguez-Millán M Marco M Olmedo A and Loya JA 2019 Ballistic performance of aramid composite combat helmet for protection against small projectiles *Composite Structures* **226** ISSN 02638223 doi:10.1016/j.compstruct.2019.111153
[4] Luz FSd Garcia Filho FdC Oliveira MS Nascimento LFC and Monteiro SN 2020 Composites with Natural Fibers and Conventional Materials Applied in a Hard Armor: A Comparison *Polymers* **12** 1920 https://doi.org/10.3390/polym12091920
[5] Struszczyk MH Dmowska-Jasek P Kubiak P Łandwijt M and Fejdyś M 2021 Multi-Criterial Analysis Tool to Design a Hybrid Ballistic Plate *Materials* **14** 4058 https://doi.org/10.3390/ma14144058
[6] Alkhatib F Mahdi E and Dean A 2021 Design and Evaluation of Hybrid Composite Plates for Ballistic Protection: Experimental and Numerical Investigations *Polymers* **13** 1450 https://doi.org/10.3390/polym13091450
[7] Haro EE Szpunar JA and Odeshi AG 2016 Ballistic impact response of laminated hybrid materials made of 5086-H32 aluminum alloy, epoxy and Kevlar® fabrics impregnated with shear thickening fluid *Composites Part A: Applied Science and Manufacturing* **87** pp 54-65 ISSN 1359-835X doi:10.1016/j.compositesa.2016.04.007
[8] Mora P Jubsilp C Bielawski CW and Rimdusit S 2021 Impact Response of Aramid Fabric-Reinforced Polybenzoxazine/Urethane Composites Containing Multiwalled Carbon Nanotubes Used as Support Panel in Hard Armor *Polymers* **13** 2779 https://doi.org/10.3390/polym13162779
[9] Nunes SG Scazzosi R Manes A Amico SC De Amorim Júnior WF and Giglio M 2019 Influence of projectile and thickness on the ballistic behavior of aramid composites: Experimental and numerical study *International Journal of Impact Engineering* **132** ISSN 0734-743X doi:10.1016/j.ijimpeng.2019.05.021
[10] Gloger M and Stempien Z Experimental Study of Soft Ballistic Packages with Embroidered Structures Fabricated by Using the Tailored Fiber Placement Technique *Materials* **15** 4208 https://doi.org/10.3390/ma15124208
[11] Skalický P Komenda J Vítek R Jedlička L and Vlhová J 2021 Evaluation of explosion resistance of ballistic polymeric materials *2021 International Conference on Military Technologies (ICMT)* pp 1-4, doi : 10.1109/ICMT52455.2021.9502804
[12] Sundaram AS Eranezhuth AA Krishna KVVR *et al* 2017 Ballistic Impact Performance Study on Thermoset and Thermoplastic Composites *J Fail. Anal. and Preven.* **17** pp 1260–1267 https://doi.org/10.1007/s11668-017-0364-6
[13] Rubin W Wen Z Feng L Yingjian L Wenting Z Yongfang L Bin D and Xiaowei W 2019

Enhancing stab resistance of thermoset–aramid composite fabrics by coating with SiC particles *Journal of Industrial Textiles* **48**(7) pp 1228-1241 ISSN 1528-0837 doi:10.1177/1528083718760804

[14] Asemani SS Liaghat G Ahmadi H Anani Y Charandabi SCh and Khodadadi A 2022 Analysis of Ballistic Impact Performance and Shear Effect on Elastomeric and Thermoset Composites *International Journal of Applied Mechanics* **14**(02) doi:10.1142/S1758825122500132

[15] Bandaru AK Chavan VV Ahmad S Alagirusamy R and Bhatnagar N 2016 Ballistic impact response of Kevlar® reinforced thermoplastic composite armors *International Journal of Impact Engineering* **89** pp 1-13 ISSN 0734743X doi:10.1016/j.ijimpeng.2015.10.014

[16] NATO AEP-55 STANAG 4569: Protection levels for occupants of logistic and light armored vehicles – Part 1-4: General – Annex A. First edition, Volume 1, 2005, 65 p.

AUTHOR INDEX